Boundary Elements and other Mesh Reduction Methods

Boundary Elements and other Mesh Reduction Methods

Editors

A. H-D. Cheng
University of Mississippi, USA

C.A. Brebbia
Wessex Institute, UK

WIT*PRESS* Southampton, Boston

Editors:

A. H-D. Cheng
University of Mississippi, USA

C.A. Brebbia
Wessex Institute, UK

Published by

WIT Press
Ashurst Lodge, Ashurst, Southampton, SO40 7AA, UK
Tel: 44 (0) 238 029 3223; Fax: 44 (0) 238 029 2853
E-Mail: witpress@witpress.com
http://www.witpress.com

For USA, Canada and Mexico

Computational Mechanics International Inc
25 Bridge Street, Billerica, MA 01821, USA
Tel: 978 667 5841; Fax: 978 667 7582
E-Mail: infousa@witpress.com
http://www.witpress.com

British Library Cataloguing-in-Publication Data

A Catalogue record for this book is available
from the British Library

Library of Congress Catalog Card Number: 2017958175

ISBN: 978-1-78466-227-1
eISBN: 978-1-78466-228-8

The texts of the papers in this volume were set individually by the authors or under their supervision. Only minor corrections to the text may have been carried out by the publisher.

Printed in Great Britain by Lightning Source, UK.

The material contained herein is reprinted from a special edition of the International Journal of Computational Methods and Experimental Measurements, Volume 6, Number 6, published by WIT Press.

Preface

This volume contains papers selected from the 40th International Conference on Boundary Elements and other Mesh Reduction Methods, held in the New Forest, UK, organised by the Wessex Institute.

The annual conference on Boundary Elements and other Mesh Reduction Methods (BEM/MRM) started in 1978 and is now in its 40th version. It is well established as the recognised international forum for the latest advances in those techniques and their application in science and engineering.

The continued success of the meeting is a result of the strength of the research on boundary elements and mesh reduction techniques being carried out all over the world. The conference has continually evolved in line with the latest developments in the field since the successful development of boundary integral techniques into BEM was reported in the first meeting held in Southampton in 1978.

The objective of the research papers presented at the meetings is the further development of techniques that reduce or eliminate the type of meshes required by first generation computational methods, such as finite differences or finite elements.

This has steadily been achieved through the development of BEM as a computational tool and continues through more recent research into advanced techniques, leading to further mesh reduction aiming to produce truly meshless methods.

Also included are papers on the use of BEM and, in particular, the description of new applications.

The Editors would like to thank all authors for the quality of their papers and other colleagues for their help in reviewing the material.

The Editors

Contents

THE COLLOCATION BOUNDARY ELEMENT METHOD REVISITED: PERFECT CODE FOR 2D PROBLEMS

NEY AUGUSTO DUMONT
Department of Civil and Environmental Engineering, Pontifical Catholic University of Rio de Janeiro, Rio de Janeiro, Brazil.

ABSTRACT
The paper reviews the collocation boundary element method (BEM) exactly as it has been originally proposed on the basis of a weighted residuals statement that leads to Somigliana's identity, but with two subtle conceptual improvements for a generally curved boundary: (a) the interpolation function for normal fluxes or traction forces (for potential or elasticity problems) must be redefined and (b) only Gauss-Legendre quadrature turns out to be required if the numerical integration issues are mathematically adequately stated. A simple, unified code is proposed – as presently shown for 2D problems – to arrive at arbitrarily high computational accuracy of the constituent matrices as well as of results at internal points independently from how convoluted a problem's topology may be (but given the representation limitations of a discretization mesh). In fact, the higher the effect of a quasi-singularity may be, as for an internal point infinitely close to the boundary, the more accurate a result is achievable with just a few number of quadrature points. A collateral, but not less relevant, outcome of the proposed developments is that regularization methods, special quadrature schemes and so many methods that intend to conceptually deviate from the originally stated BEM as an attempt to offer numerical improvements are actually unnecessary (they are in most cases just misleading). Moreover, the inaccurate, albeit popular constant element is actually not simpler to deal with than high-order elements. Owing to space restrictions, most of the detailed developments as well as the hopefully very convincing numerical results deal with potential problems, although the more general problem of elasticity is adequately posed and assessed.
Keywords: Collocation boundary element method, curved elements, high precision computation, quasi singularities

1 INTRODUCTION

The proposition of a perfect code, as provocatively stated in the title, presupposes a perfect conceptual environment of the BEM to work with. Although Brebbia *et al.* [1] should always be referred to when dealing with the basics of the BEM, a simple, sound formulation has been brought forth by Dumont [2], not only exploring some key properties of the single layer potential matrix **G** [3], but also proposing an improvement in the boundary approximate interpolation of traction forces (for elasticity problems) or normal gradients (for potential problems) that at the time was qualified as a *slight* one. In fact, this improvement, as stated in the following eqn (4), has turned out to be decisive in the direction of writing a code that numerically evaluates **G** as well as results at internal points as accurately as desired without resorting to special schemes other than the Gauss-Legendre quadrature, or unreasonable amounts of integration points, interval subdivisions or computational tricks whatsoever that unfortunately teem in the technical literature. Also decisive in the present developments are some previous studies for the accurate numerical evaluation of quasi singular integrals that only deal with the mathematical aspects of the problem in complete abstraction of any mechanical interpretation [4, 5].

2 PROBLEM FORMULATION

As applied to an elastostatics problem, the single-layer and double-layer potential matrices **G** and **H** of the BEM are obtained as

DOI: 10.2495/CMEM-V6-N6-965-975

$$\mathbf{Hd} = \mathbf{Gt}, \tag{1}$$

(body forces not considered) for nodal displacements **d** and boundary nodal traction attributes **t**. This equation turns out to be an application of Somigliana's identity that converts boundary data into domain displacements. Whenever the boundary data are inaccurate, as the result of approximated values of $\mathbf{d} \equiv d_f$ and $\mathbf{t} \equiv t_\ell$ for a given problem as well as of the piecewise boundary interpolation, an error term ε should actually be added to this equation [2, 3]. The matrices **G** and **H** are expressed in terms of the boundary integrals

$$\begin{aligned} G_{s\ell} &= \int_\Gamma u^*_{is}(\mathbf{x}-\mathbf{x}_s) t_{i\ell}(\mathbf{x}) \mathrm{d}\Gamma(\mathbf{x}) \\ H_{sf} &= \int_\Gamma \sigma^*_{jis}(\mathbf{x}-\mathbf{x}_s) n_j(\mathbf{x})\, u_{if}(\mathbf{x}) \mathrm{d}\Gamma(\mathbf{x}), \end{aligned} \tag{2}$$

where $u^*_{is}(\mathbf{x}-\mathbf{x}_s)$ and $\sigma^*_{jis}(\mathbf{x}-\mathbf{x}_s)$ are the displacement and stress (Kelvin's) fundamental solutions of the elastic problem – which have global support – and $\Gamma(\mathbf{x})$ is the integration boundary[1]. In the above equation and in the following, repeated indices mean summation. The vector $\mathbf{x} \equiv (x, y, z)$ stands for the Cartesian coordinates of a given point, in the case a *field* point, and $n_j(\mathbf{x}(\xi,\eta))$ are the Cartesian components of the unity outward vector $\vec{n}(\mathbf{x}(\xi,\eta))$ to $\Gamma(\mathbf{x}(\xi,\eta))$, in terms of parametric variables (ξ,η), for a general 3D problem. The subscript s refers to a given *source* node (at which the unit point force of the singular fundamental solution is applied) and the subscripts f (which stands for *field*) and ℓ (also a field reference) indicate, respectively, to which node or surface point the displacement-interpolation function $u_{if}(\mathbf{x}(\xi,\eta))$ or the traction-interpolation function $t_{i\ell}(\mathbf{x}(\xi,\eta))$ – both with local support – are referred. $u_{if}(\mathbf{x}(\xi,\eta))$ comes from the piecewise interpolation of displacements $u_i(\mathbf{x}(\xi,\eta))$ along the boundary, $u_i(\mathbf{x}) = u_{if}(\mathbf{x}(\xi,\eta))d_f$, where d_f are the nodal displacements. In a practical finite element or boundary element implementation, $u_{if}(\mathbf{x}(\xi,\eta))$ is actually represented by polynomial shape functions $N_f(\xi,\eta)$:

$$u_{if}(\mathbf{x}(\xi,\eta)) = \begin{cases} N_f(\xi,\eta) & \text{if } i \text{ and } f \text{ refer to the same Cartesian direction} \\ 0 & \text{otherwise.} \end{cases} \tag{3}$$

In the expression of **H**, the Jacobian used in the definition of $n_j(\mathbf{x}(\xi,\eta))$ – see eqn (9)– cancels out with the Jacobian of $\mathrm{d}\Gamma(\mathbf{x}) = |J(\xi,\eta)|\mathrm{d}\xi\mathrm{d}\eta$.

For the single-layer potential matrix **G**, it is proposed that the usual (as found in the literature) interpolation polynomials $t_{i\ell}$ of traction forces $T_i(\mathbf{x}) = t_{i\ell}(\mathbf{x}(\xi,\eta))t_\ell$ in eqn (2) be replaced with

$$t_{i\ell} = \frac{|J|_{(\text{at } \ell)}}{|J|} \begin{cases} N_\ell(\xi,\eta) & \text{if } i \text{ and } \ell \text{ refer to the same Cartesian direction} \\ 0 & \text{otherwise,} \end{cases} \tag{4}$$

where $|J|_{(\text{at } \ell)}$ is the value of the Jacobian at the point characterized by the subscript ℓ [2]. Nothing changes formally in the development of the BEM for curved boundary segments (and, of course, nothing changes numerically for the trivial cases of straight or flat boundary

[1]Since both interpolation functions $u_{if}(\mathbf{x}(\xi,\eta))$ and $t_{i\ell}(\mathbf{x}(\xi,\eta))$ formally introduced in eqn (2) have local support, there is no need of indicating that the boundary integrations are to be carried out segment by segment.

segments), except that the evaluation of **G** becomes much easier and actually more consistent as compared to proposed implementations given in the technical literature [2]. In fact, $|J|$ cancels out in the product $t_{i\ell}\mathrm{d}\Gamma$ in eqn (2) for $t_{i\ell}$ defined as suggested, and the integrand of **G** becomes a polynomial that multiplies the assumed kernel u^*_{is}.

2.1 The misconception of *continuous* versus *discontinuous* elements

In a practical implementation, the same shape functions $N_i(\xi)$ (they might be different as a corollary of the theorem to be enounced next) are used in eqns (3) and (4), although the context differs conceptually, as **G**, among other features, is in general a rectangular matrix (ℓ in general refers to a larger number of parameters than f). There is in fact a remarkable difference between the representation of boundary displacements according to $u_i(\mathbf{x}) = u_{if}(\mathbf{x}(\xi,\eta))d_f$ and the representation of boundary tractions according to $T_i(\mathbf{x}) = t_{i\ell}(\mathbf{x}(\xi,\eta))t_\ell$, namely that d_f are *nodal attributes*, whereas t_ℓ are *surface attributes*: While $u_i(\mathbf{x}) = u_{if}(\mathbf{x}(\xi,\eta))d_f$ is single valued at a node f independently from the approaching direction (discontinuities due to a cracked domain not considered, for simplicity), in the case of ℓ referring to a corner point the traction forces $T_i(\mathbf{x}) = t_{i\ell}(\mathbf{x}(\xi,\eta))t_\ell$ [or normal gradients $q(\mathbf{x}) = t_\ell(\mathbf{x}(\xi,\eta))q_\ell$, for potential problems] depend on from which direction the point is approached (that is, to which surface segment the point ℓ is meant to belong). As a result, attaching the precise meaning to the quantities at geometric locations characterized by the subscripts ℓ and f in the formulation of matrices **G** and **H** in eqn (2), so that eqn (1) becomes solvable, may not be trivial in a general mixed boundary (Cauchy) formulation. However, the dilemma *continuous* versus *discontinuous* elements should not be posed in a consistent formulation, as *continuous* elements in the definition of **G** simply stem from a misconception: while d_f and f are meant to refer to a *nodal* point, t_ℓ and ℓ should refer to just some *surface* point on a given boundary segment regardless the neighboring segments. (Observe for comparison that in a variational displacement formulation there is the concept of equivalent *nodal* forces p_f in such a way that $\mathbf{p}^{\mathrm{T}}\mathbf{d} \equiv p_f d_f$ means virtual work. This is by far not the case in the BEM.)

The conceptual advantage of using the definition of eqn (4) – and the eventual need of having the same functions $N_i(\xi)$ in both eqns (3) and (4) – becomes clear from the following theorem.

2.2 Theorem on the accurate representation of constant stress states

Theorem. For 2D problems with the boundary represented by piecewise, generally curved segments, the boundary element eqns (1) - (4) are able to exactly simulate constant stress state fields in the frame of an isoparametric formulation. For 3D problems, constant stress states are exactly simulated only if the boundary geometry is piecewise represented by up to second degree polynomials (linear and quadratic triangular as well as quadrilateral boundary elements).

Proof: One should first remark that eqn (1) only applies to a homogeneous medium, be it isotropic or not. Moreover, eqn (1) comes directly from Somigliana's identity, which is true only for exact boundary data [2, 3]. Let an exact, in principle arbitrary analytical solution in the domain Ω of a problem represented by eqn (1) be given by a displacement field u^p_i, where the superscript stands for *polynomial*, expressed as a sum of polynomial functions u^p_{im} multiplied by some constant parameters c^p_m.

Part 1. This displacement field must correspond to boundary data given in terms of the polynomial interpolation functions u_{if} introduced in eqn (2) , of sufficiently high degree to represent u^p_{im} along Γ (to adequately take the curved geometry into account) and multiplied by some nodal displacement parameters d_f, as proposed in the definition of eqn (1), that is,

$$u_i^p = u^p_{im}\left(\mathbf{x}(\xi,\eta)\right)c^p_m = u_{if}(\xi,\eta)d_f \text{ on } \Gamma. \tag{5}$$

The corresponding exact nodal displacements d_f are given by $d_f = u^p_{im}\big|_{(\text{at } f)} c^p_m$. Then,

$$u^p_{im}\left(\mathbf{x}(\xi,\eta)\right)c^p_m = u_{if}(\xi,\eta)u^p_{im}\big|_{(\text{at } f)} c^p_m \text{ on } \Gamma. \tag{6}$$

Since this must be valid for arbitrary values of c^p_m, it follows that eqn (6) only holds if $u^p_{im}\left(\mathbf{x}(\xi,\eta)\right)$ on Γ is a linear combination of $u_{if}(\xi,\eta)$. For a generally curved boundary, this is only possible if $u^p_{im}\left(\mathbf{x}(\xi,\eta)\right)$ are linear functions of the coordinates $\mathbf{x} \equiv (x,y,z)$ interpolated from nodal geometric data using a linear combination of $u_{if}(\xi,\eta)$, which constitutes an isoparametric representation. [In the case of non-homogeneous materials, only rigid body displacements are generally represented by eqn (5), as the field functions $u^p_{im}(\mathbf{x})$ of a non-trivial analytical solution, if available, are no longer necessarily polynomials. However, eqn (1) only applies to a homogeneous medium.] This proves the first part of the theorem, namely, that the domain solutions $u^p_{im}(\mathbf{x})$ must be linear polynomial functions in order for eqn (1) to hold and in an isoparametric formulation.

Part 2. The constant stress state $\sigma^p_{ij} = \sigma^p_{ijm}(\mathbf{x})c^p_m$ corresponding to the analytical, linear displacement solution u^p_i of the first part of this proof leads to boundary traction forces (from now on dropping arguments for simplicity)

$$\sigma^p_{ij}n_j = \sigma^p_{ijm}\, n_j\, c^p_m = t_{i\ell}\, t_\ell \text{ on } \Gamma. \tag{7}$$

The corresponding exact traction force attributes t_ℓ are given by $t_\ell = \left(\sigma^p_{ijm}\, n_j\right)_{(\text{at } \ell)} c^p_m$. Then,

$$\sigma^p_{ijm}\, n_j c^p_m = t_{i\ell}\left(\sigma^p_{ijm}\, n_j\right)_{(\text{at } \ell)} c^p_m \text{ on } \Gamma. \tag{8}$$

Since this should be valid for arbitrary values of c^p_m as well as for any constant values σ^p_{ijm}, and making use of eqn (4), one should investigate the conditions for $n_j = N_\ell\left(|J|n_j\right)_{(\text{at } \ell)}\big/|J|$ to hold. Recall that

$$\vec{n} = \frac{y'\vec{i} - x'\vec{j}}{|J|} \text{ for 2D, } \vec{n} = \frac{1}{|J|}\begin{vmatrix} \partial x/\partial\xi & \partial y/\partial\xi & \partial z/\partial\xi \\ \partial x/\partial\eta & \partial y/\partial\eta & \partial z/\partial\eta \\ \vec{i} & \vec{j} & \vec{k} \end{vmatrix} \text{ for 3D problems.} \tag{9}$$

Then, expressing $n_j = \tilde{n}_j/|J|$, eqn (8) holds only if

$$\tilde{n}_j = N_\ell\, \tilde{n}_\ell \tag{10}$$

also holds (j and ℓ referring to the same Cartesian directions). This is generally true for two-dimensional problems in an isoparametric representation. For the three-dimensional case, as $\tilde{n}_j$ involves the multiplication of two polynomial derivatives, according to eqn (9), eqn (10) is attained only if the boundary geometry is piecewise represented by up to complete quadratic polynomials (linear and quadratic triangular as well as linear quadrilateral elements).

In spite of the lengthy proof of this theorem, the computational implementation of the matrix **G** in eqn (2) with $t_{i\ell}(\xi,\eta)$ given by eqn (4) is actually simpler than in the developments hitherto presented in the technical literature for curved elements. As a side remark or, more formally, as a corollary from the above theorem, one obtains that the shape functions $N_f(\xi,\eta)$ of eqn (3), used to interpolate both displacements and geometry in the frame of an isoparametric formulation, should be able to represent the shape functions $N_\ell(\xi,\eta)$ of eqn (4), which justifies using the same functions in both equations.

3 ACCURATE NUMERICAL EVALUATION OF MATRIX G

Although the present developments are seamlessly applicable to two-dimensional elasticity problems, owing to space restrictions only the simpler case of a potential problem (solution of the Laplace equation) is outlined.

3.1 Evaluation of G for the improper integral

The matrix **G** is to be evaluated as in eqn (2) along a boundary segment Γ_{seg} for an interpolation function given according to eqn (4) by $|J|_{(\text{at } n)} N_n^{o_e}(\xi)/|J(\xi)|$, where $n = 1\ldots o_e+1$ is the local numbering ℓ of the boundary segment and o_e is the degree of the interpolation polynomial $N_n^{o_e}(\xi)$. One uses $u_m^*(\mathbf{x}-\mathbf{x}_m) = -\ln|\mathbf{x}(\xi)-\mathbf{x}(\xi_m)|/2\pi \equiv -\ln|r_m(\xi-\xi_m)|/2\pi \equiv -\ln r_m/2\pi$, where $0 \le \xi_m \le 1$ is the parametric coordinate of the source point s, which is locally numbered $m = 1\ldots o_e+1$, for o_e+1 source points along Γ_{seg}. (Implementations using $0 \le \xi_m \le 1$ turn out to be simpler than for $-1 \le \xi_m \le 1$.) Then, a submatrix G_{mn} of $G_{s\ell}$ is evaluated as

$$2\pi G_{mn} = -\int_{\Gamma_{seg}} \ln r_m \frac{|J|_{(\text{at } n)}}{|J(\xi)|} N_n^{o_e}(\xi)\,\mathrm{d}\Gamma(\xi) = -|J|_{(\text{at } n)} \int_0^1 \ln r_m\, N_n^{o_e}\,\mathrm{d}\xi, \tag{11}$$

which is the same as

$$\begin{aligned}
\frac{-2\pi}{|J|_{(\text{at } n)}} G_{mn} &= \left(\int_0^1 \ln\left|\frac{r_m}{\xi-\xi_m}\right| N_n^{o_e}\,\mathrm{d}\xi + \int_0^1 \ln|\xi-\xi_m|\, N_n^{o_e}\,\mathrm{d}\xi \right) \\
&= \left(GL \int_0^1 \ln\left|\frac{r_m}{\xi-\xi_m}\right| N_n^{o_e}\,\mathrm{d}\xi + \int_0^1 \ln|\xi-\xi_m|\, N_n^{o_e}\,\mathrm{d}\xi \right) \\
&= \left[GL \int_0^1 \ln r_m\, N_n^{o_e}\,\mathrm{d}\xi + \left(\int_0^1 \ln|\xi-\xi_m|\, N_n^{o_e}\,\mathrm{d}\xi - GL \int_0^1 \ln|\xi-\xi_m|\, N_n^{o_e}\,\mathrm{d}\xi \right) \right].
\end{aligned} \tag{12}$$

In this equation, the symbol GL is introduced to mean a Gauss-Legendre quadrature using a given number n_g of evaluation points, which is applicable to the indicated evaluation in the second row up to a desired accuracy, and then split in the third row. Observe that, in this third row, one obtains the intended evaluation of the integral of eqn (11) as if carried out in the frame of a Gauss-Legendre quadrature plus the correction term given in brackets. This correction term is an array with o_e+1 constants that only depend on the shape function $N_n^{o_e}$ and can be previously evaluated for a given number n_g of Gauss-Legendre points and stored. The analytical part of this correction term can be already expressed as an array of $(o_e+1)\times(o_e+1)$

matrices, whose rows and columns correspond to the shape functions $N_n^{o_e}$ and singularity poles ξ_m, respectively, as illustrated for linear, quadratic and cubic elements:

$$\int_0^1 \ln|\xi-\xi_m|N_n^{o_e}\,\mathrm{d}\xi := \begin{cases} 1=\begin{bmatrix} \frac{-3}{4} & \frac{-1}{4} \\ \frac{-1}{4} & \frac{-3}{4} \end{bmatrix}, \quad 2=\begin{bmatrix} \frac{-17}{36} & \frac{-\ln 2}{6}-\frac{1}{18} & \frac{1}{36} \\ \frac{-5}{9} & \frac{-2\ln 2}{3}-\frac{8}{9} & \frac{-5}{9} \\ \frac{1}{36} & \frac{-\ln 2}{6}-\frac{1}{18} & \frac{-17}{36} \end{bmatrix}, \\ 3=-\begin{bmatrix} \frac{11}{32} & \frac{\ln 3}{8}+\frac{1}{32} & \frac{\ln 3}{8}+\frac{1}{96}-\frac{\ln 2}{9} & \frac{1}{32} \\ \frac{19}{32} & \frac{3\ln 3}{8}+\frac{67}{96}-\frac{\ln 2}{9} & \frac{3\ln 3}{8}+\frac{25}{96}-\frac{4\ln 2}{9} & \frac{1}{32} \\ \frac{1}{32} & \frac{3\ln 3}{8}+\frac{25}{96}-\frac{4\ln 2}{9} & \frac{3\ln 3}{8}+\frac{67}{96}-\frac{\ln 2}{9} & \frac{19}{32} \\ \frac{1}{32} & \frac{\ln 3}{8}+\frac{1}{96}-\frac{\ln 2}{9} & \frac{\ln 3}{8}+\frac{1}{32} & \frac{11}{32} \end{bmatrix} \end{cases} \tag{13}$$

3.2 Evaluation of **G** for complex and real quasi-singularities

When the source point $\mathbf{x}_s$ is not on the integration segment Γ_{seg} but sufficiently close to configure a quasi-singularity, one evaluates ξ_s such that $r^2(\xi-\xi_s)\equiv r^2=0$ and obtains

$$r^2=\left(x(\xi_s)-x_s\right)^2+\left(y(\xi_s)-y_s\right)^2=\rho(\xi)w(\xi_s)\equiv\rho(\xi)\left[(\xi_s-a)^2+b^2\right]=0, \tag{14}$$

where $w(\xi)\equiv w=(\xi-a)^2+b^2$ has roots $\xi_s=a\pm bi$, with $i=\sqrt{-1}$, and $\rho(\xi)$ is a polynomial whose degree depends on o_e and has only complex roots of large values. Then, $\xi_s=a\pm bi$ characterizes the complex quasi-singularity one must deal with. The evaluation of *(a,b)* is not trivial and must be carried out iteratively using a Newton-Raphson algorithm, for instance, as given by Dumont [4] for quadratic elements, and generalized in the present developments for an element of arbitrary order o_e. The explicit evaluation of $\rho(\xi)$ in $r^2(\xi)=\rho(\xi)w(\xi)$ can be circumvented by simply expressing $\rho=r^2/w$.

Complex quasi-singularity. In such a case, one may write from eqn (11)

$$\begin{aligned}\frac{-4\pi}{|J|_{(\text{at } n)}}G_{sn} &= \left(\int_0^1 \ln\left(\frac{r^2}{w}\right)N_n^{o_e}\,\mathrm{d}\xi+\int_0^1\ln(w)N_n^{o_e}\,\mathrm{d}\xi\right)\\ &= GL\int_0^1\ln r^2 N_n^{o_e}\,\mathrm{d}\xi+\left(\int_0^1\ln(w)N_n^{o_e}\,\mathrm{d}\xi-GL\int_0^1\ln(w)N_n^{o_e}\,\mathrm{d}\xi\right).\end{aligned} \tag{15}$$

The definite integral $\int_0^1\ln(w)N_n^{o_e}\mathrm{d}\xi$, $n=1\ldots o_e+1$, for shape functions $N_n^{o_e}$ of order o_e, may be expressed as

$$\int_0^1 \ln(w) N_n^{o_e} \mathrm{d}\xi = \mathbf{L}_0^w \ln\left(a^2 + b^2\right) + \mathbf{L}_1^w \ln\left((a-1)^2 + b^2\right)$$
$$+\mathbf{A}^w \left(\arctan\frac{a-1}{b} - \arctan\frac{a}{b}\right) + \mathbf{P}^w, \tag{16}$$

where $\mathbf{L}_0^w$, $\mathbf{L}_1^w$, $\mathbf{A}^w$ and $\mathbf{P}^w$ are arrays with $o_e + 1$ low order polynomial terms in a and b that fulfill the properties:

$$\sum_{n=1}^{o_e+1} P^w[n] = -2, \quad \sum_{n=1}^{o_e+1} L_0^w[n] = a, \quad \sum_{n=1}^{o_e+1} L_1^w[n] = 1 - a, \quad \sum_{n=1}^{o_e+1} A^w[n] = 2b. \tag{17}$$

Real quasi-singularity. This occurs as a particular case of the previous section, when $b = 0$, that is, $\xi_s = a$. Then, equation becomes

$$\frac{-4\pi}{|J|_{(\text{at } n)}} G_{sn} = \left(\int_0^1 \ln\left(\frac{r^2}{|\xi - a|}\right) N_n^{o_e} \,\mathrm{d}\xi + \int_0^1 \ln|\xi - a| N_n^{o_e} \,\mathrm{d}\xi\right)$$
$$= GL\int_0^1 \ln r^2 N_n^{o_e} \,\mathrm{d}\xi + \left(\int_0^1 \ln|\xi - a| N_n^{o_e} \,\mathrm{d}\xi - GL\int_0^1 \ln|\xi - a| N_n^{o_e} \,\mathrm{d}\xi\right) \tag{18}$$

and one may express the definite integral $\int_0^1 \ln|\xi - a| N_n^{o_e} \mathrm{d}\xi$, $n = 1 \ldots o_e + 1$, for shape functions $N_i^{o_e}$ of order o_e, as

$$\int_0^1 \ln|\xi - a| N_n^{o_e} \mathrm{d}\xi = \left(a\mathbf{L}_0^a \ln|a| + (1-a)\mathbf{L}_1^a \ln|1 - a| + \mathbf{P}^a\right), \tag{19}$$

where $\mathbf{L}_0^a$, $\mathbf{L}_1^a$ and $\mathbf{P}^a$ are arrays with $o_e + 1$ low order polynomial terms in a that fulfill the properties:

$$\sum_{n=1}^{o_e+1} P^a[n] = -1, \quad \sum_{n=1}^{o_e+1} L_0^a[n] = \sum_{n=1}^{o_e+1} L_1^a[n] = 1. \tag{20}$$

4 ACCURATE NUMERICAL EVALUATION OF MATRIX H

4.1 Evaluation of H for a $1/r$ singularity

The $1/r$ singularity related to the double-layer matrix **H** is already adequately dealt with in the technical literature. For $r_m^2 = \left(x(\xi) - x(\xi_m)\right)^2 + \left(y(\xi) - y(\xi_m)\right)^2$, where $0 \le \xi_m \le 1$ is the parametric coordinate of the source point s, locally numbered $m = 1 \ldots o_e + 1$, one expresses the normal gradient to the boundary as $q_m^*(\mathbf{x} - \mathbf{x}_m) = -\left(\left(x(\xi) - x(\xi_m)\right) y'(\xi) - \left(y(\xi) - y(\xi_m)\right) x'(\xi)\right) / 2\pi r_m^2$. Then, a submatrix H_{mn} of H_{sf} is evaluated for $o_e + 1$ shape functions along Γ_{seg} as

$$H_{mn} = GL\int_{\Gamma_{seg}} q_{jm}^* n_j N_n^{o_e} \mathrm{d}\Gamma + \left(1 - \frac{\Delta\theta_m}{2\pi}\right)\delta_{mn}, \tag{21}$$

where $\Delta\theta_m$ is the angular jump from the left to right tangent to node m.

4.2 Evaluation of **H** for complex and real quasi-singularities

Referring to the expression of $r^2(\xi) = \rho(\xi)w(\xi)$ in Section 3.2 and the above definitions for the matrix $\mathbf{H}$, one may write for a submatrix H_{mn} of H_{sf}

$$\begin{aligned} H_{mn} &= \frac{-1}{2\pi}\int_{\Gamma_{seg}} \frac{1}{w}\left[(xy' - yx')\frac{uw}{r^2}\right]\mathrm{d}\xi \\ &= \frac{-1}{2\pi}\int_0^1 \frac{1}{w}\left[(xy' - yx')\frac{uw}{r^2}\right]\mathrm{d}\xi \equiv \int_0^1 \frac{g}{w}\mathrm{d}\xi \equiv \int_0^1 \frac{1}{w}\left(\frac{fw}{r^2}\right)\mathrm{d}\xi. \end{aligned} \tag{22}$$

According to eqn (25) in Dumont [4], one obtains for the integral

$$\int_0^1 \frac{g}{w}\mathrm{d}\xi = GL\int_0^1 \frac{g}{w}\mathrm{d}\xi + R_1C_1 + R_2C_2, \tag{23}$$

where $\xi_m = a \pm bi$ are the roots of $w \equiv w(\xi) = (\xi - a)^2 + b^2$, with

$$R_1 = b_0/b, \quad R_2 = a_0 - R_1 a, \quad \text{for } b \neq 0. \tag{24}$$

In these expressions, $a_0 = \Re\left(g(\xi_0)\right)$ and $b_0 = \Im\left(g(\xi_0)\right)$, and

$$C_1 = \int_0^1 \frac{\xi \mathrm{d}\xi}{w} - \sum_{i=1}^{n} \frac{\xi_i h_i}{w_i}, \quad C_2 = \int_0^1 \frac{\mathrm{d}\xi}{w} - \sum_{i=1}^{n} \frac{h_i}{w_i}. \tag{25}$$

The integrals in the above equations are defined analytically as

$$\int_0^1 \frac{\xi \mathrm{d}\xi}{(\xi - a)^2 + b^2} = \frac{1}{2}\ln\frac{(1-a)^2 + b^2}{a^2 + b^2} + \frac{a}{b}\left(\arctan\frac{1-a}{b} + \arctan\frac{a}{b}\right), \quad \text{for } b \neq 0 \tag{26}$$

$$\int_0^1 \frac{\mathrm{d}\xi}{(\xi - a)^2 + b^2} = \frac{1}{b}\left(\arctan\frac{1-a}{b} + \arctan\frac{a}{b}\right), \quad \text{for } b \neq 0. \tag{27}$$

For a real quasi-singularity $\xi_m = a$, with $b = 0$, the integral to be obtained is

$$\int_0^1 \frac{g}{\xi - a}\mathrm{d}\xi = GL\int_0^1 \frac{g}{\xi - a}\mathrm{d}\xi + R_1C_1, \text{ for either } a < 0 \text{ or } a > 1, \tag{28}$$

where $R_1 = g(\xi_0)$ and

$$C_1 = \int_0^1 \frac{\mathrm{d}\xi}{\xi - a} - \sum_{i=1}^{n} \frac{h_i}{\xi_i - a} = a\ln\left|\frac{a-1}{a}\right| + 1 - \sum_{i=1}^{n} \frac{h_i}{\xi_i - a}. \tag{29}$$

5 ACCURATE EVALUATION OF RESULTS AT INTERNAL POINTS

Owing to space restriction, the complete algorithm for the evaluation of results at internal points cannot be outlined in this paper. Potential and displacement results are evaluated using the same procedures applied to matrices **G** and **H**. For internal gradients, a procedure to deal with $\int_0^1 g/w^2 d\xi$ must be developed, whereas stress results also demand the development of an algorithm to deal with the more elaborate case $\int_0^1 g/w^3 d\xi$. Such general developments have already been outlined by Dumont [4] and were especially coded in the frame of the present research work. They have resulted into black boxes that are straight forward to implement, as shown in the full manuscript that is being prepared.

6 A SIMPLE NUMERICAL ILLUSTRATION

Figure 1 shows a 2D domain discretized externally with 11 quadratic elements and four more elements to configure a hole. Node 27 of the hole is almost coincident with the external boundary. Moreover, nodes 12 and 29 are very close. Several internal (and some external) points are placed along the indicated dash line. Some of the points are just inside the domain,

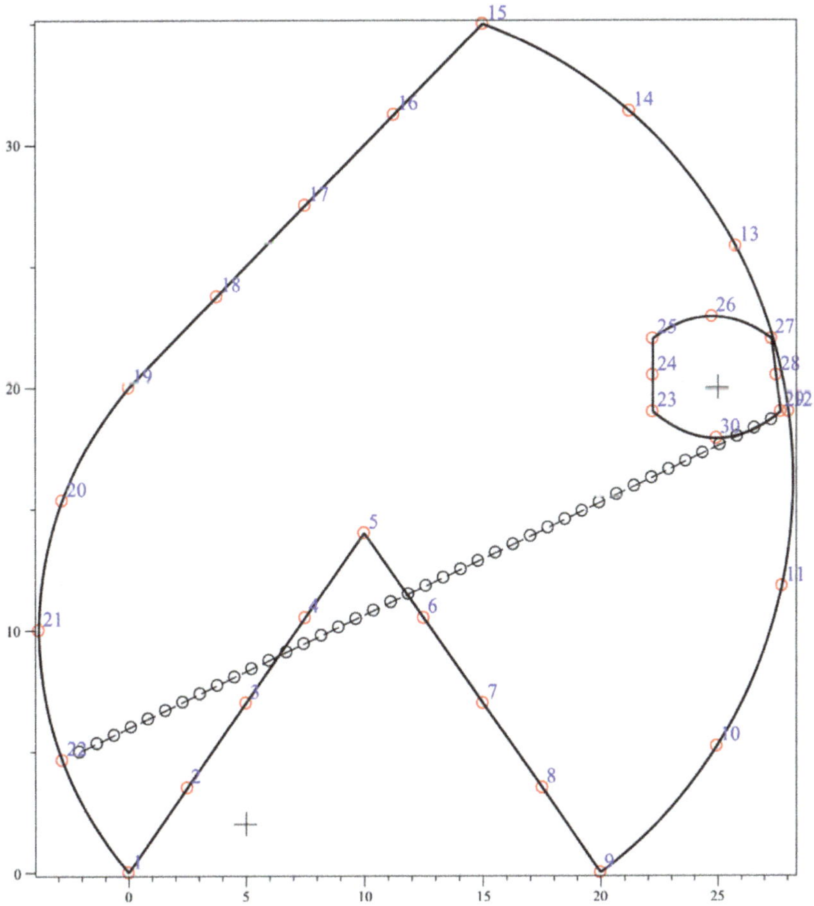

Figure 1. Multiply-connected domain discretized with a very coarse mesh of 15 quadratic elements and dash line along which internal results are evaluated.

whereas some others are just outside. The matrices **G** and **H** were evaluated using in a first application four Gaussian points for all calculations and in a second case ten Gaussian points. The orthogonality of **H** with respect to a constant potential field was checked with global Euclidean errors of 10^{-7} and 10^{-15}, respectively. These same errors affected the evaluation of eqn (1) for potential fields that vary linearly in the Cartesian directions, which attests the validity of the theorem of Section 2.2 within Gauss-Legendre quadrature accuracy. For these linear potential fields results at internal points could be obtained with the same previous accuracies all along the dash line independently from how close just inside or just outside the domain a point was placed. Results for higher order potential fields as well for two logarithmic fields going to infinity at the points marked with a cross were obtained with relatively good accuracy for the applied coarse mesh.

7 CONCLUDING REMARKS

This paper reviews the collocation boundary element method exactly as it may be obtained from Somigliana's identity, but with a decisive improvement for generally curved boundaries, which leads to a theorem akin to the basic one of the displacement finite element method, namely that constant stress states must be exactly represented for an elastic isotropic medium. A practical outcome of the proposed improvement is that the logarithmic singularity of the single-layer potential matrix, for 2D problems, can always be dealt with analytically, with a remaining regular part that only requires a Gauss-Legendre quadrature. Such analytical results are actually independent from problem geometry and can be pre-evaluated and stored in tables for different element orders and numbers of quadrature points. General quasi-singularities – both for matrix assemblages and evaluation of results at internal points – are dealt with mathematically to an arbitrarily large accuracy, which completely dispenses with regularization techniques (including not always legitimate mechanical justifications) and artifacts such as interval subdivision and special quadrature schemes. It is in passing discussed that the qualifications *continuous* and *discontinuous*, as applied to boundary elements, are actually the result of misleading concepts and should not even be mentioned in relation to a consistent formulation, as presently proposed. A simple – however convincing – numerical example is displayed to just illustrate how remarkable the results can be in the present context, for instance with global or local errors never larger than 10^{-7} when using just four Gauss-Legendre quadrature points along curved quadratic elements independently from how high a singularity or quasi-singularity may occur.

ACKNOWLEDGEMENTS

This work was supported by the Brazilian agencies CAPES, CNPq and FAPERJ.

REFERENCES

[1] Brebbia, C.A., Telles, J.C.F. & Wrobel, L.C., *Boundary Element Techniques*, Springer-Verlag: Berlin and New York, 1984.

[2] Dumont, N.A., The boundary element method revisited, *Boundary Elements and Other Mesh Reduction Methods XXXII*, ed C.A. Brebbia, WITPress, Southampton, pp. 227–238, 2010.
https://doi.org/10.2495/be100201

[3] Dumont, N.A., An assessment of the spectral properties of the matrix **G** used in the boundary element methods. *Computational Mechanics,* **22**(1), pp. 32–41, 1998.
https://doi.org/10.1007/s004660050336

[4] Dumont, N.A., On the efficient numerical evaluation of integrals with complex singularity poles. *Engineering Analysis with Boundary Elements*, **13**, pp. 155–168, 1994. https://doi.org/10.1016/0955-7997(94)90018-3

[5] Dumont, N.A. & Noronha, M., A simple, accurate scheme for the numerical evaluation of integrals with complex singularity poles. *Computational Mechanics,* **22**(1), pp. 42–49, 1998.
https://doi.org/10.1007/s004660050337

COUPLED BEM AND FEM ANALYSIS OF FLUID-STRUCTURE INTERACTION IN DUAL COMPARTMENT TANKS

VASYL V. GNITKO, KYRYL G. DEGTYARIOV, VITALY V. NAUMENKO & ELENA A. STRELNIKOVA*
A.N. Podgorny Institute for Mechanical Engineering Problems of the Ukrainian Academy of Sciences, Ukraine.

ABSTRACT

The paper presents a fluid-structure interaction analysis of fuel tanks with cylindrical and spherical compartments partially filled with a liquid. The compound shell of revolution is considered as a container model. The shell is supposed to be thin, so the Kirchhoff–Love linear theory hypotheses are applied. The liquid is an ideal and incompressible one. Its properties and filling levels may be different within each compartment. The shell vibrations coupled with liquid sloshing under the force of gravity have been considered. The tank structure is modelled by a finite element method, whereas liquid sloshing in the compartments is described by a boundary element method. A system of singular integral equations is obtained for evaluating the fluid pressure. At the first stage, both spherical and cylindrical fluid-filled unconnected rigid shells are considered. Different filling levels as well as small radii of free surfaces are taken into account in problems of liquid sloshing in spherical shells. The sloshing frequencies in the presence of complete or partially covered free surfaces are determined for cylindrical shells. The boundary element method has proven to be effective and accurate in all the problems considered. At the second stage, the natural frequencies and modes of the dual compartment tank are obtained including sloshing, elasticity, and gravity effects.

Keywords: baffles, boundary and finite element methods, fluid-structure interaction, free vibrations, fuel tanks, sloshing

1 INTRODUCTION

The low frequency oscillations of free surfaces in partially filled containers significantly increase dynamic responses of structures containing liquids. This phenomenon is known as sloshing. It can result in severe structural damage, loss of stability, and failure.

In order to suppress sloshing, a variety of methods have been proposed, simulated, and tested. The boundary element method (BEM) has been successfully used in solution of both linear and non-linear sloshing problems. In order to solve the large-scale problems, researchers developed the multi-domain boundary element method (MBEM). The multi-domain collocation strategy was briefly introduced by Brebbia *et al.* [1] and then developed by Wang and Gao [2]. The MBEM is especially effective if the computational domain has a complicated structure. It is successfully applied for numerical simulation of sloshing in multi-compartment fuel tanks with different types of baffles.

The effect of baffles on sloshing frequencies was studied by Biswal *et al.* [3]. The numerical method using a finite element formulation was developed by Kumar and Sinhamahapatra [4] to analyze dynamic effects of perforated vertical baffles. Sloshing in spherical tanks for liquefied natural gas carriers was studied by Faltinsen and Timokha [5] and for water supply towers by Curadelli *et al.* [6]. Ravnik *et al.* [7] presented a fluid-structure interaction analysis of the cylindrical fuel tank with two compartments partially filled with a liquid. But fluid-structure interaction problems for fuel tanks of more complicated shapes are still not fully described in the literature.

**ORCID: http://orcid.org/0000-0003-0707-7214*

DOI: 10.2495/CMEM-V6-N6-976-988

2 PROBLEM STATEMENT

In this paper, we describe applications of multi-domain and single-domain boundary element methods to the fluid-structure interaction analysis of compound fuel tanks.

Consider a fuel tank with cylindrical and spherical compartments. The internal baffles can be installed in both compartments for slosh suppression. The structure and its sketch are shown in Fig. 1.

Assume that both compartments are partly filled with an incompressible ideal liquid. Fluid motion is considered to be irrotational.

Suppose σ is a wetted part of the structure surfaces, S_{01} and S_{02} are free surfaces of the liquid in spherical and cylindrical compartments, respectively. Let S_0 be $S_0 = S_{01} \cup S_{02}$. The wetted surface σ consists of four parts, namely, $\sigma = S_{w1} \cup S_{w2} \cup S_{\text{bot}} \cup S_{\text{baf}}$. Here S_{w1} and S_{w2} are the wetted surfaces of cylindrical and spherical parts, S_{bot} is a bottom surface of the tank and S_{baf} is a baffle surface. The domains occupied with the liquid are denoted by Σ_1 and Σ_2 for cylindrical and spherical compartments, respectively. Let $\Sigma_{12}=\partial\Sigma_1 \cap \partial\Sigma_2$ be the common part of wetted surface areas. The liquid densities are ρ_1 and ρ_2, and filling levels are $h_1$and h_2 in Σ_1 and Σ_2, respectively. Let R_{ir}be the inner radius of the ring free surface (Fig. 1).

Consider at first free vibrations of the empty elastic shell structure. Assume that a time dependent vector of the shell displacements $\tilde{\mathbf{U}}$is given by

$$\tilde{\mathbf{U}} = \tilde{\mathbf{u}}\exp(i\Omega t); \quad \tilde{\mathbf{u}} = \left(u_1, u_2, u_3\right),$$

where Ω is a vibration frequency. The time factor $\exp(i\Omega)$ will be omitted further on. After separation of the time factor, the shell vibrations are described by the system of three partial differential equations

$$\sum_{i=1}^{3} L_{ij}u_i = \Omega^2 u_j, \quad j = 1,2,3,$$

where L_{ij} are linear differential operators of the Kirchhoff–Love shell theory described by Wan *et al.* [8]. To obtain the natural frequencies Ω_k and modes $\mathbf{u}_k\left(k = \overline{1,N}\right)$ of the shell structure, the finite element method is applied [7].

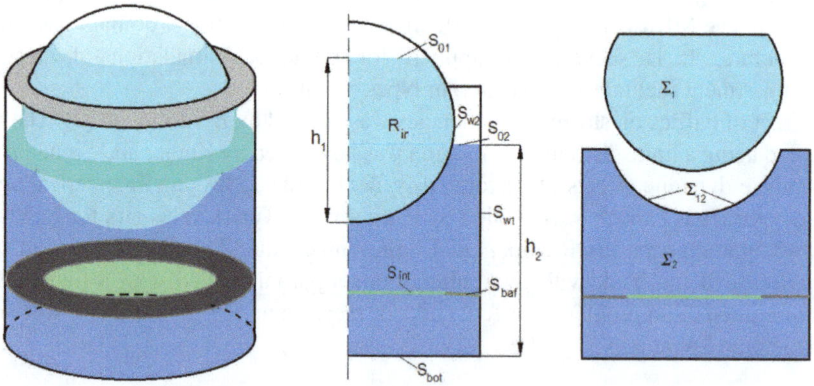

Figure 1: Shell structure with an internal baffle, its sketch and fluid subdomains.

Let $\mathbf{U}=(U_1,U_2,U_3)$ be displacements of the fluid-filled elastic shell structure. After forming the global stiffness and mass matrices **L** and **M**, the following equation of motion for the tank filled with the liquid has been obtained [7]:

$$\mathbf{LU}+\mathbf{M}\ddot{\mathbf{U}}=\mathbf{P}. \tag{1}$$

Here **P** gives the fluid dynamical pressure onto the shell structure normal to its surface.

Assuming the flow to be inviscid and irrotational, the incompressible fluid motion in the 3D tank is described by the Laplace equation

$$\nabla^2\Phi=0, \tag{2}$$

where Φ is a velocity potential.

To determine this potential a mixed boundary value problem for the Laplace equation is formulated in the double domain $\Sigma_1\cup\Sigma_2$. The non-penetration condition on the wetted tank surfaces σ is following:

$$\left.\frac{\partial\Phi}{\partial\mathbf{n}}\right|_{\sigma}=\frac{\partial w}{\partial t}. \tag{3}$$

Here **n** is an external unit normal to the tank wetted surfaces, w denotes a normal component of the displacement vector **U**, namely, $w=(\mathbf{U},\mathbf{n})$. Let functions $\varsigma_1(t,x,y)$ and $\varsigma_2(t,x,y)$ be free surface elevations in the first and second compartments (Fig. 1). By p_i denote the pressure on the free surfaces $S_{0i}(i=1,2)$. Let p_0 be an atmospheric pressure. Then the kinematic and dynamic boundary conditions on S_{01} and S_{02} can be expressed as follows:

$$\left.\frac{\partial\Phi}{\partial\mathbf{n}}\right|_{S_{0i}}=\frac{\partial\zeta_i}{\partial t};\quad \left.p_i-p_0\right|_{S_{0i}}=0;\quad i=1,2, \tag{4}$$

On the free surfaces, the following formulae for p_i $(i=1,2)$ are valid:

$$p_1-p_0=-\rho_1\left.\left(\frac{\partial\Phi}{\partial t}+g\zeta_1\right)\right|_{S_{01}};\quad p_2-p_0=-\rho_2\left.\left(\frac{\partial\Phi}{\partial t}+g\zeta_2\right)\right|_{S_{02}}.$$

Here g is the gravity acceleration.

Denote by $\Phi^{(i)}$ values of the velocity potential Φ at points $P\in\partial\Sigma_i\ (i=1,2)$.

Equations (1) and (2) are solved simultaneously using the shell fixation conditions relative to **U**, boundary conditions (3) and (4), and the following expressions for dynamical components of the liquid pressure on elastic walls:

$$p_l=(\mathbf{P},\mathbf{n})=\begin{cases}-\rho_1\dfrac{\partial\Phi^{(1)}}{\partial t}; & P\in S_{w1}\cup S_{\text{bot}};\\[2mm] -\rho_2\dfrac{\partial\Phi^{(2)}}{\partial t}; & P\in S_{w2}\setminus\Sigma_{12};\\[2mm] -\left(\rho_1\dfrac{\partial\Phi^{(1)}}{\partial t}-\rho_2\dfrac{\partial\Phi^{(2)}}{\partial t}\right); & P\in\Sigma_{12}.\end{cases}$$

To define modes of shell vibrations coupled with liquid sloshing, we will represent displacements of the fluid-filled tank as $\mathbf{U}=\mathbf{u_f}\times\exp(i\omega t)$. Here ω and $\mathbf{u_f}$ are natural frequencies and vibration modes of the fluid-filled shell structure.

3 THE MODE SUPERPOSITION METHOD FOR COUPLED DYNAMIC PROBLEMS

Consider the vibration modes of the fluid-filled tank in a form

$$\mathbf{U}=\sum_{k=1}^{N} c_k \mathbf{u}_k, \tag{5}$$

where $c_k = c_k(t)$ are unknown coefficients, and $\mathbf{u}_k$ are eigenmodes of the empty tank. In other words, the mode of vibration of the fluid-filled tank is determined as a linear combination of eigenmodes of the empty shell structure. Note that the following relationships are fulfilled [7]

$$\mathbf{L}(\mathbf{u}_k)=\Omega_k^2\mathbf{M}(\mathbf{u}_k), \quad (\mathbf{M}(\mathbf{u}_k),\mathbf{u}_j)=\delta_{kj} \ . \tag{6}$$

Hence

$$(\mathbf{L}(\mathbf{u}_k),\mathbf{u}_j)=\Omega_k^2\delta_{kj} \ , \tag{7}$$

where Ω_k is the k-th frequency of the empty tank vibrations. Equations (6) and (7) show that the abovementioned vibration modes have to be orthonormalized with respect to the mass matrix.

Let also introduce $\mathbf{u}_k^i$ as follows:

$$\mathbf{u}_k = \begin{cases} \mathbf{u}_k^1; & P\in\Sigma_1, \\ \mathbf{u}_k^2; & P\in\Sigma_2. \end{cases}$$

Consider $\mathbf{\Phi}$ as a sum of two potentials $\Phi=\Phi_1+\Phi_2$as it was done by Degtyarev *et al.* [9]. Represent potential $\mathbf{\Phi}_1$ as the following series:

$$\Phi_1=\sum_{k=1}^{N}\dot{c}_k(t)\phi_{1k}. \tag{8}$$

Here the time-dependant coefficients $c_k(t)$ are defined in eqn (5). To determine functions φ_{1k} $\left(k=\overline{1,N}\right)$, we have the following boundary value problems:

$$\nabla^2\phi_{1k}^i=0; \quad P\in\Sigma_i; \quad \left.\frac{\partial\phi_{1k}^i}{\partial\mathbf{n}}\right|_{S_{wi}}=w_k^i; \quad \left.\frac{\partial\phi_{1k}^1}{\partial\mathbf{n}}\right|_{S_{\text{bot}}}=w_k^i; \quad \phi_{1k}^i\Big|_{S_{i0}}=0. \tag{9}$$

Here, w_k^1, w_k^2 are normal components of the mode $\mathbf{u}_k$ in the first and second compartments, $w_k^i=\left(\mathbf{u}_k^i,\mathbf{n}\right)$. So functions ϕ_{1k}^1 and ϕ_{1k}^2 are solutions of problems (9) for $i=1$ and $i=2$, accordingly, as in Ref. [10].

To determine potential $\mathbf{\Phi}_2$, we must solve the boundary value problems of fluid vibrations in two disconnected compartments with rigid walls. These problems are solved separately for spherical and cylindrical parts. To obtain the sloshing modes, the following sequence of boundary value problems for auxiliary functions ψ_{2k}^i $(i=1,2)$ is formulated:

$$\Delta\psi^i_{2k}=0;\quad \left.\frac{\partial\psi^i_{2k}}{\partial\mathbf{n}}\right|_{S_{wi}}=0;\quad \left.\frac{\partial\psi^1_{2k}}{\partial\mathbf{n}}\right|_{S_{\text{bot}}}=0;$$

$$\left.\frac{\partial\psi^i_{2k}}{\partial\mathbf{n}}\right|_{S_{0i}}=\frac{\partial\varsigma_i}{\partial t};\quad \left.\frac{\partial\psi^i_{2k}}{\partial t}+g\zeta_i\right|_{S_{0i}}=0,\ k=\overline{1,N};\quad i=1,2.$$

Suppose hereinafter $\psi_{2k}(t,x,y,z)=e^{i\chi_k t}\phi_{2k}(x,y,z)$. Then we obtain the following eigenvalue problems [9, 10] for each $\phi^i_{2k}\ (i=1,2)$ in the liquid domains $\Sigma_i(i=1,2)$:

$$\Delta\phi^i_{2k}=0\Big|_{\Sigma_i};\ \left.\frac{\partial\phi^i_{2k}}{\partial\mathbf{n}}\right|_{S_{wi}}=0;\quad \left.\frac{\partial\phi^1_{2k}}{\partial\mathbf{n}}\right|_{S_{\text{bot}}}=0;\quad \left.\frac{\partial\phi^i_{2k}}{\partial\mathbf{n}}=\frac{\chi_k^2}{g}\phi^i_{2k}\right|_{S_{0i}};\quad \int_{S_{0i}}\phi^i_{2k}dS=0. \tag{10}$$

So for potentials Φ_1 and Φ_2 we obtain the next representations:

$$\Phi_1=\begin{cases}\sum_{k=1}^{M_1}c_k\phi^1_{1k}, & P\in\Sigma_1,\\ \sum_{k=1}^{M_2}c_k\phi^2_{1k}, & P\in\Sigma_2,\end{cases};\Phi_2=\begin{cases}\sum_{k=1}^{L_1}\dot{d}_k\phi^1_{2k}, & P\in\Sigma_1,\\ \sum_{k=1}^{L_2}\dot{d}_k\phi^2_{2k}, & P\in\Sigma_2.\end{cases} \tag{11}$$

Here $c_k(t), d_k(t)$ are unknown time-dependant coefficients.

The effective numerical procedure for solution of eigenvalue problems (10) using the single and multi-domain boundary element methods has been introduced in Refs. [7, 10].

Thus, the problem under consideration involves the following steps. First, it is necessary to obtain the sloshing frequencies and modes ϕ^i_{2k} using rigid wall assumption. Second, we obtain the natural frequencies Ω_k and modes $\mathbf{u}_k$ of the empty elastic tank. Then we define the free vibration frequencies and modes ϕ^i_{1k} of the elastic tank without considering effects of sloshing. When functions ϕ^i_{1k} and ϕ^i_{2k} are defined, we substitute them in eqns (11) and (1) and obtain the system of ordinary differential equations as it was done in Ref. [10]. Finally, the flow-induced vibrations of elastic structures are studied.

To define functions ϕ^i_{1k} and ϕ^i_{2k} we use the boundary element method in its direct formulation [1]. Dropping indices i, $1k$ and $2k$ one can obtain the main integral equation in the following form:

$$2\pi\phi(P_0)=\iint_S q\frac{1}{|P-P_0|}dS-\iint_S\phi\frac{\partial}{\partial\mathbf{n}}\frac{1}{|P-P_0|}dS. \tag{12}$$

Here, $S=\sigma\cup S_0$, points P and P_0 belong to the surface S. The value $|P-P_0|$ represents Cartesian distance between the points P and P_0. In doing so, the function φ defined on the wetted tank surface σ presents the pressure, and the function q defined on the free surface S_0, is the flux, $q=\partial\phi/\partial\mathbf{n}$. To apply the MBEM, we introduce the artificial interface surface S_{int} [11]–[12]. In MBEM, the computational domain is divided into a number of subdomains, and the BEM algebraic equations are established for each subdomain. Then the global system of algebraic equations is formed by assembling results of all subdomains in terms of the equilibrium and matching conditions over common interface nodes. This system has a blocked and sparse character.

The basic procedure is to start with the standard boundary integral equation for potential (12), replace Cartesian coordinates (*x, y, z*) with cylindrical ones (*r, θ, z*), and integrate with respect to z and θ. We use furthermore the cylindrical coordinate system, and represent unknown functions as Fourier series by the circumferential coordinate θ

$$w_k^i(r,z,\theta)=w_k^i(r,z)\cos\alpha\theta;\quad \phi_{jk}^i(r,z,\theta)=\phi_{jk}^i(r,z)\cos\alpha\theta;\, i=1,2;\, j=1,2;\, k=1,2,..., \tag{13}$$

where α is a given integer (the number of nodal diameters). In this case, the solution is independent of the angular coordinate θ, and the three-dimensional problem is reduced to a two-dimensional one in the radial coordinate r and the axial coordinate z.

Let Γ be a generator of the surface σ. Using (12), (13) we have obtained the following system of singular integral equations for unknown functions φ and q in problem (9):

$$2\pi\phi(z_0)+\int_\Gamma\phi(z)Q(z,z_0)r(z)d\Gamma-\int_0^R q(\rho)\Psi(P,P_0)\rho d\rho=\int_\Gamma w(z)\Psi(P,P_0)r(z)d\Gamma_1;P_0\in\sigma;$$

$$\int_\Gamma\phi(z)Q(z,z_0)r(z)d\Gamma-\int_0^R q(\rho)\Psi(P,P_0)\rho d\rho=\int_\Gamma w(z)\Psi(P,P_0)r(z)d\Gamma_1;\; P_0\in S_0.$$

Here

$$Q(z,z_0)=\frac{4}{\sqrt{a+b}}\left\{\frac{1}{2r}\left[\frac{r^2-r_0^2+(z_0-z)^2}{a-b}\mathsf{E}_\alpha(k)-\mathsf{F}_\alpha(k)\right]n_r+\frac{z_0-z}{a-b}\mathsf{E}_\alpha(k)n_z\right\};$$

$$\Psi(P,P_0)=\frac{4}{\sqrt{a+b}}\mathsf{F}_\alpha(k);\,\mathsf{E}_\alpha(k)=(-1)^\alpha(1-4\alpha^2)\int_0^{\pi/2}\cos2\alpha\psi\sqrt{1-k^2\sin^2\psi}\,d\psi;$$

$$\mathsf{F}_\alpha(k)=(-1)^\alpha\int_0^{\pi/2}\frac{\cos2\alpha\psi d\psi}{\sqrt{1-k^2\sin^2\psi}};\, a=\rho^2+\rho_0^2+(z-z_0)^2;\;\; b=2\rho\rho_0;k^2=\frac{2b}{a+b}.$$

The system of singular integral equations for mixed boundary value problem (10) has been obtained in Ref. [10]. Numerical procedures to solve these problems based on the boundary element method are described in details in the papers [7, 9, 11].

4 SOME NUMERICAL RESULTS

4.1 Low frequency sloshing modes for partially filled spherical shells

4.1.1 Spherical shells without baffles

Now we turn to the problem of sloshing in the spherical container without baffles. Radius of the sphere is R_1=1m; the filling level is h_1.

The numerical simulation has been provided for different values of h_1 $(0.2<h_1/R_1<1.99)$ and different modes α $(\alpha=\overline{0,3})$. Both SBEM and MBEM are applied here. The boundary elements with constant approximation of unknowns inside elements are used. In SBEM there are 200 elements along the spherical surface and 150 elements along the free surface. In MBEM we divide the computational domain into two parts by an artificial interface surface at $h_{\text{int}}=0.5h_1$ using 100 boundary elements in each subdomain along the spherical surface and 150 elements along the free surface.

We use practically the same mesh to find a numerical approximation of low eigenvalues for the so-called 'ice-fishing problem'. In this problem, formally, we should consider an infinitely wide and deep ocean covered with ice, with a small round fishing hole. Sloshing in such 'containers' was studied by McIver [12]. We approximate this infinite case using the spherical tank with the small round hole on its top. It allows us to compare our numerical results with those obtained in the papers [12–14].

Proposed boundary element approach gives nearly the same values as the other authors. In Tables 1 and 2 we compare our results obtained by using SBEM and MBEM with those obtained in Refs. [12–14] for axisymmetric ($\alpha = 0$) and non-axisymmetric ($\alpha = 1$) modes. Four first frequencies ($m = \overline{1,4}$) are evaluated for each α. Here we consider different filling levels h_1. The value $h_1/R_1 = 1.99$ corresponds to the ice-fishing problem. The results of Faltinsen and Timokha [13], and Kulczycki *et al.* [14] are very close. Both Refs. [13–14] and [12] results analytically approximate the sloshing eigenvalues of the spherical tank.

It should be noted that results obtained by SBEM are more precise than MBEM ones, but the matrix size in SBEM is twice larger compared with MBEM. If un-baffled tanks are at low filling levels, it is preferable to use SBEM. The numerical analysis demonstrates that the lowest liquid sloshing frequency occurs for mode $\alpha = 1$.

Considering our approximate natural sloshing modes one can observe how free surface profiles change with the liquid depth. These results are illustrated in Fig. 2 for the three lowest eigenvalues of the mode $\alpha = 1$. Here numbers 1, 2, 3, 4 correspond to the different non-dimensional filling levels: $h_1/R_1 = 1.0$; 0.2; 1.8; 1.9, respectively.

In the spherical tank with $0 < h_1/R_1 < 0.5$ the lowest mode presents a spatial wave pattern that look like inclination of an almost flat free surface. Increasing the liquid depth yields more

Table 1: Axisymmetric slosh frequencies of the fluid-filled spherical shell, Hz.

		Filling level h_1, m				
m	Method	$h_1 = 0.2$	$h_1 = 0.6$	$h_1 = 1.0$	$h_1 = 1.8$	$h_1 = 1.99$
1	[13,14],	3.8261	3.6501	3.7451	6.7641	29.0500
	[12]	3.8261	3.6501	3.7451	6.7641	29.2151
	MBEM	3.4034	3.5455	3.7294	6.6098	30.7081
	SBEM	3.8314	3.6510	3.7456	6.7665	29.1811
2	[13,14],	9.2561	7.2659	6.9763	12.1139	51.8122
	[12]	9.2561	7.2659	6.9763	12.1139	52.0467
	MBEM	9.2636	7.2893	6.9796	12.0008	52.9393
	SBEM	9.2686	7.2684	6.9780	12.1205	52.0255
3	[13,14],	14.7556	10.7443	10.1474	17.3960	74.2909
	[12]	14.7556	10.7443	10.1474	17.3960	74.5537
	MBEM	14.9214	10.7483	10.1496	17.3136	75.3139
	SBEM	14.7763	10.7502	10.1512	17.4086	74.5547
4	[13,14],	20.1187	14.1964	13.3041	22.6579	96.6207
	[12]	20.1187	14.1964	13.3041	22.6570	96.9560
	MBEM	20.2066	14.2023	13.3083	22.5962	97.7771
	SBEM	20.1498	14.2056	13.3110	22.6777	96.9021

Table 2: Non-axisymmetric slosh frequencies of the fluid-filled spherical shell, Hz.

		Filling level h_1, m				
m	Method	$h_1 = 0.2$	$h_1 = 0.6$	$h_1 = 1.0$	$h_1 = 1.8$	$h_1 = 1.99$
1	[13,14],	1.0723	1.2625	1.5601	3.9593	18.9838
	[12]	1.0723	1.2625	1.5601	3.9593	19.1582
	MBEM	1.1034	1.2777	1.5638	3.9606	19.1603
	SBEM	1.0723	1.2626	1.5603	3.9508	19.1130
2	[13,14],	6.2008	5.3860	5.2755	9.4534	41.3491
	[12]	6.2008	5.3860	5.2755	9.4534	41.7683
	MBEM	6.1227	5.3534	5.2749	9.4582	41.5327
	SBEM	6.2090	5.3697	5.2764	9.4538	41.5333
3	[13,14],	11.8821	8.9418	8.5044	14.7548	63.5354
	[12]	11.8821	8.9418	8.5044	14.7548	64.0323
	MBEM	11.9650	8.9529	8.5062	14.7648	63.9483
	SBEM	11.8981	8.9429	8.5069	14.7574	63.8783
4	[13,14],	17.3581	12.4234	11.6835	20.0224	85.9166
	[12]	17.3584	12.4234	11.6835	20.0224	86.3001
	MBEM	17.4540	12.4276	11.6863	20.0394	86.2972
	SBEM	17.3842	12.4291	11.6884	20.0278	86.2034

complicated free surface profiles. Figure 3 demonstrates the spatial wave patterns for $\alpha = 1$, $m = 1, 2, 3$ at $h_1/R_1 = 1.8$.

4.1.2 Spherical shells with baffles

Consider the rigid spherical tank of radius $R_1 = 1$ m filled to the depth $h_1 = 1.4$ m. The inner periphery of the tank contains a thin rigid-ring baffle. The baffle position is $h_{\text{baf}} = 1$ m. The different annular orifices in the baffle are considered. Radii of these orifices are radii R_{int} of the interface surfaces. The first four frequencies for mode $\alpha = 1$ are evaluated for radii $R_{\text{int}} = 1.0$ m, $R_{\text{int}} = 0.7$ m, and $R_{\text{int}} = 0.2$ m. Note that $R_{\text{int}} = 1.0$ m correspond to the un-baffled tank. The frequencies are presented in Table 3.

The results show that frequencies decrease as radius of the baffle orifice is decreased. Only the lowest eigenvalue is essentially influenced by installing the baffles.

Figure 4 demonstrates the first non-axisymmetric modes of liquid vibrations in spherical tanks with and without baffles. When the baffle is installed, the mode shape becomes almost flat.

4.2 Sloshing frequencies for partially filled cylindrical tanks with covered free surfaces

Consider eigenvalue problem (10) for the second fluid domain Σ_2 bounded by cylindrical and spherical surfaces (Fig. 1). By H denote a height of the cylinder, by R_1, R_2 radii of the spherical and cylindrical parts, and by h_2 the filling level in Σ_2. Let $(0, z_s)$ be centre coordinates of the sphere. The shell structure has following dimensions: $R_1 = 1$ m; $R_2 = 1.2$ m; $z_s = 1.5$ m; $H = 2$.m. The modes and frequencies are evaluated for different h_2 in the range of

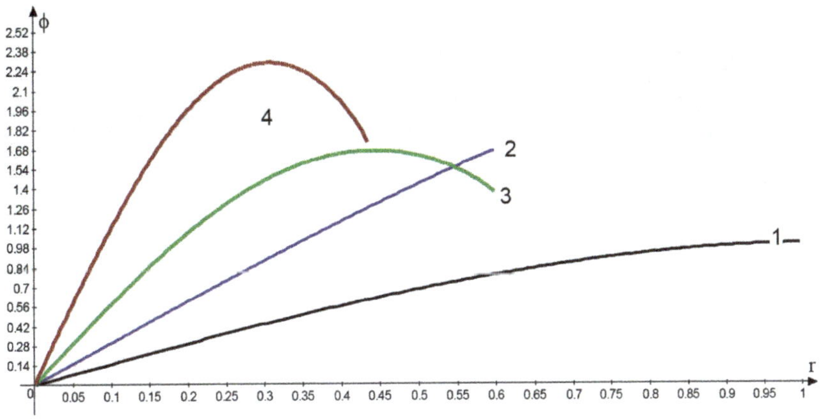

a) $\alpha = 1$, m=1

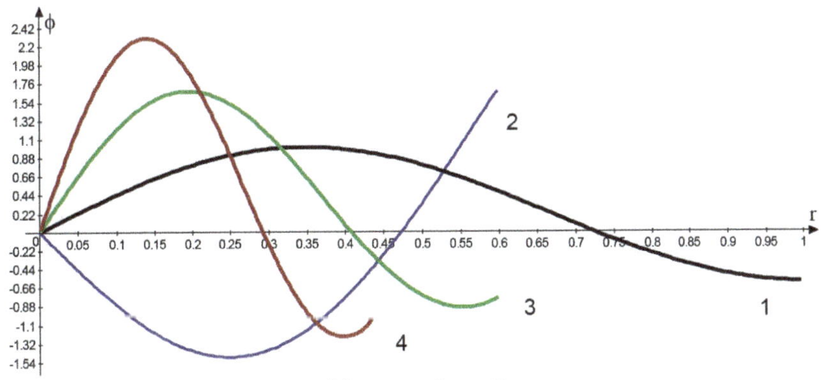

b) $\alpha = 1$, m=2

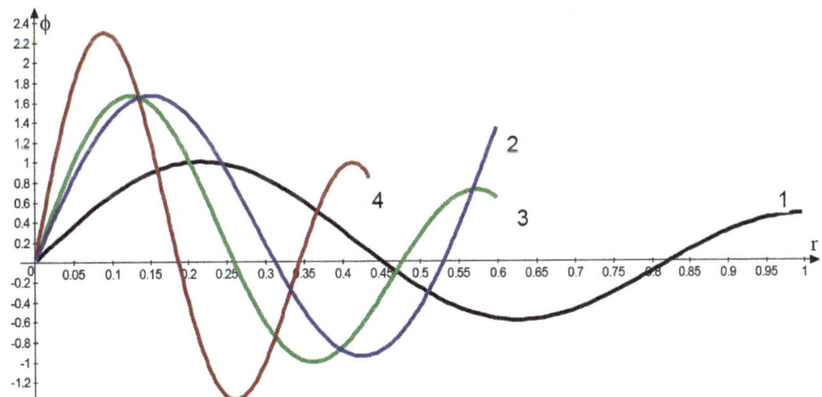

c) $\alpha = 1.m$=3

Figure 2: The radial wave profiles m = 1, 2, 3 for different non-dimensional liquid depths h_1 / R_1.

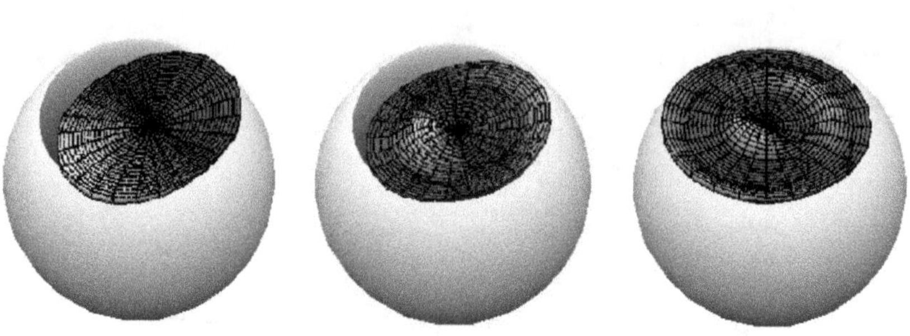

Figure 3: Spatial wave patterns for $\alpha = 1$; $m = 1, 2, 3$.

Table 3: Sloshing frequencies of the fluid-filled spherical shell with baffles, Hz.

	ω^2/g		
m	$R_{int} = 1.0$ m	$R_{int} = 0.7$ m	$R_{int} = 0.2$ m
1	2.1232	2.0435	1.4234
2	5.9800	5.9723	5.8405
3	9.4789	9.4785	9.4567
4	12.9431	12.9430	12.9358

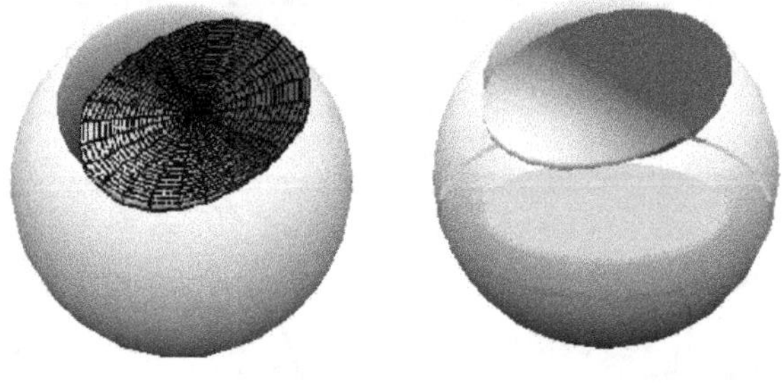

Figure 4: Spatial mode shapes for $\alpha = 1$; $m = 1$ of un-baffled a) and baffled tanks b).

$0.5 < h_2 / R_1 < 2.39$. The inner radius R_{ir} of the ring free surface is defined by the dive level of the spherical shell z_s and the filling level h_2 as follows: $R_{ir} = \sqrt{R_1^2 - (h_2 - z_s)^2}$.

The results are illustrated in Fig. 5 for two lowest eigenvalues of the mode $\alpha = 1$. Here numbers 1, 2 correspond to frequencies of cylindrical tanks with ring and circular free surfaces, respectively. If $h_2 = 0.5$ m, then both tanks have the complete circular free surfaces.

Figure 5a) and b) demonstrates changes in the first and second frequencies of the mode $\alpha = 1$ via the filling level h_2. The values of frequencies increase significantly with increasing the inner radius of the ring free surface. The maximum values of frequencies correspond to $h_2 = z_s$; $R_{ir} = R_1$. Here we get the analogy with the "ice-fishing problem" for the spherical shell with $h_1/R_1 = 1.99$ (see Tables 1 and 2). Sloshing frequencies of cylindrical shell with circular free surface are stabilized when $h_2 > z_s$; one can observe here the typical monotonic

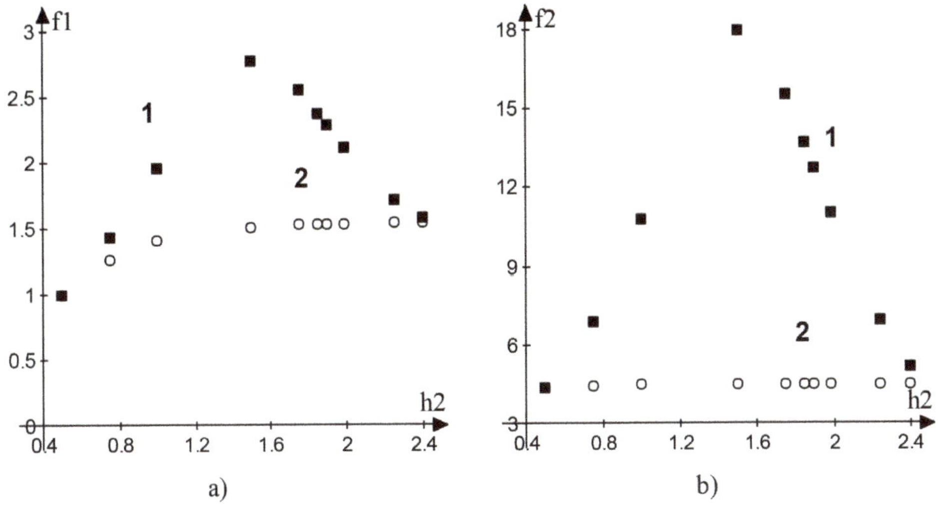

Figure 5: Frequencies of cylindrical shell with ring and circular free surfaces.

dependences. The benefit of a partly covered free surface is that it increases the natural frequencies.

4.3 Modes and frequencies of the empty cylindrical–spherical tank

Consider the elastic spherical-cylindrical tank without baffles (Fig. 1). The shell structure is clamped at the bottom edge. Natural frequencies Ω_k and modes $\mathbf{u}_k$, $k = \overline{1,N}$ of the shell structure without the liquid are calculated using FEM as it is proposed in Ref. [7]. The shell structure has the following dimensions: $R_1 = 1$ m, $R_2 = 1.2$ m, $z_s = 1.5$ m, $H = 2.0$ m, wall thickness $h = 0.025$ m, and material properties: Young's modulus $E = 2{\cdot}10^5$ MPa, Poisson's ratio $\nu = 0.3$, density $\rho_s = 2700$ kg/m^3. Four modes of vibration are shown in Fig. 6.

The modes 6a) and 6b) correspond to the duplicate frequency $\Omega_1 = \Omega_2 = 4.6359$ Hz. This frequency is the lowest one of the mode $\alpha = 1$. The axisymmetric mode 6c) corresponds to the frequency $\Omega_3 = 5.9312$ Hz, that is the lowest one of the mode $\alpha = 0$. The axisymmetric mode 6d) corresponds to the frequency $\Omega_4 = 50.9721$ Hz.

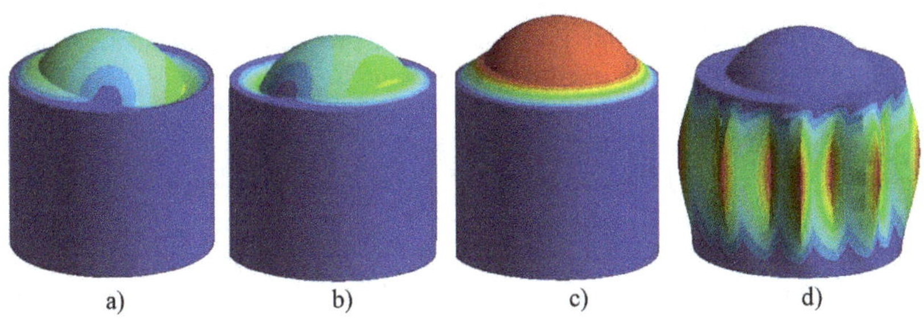

Figure 6: Modes of the empty cylindrical–spherical tank.

Table 4: Frequencies of the fluid-filled double tank, Hz.

m	ω	Dominant vibration	a
1	0.6223	Sphere sloshing	1
2,3	0.6359	Sphere wall	1
4	0.8294	Cylinder sloshing	1
5	0.9635	Sphere sloshing	0
22	5.0785	Cylinder wall	0

4.4 Modes and frequencies of the fluid-filled cylindrical–spherical tank

To define coupled modes of harmonic vibrations of the elastic tank, we represent the time-dependant unknown coefficients in series for potentials Φ_1,Φ_2, eqn (11), and for vibration modes **u** of the fluid-filled shell, eqn (5), in the form

$$c_k(t) = C_k e^{i\omega t}; \quad d_k(t) = D_k e^{i\omega t}, \tag{14}$$

where ω is an own frequency, C_k and D_k are unknown constants. Using (14) we obtain the eigenvalue problem for evaluating natural frequencies and modes of the fluid-filled cylindrical-spherical tank [9].

The filling levels are h_1 =1 m and h_2 = 1.5 m. The lowest frequencies of the double tank are presented in Table 4.

The first frequencies are very close in value. For example, modes 1–5 have frequencies with difference within 1 Hz, and these modes are associated with different vibration types. The first twenty modes are predominantly sloshing ones.

5 CONCLUSIONS

The numerical procedure based on the coupling finite and boundary element methods is developed for the fluid-structure interaction analysis of the dual compartment tank. The considered problem has been solved using the multi-domain and single-domain boundary element approaches. The analysis demonstrates that sloshing and shell vibrations can not be considered separately. The lower frequencies of the fluid-filled tanks decrease to 0.2–0.4 and 0.1–0.6 of those of empty tanks for spherical and cylindrical compartments, respectively. The fluid-stricture interaction effects are more significant for the outer cylindrical part.

ACKNOWLEDGMENTS

The authors gratefully acknowledge Professor Carlos Brebbia, Wessex Institute of Technology, for his constant support and interest to our research.

REFERENCES

[1] Brebbia, C.A, Telles, J.C.F. & Wrobel, L.C., *Boundary Element Techniques: Theory and Applications in Engineering*. Springer-Verlag: Berlin and New York, 1984.

[2] Wang, J. & Gao, X.W., Structural multi-scale boundary element method based on element subdivision technique. *Chinese Journal of Computational Mechanics*, **27**(2), pp. 258–263, 2010.

[3] Biswal, K.C., Bhattacharyya, S.K. & Sinha, P.K., Dynamic characteristics of liquid filled rectangular tank with baffles. *IE (I) Journal-CV,* **84**, pp. 145–148, 2004.
[4] Kumar, A. & Sinhamahapatra K.P., Dynamics of rectangular tank with perforated vertical baffle. *Ocean Engineering,* **126**(1), pp. 384–401, 2016. https://doi.org/10.1016/j.oceaneng.2016.09.012
[5] Faltinsen, O.M. & Timokha, A.N., *Sloshing*. Cambridge University Press, Cambridge, 2009.
[6] Curadelli, O., Ambrosini, D., Mirasso, A. & Amani, M., Resonant frequencies in an elevated spherical container partially filled with water: FEM and measurement. *Journal of Fluids and Structures*, **26**(1), pp. 148–159, 2010. https://doi.org/10.1016/j.jfluidstructs.2009.10.002
[7] Ravnik, J., Strelnikova, E., Gnitko V., Degtyarev, K. & Ogorodnyk,U., BEM and FEM analysis of fluid-structure interaction in a double tank. *Engineering Analysis with Boundary Elements*, **67**, pp. 13–25, 2016. https://doi.org/10.1016/j.enganabound.2016.02.006
[8] Wan, F.Y.M. & Weinitschke, H.J., On shells of revolution with Love-Kirchoff hypothesis. *Journal of Engineering Mathematics*, **22**, pp. 285–334, 1988. https://doi.org/10.1007/bf00058512
[9] Degtyarev, K., Glushich, P., Gnitko, V. & Strelnikova, E., Numerical simulation of free liquid-induced vibrations in elastic shells. *International Journal of Modern Physics and Applications,* **1**(4), pp. 159–168, 2015.
[10] Gnitko, V., Naumenko, V., Rozova, L. & Strelnikova, E., Multi-domain boundary element method for liquid sloshing analysis of tanks with baffles. *Journal of Basic and Applied Research International*, **17**(1), pp. 75–87, 2016.
[11] Gnitko, V., Degtyariov, K., Naumenko, V. & Strelnikova, E., BEM and FEM analysis of the fluid-structure Interaction in tanks with baffles. *International Journal of Computational Methods and Experimental Measurements*, **5**(3), pp. 317–328, 2017. https://doi.org/10.2495/cmem-v5-n3-317-328
[12] McIver, P., Sloshing frequencies for cylindrical and spherical containers filled to an arbitrary depth. *Journal of Fluid Mechanics*, **201**, pp. 243, 1989. https://doi.org/10.1017/s0022112089000923
[13] Faltinsen, O.M. & Timokha, A.N., Analytically approximate natural sloshing modes for a spherical tank shape. *Journal of Fluid Mechanics*, **703**, pp. 391–401, 2012. https://doi.org/10.1017/jfm.2012.237
[14] Kulczycki, T., Kwaśnicki, M. & Siudeja, B., The shape of the fundamental sloshing mode in axisymmetric containers. *Journal of Engineering Mathematics*, **99**(1), pp. 157–193, 2016. https://doi.org/10.1007/s10665-015-9826-6

MESHLESS LARGE PLASTIC DEFORMATION ANALYSIS CONSIDERING WITH A FRICTION COEFFICIENT BY TRIPLE-RECIPROCITY BOUNDARY ELEMENT METHOD

YOSHIHIRO OCHIAI
Department of Mechanical Engineering, Kindai University, Japan.

ABSTRACT
In general, internal cells are required to solve large deformation problems using a conventional boundary element method (BEM). However, in this case, the merit of BEM, which is the ease of data preparation, is lost. Triple-reciprocity BEM enables us to solve elastoplasticity problems with a small plastic deformation. In this study, it is shown that two-dimensional large plastic deformation problems with a friction coefficient can be solved without the use of internal cells, by the triple-reciprocity BEM. Initial stress and strain formulations are adopted and the initial stress or strain distribution is interpolated using boundary integral equations. In this method, only boundary elements are remeshed. A new computer program is developed and used to solve several problems.
Key words: BEM; large plastic deformation; initial stress method; numerical analysis; strain hardening; thin plate spline

1 INTRODUCTION

The finite element method (FEM) requires remeshing several times during large-plastic-deformation analysis. Elastoplastic problems can be solved by a conventional boundary element method (BEM) using internal cells for domain integrals [1, 2]. In this case, however, the merit of BEM, which is ease of data preparation, is lost. On the other hand, several countermeasures have been considered. Ochiai has proposed the triple-reciprocity BEM without the use of internal cells for elastoplastic problems [3]. With this method, a highly accurate solution can be obtained using only fundamental solutions of a low order and by reducing the need for data preparation. Ochiai and Kobayashi applied triple-reciprocity BEM without internal cells to two-dimensional elastoplastic problems using initial stress and strain formulations [3, 4]. However, the weak point in our previous studies was that the numerical examples are elastoplasticity problems with a small plastic deformation and too simple to substantiate the theory.

In this study, triple-reciprocity BEM is applied to large-plastic-deformation problems. The initial stress and strain formulations are adopted and the theory is expressed using a few fundamental solutions. In this method, only boundary elements are remeshed. Arbitrary distributions of the initial stress or strain for elastoplastic analysis are interpolated using boundary integral equations and internal points. This interpolation corresponds to a thin plate spline. In this method, strong singularities in the calculation of stresses at internal points become weak. A new computer program was developed and applied to several elastoplastic problems to clearly demonstrate the theory. A constrained upsetting problem can be easily solved without locking and mesh rezoning by the present method.

2 THEORY

2.1 Initial stress formulation

To analyze large-plastic-deformation problems using the initial stress formulation, the following boundary integral equation must be solved [1, 2].

DOI: 10.2495/CMEM-V6-N6-989-999

$$c_{ij}\dot{u}_j = \int_\Gamma [u_{ij}^{[1]}\dot{p}_j - p_{ij}\dot{u}_j]d\Gamma + \int_\Omega \varepsilon_{ijk}^{[1]}\dot{\sigma}_{ljk}^{[1]}d\Omega \tag{1}$$

Here, $\dot{\sigma}_{ljk}^{[1]}$ is the initial stress rate and c_{ij} is the free coefficient. u_j and p_j are the jth component of the displacement rate and surface traction rate, respectively. Upper notation [1] is a conventional term. Γ and Ω are the boundary and domain, respectively. As shown in eqn (1), when there is an arbitrary initial stress rate, a domain integral becomes necessary. Denoting the distance between the observation point and the loading point as r, Kelvin's solution $u_{ij}^{[1]}$ and p_{ij} are given by

$$u_{ij}^{[1]} = \frac{1}{8\pi(1-\nu)G}[(3-4\nu)\delta_{ij}\ln(\frac{1}{r}) + r,_i r,_j] \tag{2}$$

$$p_{ij} = \frac{-1}{4\pi(1-\nu)r}\{[(1-2\nu)\delta_{ij} + 2r,_i r,_j]\frac{\partial r}{\partial n} - (1-2\nu)(r,_i n_j - r,_j n_i)\}, \tag{3}$$

where ν is Poisson's ratio and G is the shear modulus. The ith component of a unit normal vector is denoted by n_i. Moreover, let us set $\nu'=\nu/(1+\nu)$ for the plane stress and $r,_i = fr/fx_i$. For large deformation Poisson's ratio ν approximately 0.5 is assumed. The function $\varepsilon_{ijk}^{[1]}$ in eqn (1) is given by [1]

$$\varepsilon_{ijk}^{[1]} = [(1-2\nu)(\delta_{ij}r,_k + \delta_{ik}r,_j) - \delta_{jk}r,_i + 2r,_i r,_j r,_k]\frac{-1}{8\pi(1-\nu)Gr}. \tag{4}$$

2.2 Interpolation of initial stress

Interpolation using boundary integrals is necessary to avoid the domain integral in eqn (1). The distribution of the initial stress $\dot{\sigma}_{ljk}^{[1]}$ in the case of a two-dimensional problem is interpolated using the integral equation to transform the domain integral into a boundary integral. The following equations are used for interpolation [5–10]:

$$\nabla^2\dot{\sigma}_{ljk}^{[1]S} = -\dot{\sigma}_{ljk}^{[2]S}, \tag{5}$$

$$\nabla^2\dot{\sigma}_{ljk}^{[2]S} = -\sum_{m=1}^{M}\dot{\sigma}_{ljk(m)}^{[3]P}, \tag{6}$$

where $\nabla^2 = \partial^2/\partial x^2 + \partial^2/\partial y^2$. From eqns (5) and (6), we obtain

$$\nabla^4\dot{\sigma}_{ljk}^{[1]S} = \sum_{m=1}^{M}\dot{\sigma}_{ljk(m)}^{[3]P}. \tag{7}$$

This equation is the same type of equation as for the displacement ($\dot{\sigma}_{ljk}^{[1]}$) of a fictitious thin plate with point loads ($\dot{\sigma}_{ljk}^{[3]P}$), as shown in Fig. 1. The $\dot{\sigma}_{ljk}^{[1]}$ values at several points and on the boundary are given, and the unknown point load P is obtained inversely [4, 7]. The term $\dot{\sigma}_{ljk}^{[2]S}$ corresponds to the sum of the curvatures of $\dot{\sigma}_{ljk}^{[1]}$. This fictitious thin plate is simply supported, and the $\dot{\sigma}_{ljk}^{[2]S}$ values on the boundary are zero. We emphasize that eqns (5) and (6) can be used to interpolate the complicated distribution of the initial stress $\dot{\sigma}_{ljk}^{[1]}$. These equations are the same as those used to generate a free-form surface using an integral equation [6, 7]. In this study, the distribution of the initial stress is assumed to be represented by a 2.5-dimensional free-form surface. In this method, each component of the initial stress $\dot{\sigma}_{ljk}^{[1]}$ (j,k=1,2) is interpolated.

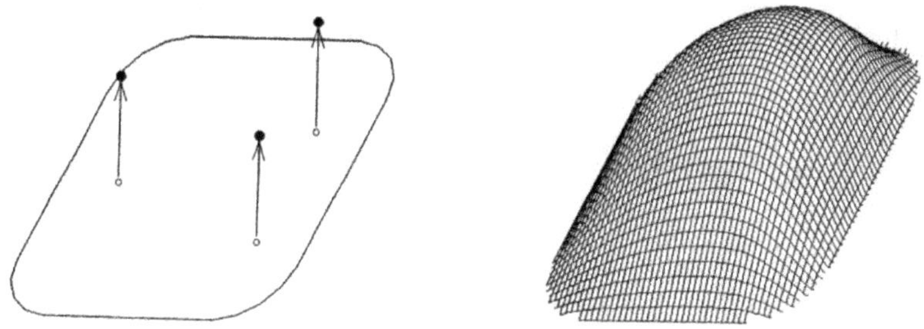

(a) Given internal points and boundary (b) Obtained surface $\dot{\sigma}_{Ijk}^{[1]}$

Figure 1: Fictitious thin plate with unknown point loads used for interpolation.

2.3 Representation of initial stress by integral equation

The distribution of the initial stress is represented by an integral equation. The harmonic function $T^{[1]}$ and biharmonic function $T^{[2]}$ are generally given by

$$T^{[1]} = \frac{1}{2\pi}[\ln(\frac{1}{r}) + B]. \tag{8}$$

$$T^{[2]} = \frac{r^2}{8\pi}[\ln(\frac{1}{r}) + B + 1], \tag{9}$$

where B is an arbitrary constant. Denoting the number of points $\dot{\sigma}_{Ijk}^{[3]P}$ as M, the curvature of the initial stress rate $\dot{\sigma}_{Ijk}^{[2]S}$ is given by Green's second identity and eqn (8) as [4–7, 11–13]

$$c\dot{\sigma}_{Ijk}^{[2]S}(P) = \int_\Gamma \{T^{[1]}(P,Q)\frac{\partial \dot{\sigma}_{Ijk}^{[2]S}(Q)}{\partial n} - \frac{\partial T^{[1]}(P,Q)}{\partial n}\dot{\sigma}_{Ijk}^{[2]S}(Q)\}d\Gamma + \sum_{m=1}^{M} T^{[1]}\dot{\sigma}_{Ijk(m)}^{[3]P}. \tag{10}$$

The initial stress rate $\dot{\sigma}_{Ijk}^{[1]}$ is given by Green's theorem and eqns (7) and (8) as [4–7]

$$c\dot{\sigma}_{Ijk}^{[1]S}(P) = -\sum_{f=1}^{2}(-1)^f \int_\Gamma \{T^{[f]}(P,Q)\frac{\partial \dot{\sigma}_{Ijk}^{[f]S}(Q)}{\partial n} - \frac{\partial T^{[f]}(P,Q)}{\partial n}\dot{\sigma}_{Ijk}^{[f]S}(Q)\}d\Gamma$$

$$-\sum_{m=1}^{M} T^{[2]}\dot{\sigma}_{Ijk(m)}^{[3]P} \tag{11}$$

where c =0.5 on the smooth boundary and c =1 in the domain. $\dot{\sigma}_{Ijk}^{[2]S}$(Q) is zero because the fictitious thin plate is simply supported. $\partial\dot{\sigma}_{Ijk}^{[2]S} / \partial n$ is zero for symmetric boundary. From eqns (10) and (11),

$$c\dot{\sigma}_{Ijk}^{[1]S}(P) = \int_\Gamma \{T^{[1]}(P,Q)\frac{\partial \dot{\sigma}_{Ijk}^{[1]S}(Q)}{\partial n} - \frac{\partial T^{[1]}(P,Q)}{\partial n}\dot{\sigma}_{Ijk}^{[1]S}(Q)\}d\Gamma$$

$$-\int_\Gamma \{T^{[2]}(P,Q)\frac{\partial \dot{\sigma}_{Ijk}^{[2]S}(Q)}{\partial n}d\Gamma - \sum_{m=1}^{M} T^{[2]}(P,q)\dot{\sigma}_{Ijk(m)}^{[3]P}(q), \tag{12}$$

$$0 = \int_\Gamma T^{[1]}(P,Q)\frac{\partial \dot{\sigma}_{Ijk}^{[2]S}(Q)}{\partial n}d\Gamma + \sum_{m=1}^{M} T^{[1]}(P,q)\dot{\sigma}_{Ijk(m)}^{[3]P}(q). \tag{13}$$

On the other hand, $\partial\dot{\sigma}_{ijk}^{[2]S}/\partial n$ can be assumed to be zero on the symmetric boundary. For internal points, the following equation is obtained:

$$c\dot{\sigma}_{ijk}^{[1]S}(p)=\int_{\Gamma}\{T^{[1]}(p,Q)\frac{\partial\dot{\sigma}_{ijk}^{[1]S}(Q)}{\partial n}-\frac{\partial T^{[1]}(p,Q)}{\partial n}\dot{\sigma}_{ijk}^{[1]S}(Q)\}d\Gamma$$

$$-\int_{\Gamma}\frac{\partial T^{[2]}(p,Q)}{\partial n}\dot{\sigma}_{ijk}^{[2]S}(Q)\}d\Gamma-\sum_{m=1}^{M}T^{[2]}(p,q)\dot{\sigma}_{ijk(m)}^{[3]P}(q). \tag{14}$$

In practice, unknown $\partial\dot{\sigma}_{ijk}^{[1]S}(Q)/\partial n$, $\partial\dot{\sigma}_{ijk}^{[2]S}(Q)/\partial n$ and $\dot{\sigma}_{ijk(m)}^{[3]P}$ must be obtained from $\dot{\sigma}_{ijk}^{[1]}(P)$ and $\dot{\sigma}_{ijk}^{[1]}(p)$. Replacing $\dot{\sigma}_{ijk}^{[f]S}(Q)$, $\partial\dot{\sigma}_{ijk}^{[f]S}(Q)/\partial n$ and $\dot{\sigma}_{ijk(m)}^{[3]P}$ with vectors $E^{[f]}$, $V^{[f]}$ and $E^{[3]P}(q)$, respectively, we obtain the following equations from eqns (12)–(14) [4].

$$E^{[1]}(P)=G_1V^{[1]}-H_1E^{[1]}(P)-G_2V^{[2]}-G_1{}^{P}E^{[3]P}(q) \tag{15}$$

$$0=G_1V^{[1]}+G_1{}^{P}E^{[3]P}(q) \tag{16}$$

$$E^{[1]}(p)=G_3V^{[1]}-H_3E^{[1]}(P)-G_4V^{[2]}-G_4{}^{P}E^{[3]P}(q) \tag{17}$$

The details of this procedure are given in reference [4]. From eqns (15)–(17),

$$\begin{bmatrix}\mathbf{V}^{[1]}\\ \mathbf{V}^{[2]}\\ \mathbf{E}^{[3]P}(q)\end{bmatrix}=\mathbf{A}^{-1}\begin{bmatrix}\mathbf{H}_1\mathbf{E}^{[1]}(P)\\ \mathbf{0}\\ \mathbf{H}_3\mathbf{E}^{[1]}+\mathbf{G}_3\mathbf{E}^{[1]}(\mathrm{p})\end{bmatrix} \tag{18}$$

where $\mathbf{A}^{-1}$ is the inverse matrix. If the boundary is divided into N_0 constant elements, and N_1 internal points are used, the simultaneous linear algebraic equations with $(2N_0+N_1)$ as unknowns must be solved.

2.4 Triple-reciprocity BEM

The function $\varepsilon_{ijk}^{[f]}$ is defined as

$$\nabla^2\varepsilon_{ijk}^{[f+1]}=\varepsilon_{ijk}^{[f]}. \tag{19}$$

Using eqns (5), (6), (18), (19) and Green's second identity, eqn (1) becomes

$$c_{ij}\dot{u}_j=\int_{\Gamma}[u_{ij}^{[1]}\dot{p}_j-p_{ij}\dot{u}_j]d\Gamma$$

$$-\sum_{f=1}^{2}(-1)^f\int_{\Gamma}\{\frac{\partial\varepsilon_{ijk}^{[f+1]}}{\partial n}\dot{\sigma}_{ijk}^{[f]S}-\varepsilon_{ijk}^{[f+1]}\frac{\partial\dot{\sigma}_{ijk}^{[f]S}}{\partial n}\}d\Gamma+\sum_{m=1}^{M}\varepsilon_{ijk}^{[3]}\dot{\sigma}_{ijk(m)}^{[3]P}. \tag{20}$$

Next, the function $\varepsilon_{ijk}^{[f]}$ is obtained, and the relationship between the biharmonic function $A^{[1]}$ and Kelvin's solution $u_{ij}^{[1]}$ is given by

$$u_{ij}^{[1]}=\frac{-1}{2(1-\nu)G}A_{,ij}^{[1]}+\frac{\delta_{ij}A_{,kk}^{[1]}}{G}. \tag{21}$$

The biharmonic function $A^{[1]}$ is given as

$$A^{[1]} = \frac{r^2}{8\pi}[\ln(\frac{1}{r}) + C_q], \tag{22}$$

where

$$C_q = \frac{7-8\nu}{2(3-4\nu)}. \tag{23}$$

The displacement-strain relationship is given by

$$\varepsilon_{ijk}^{[1]} = \frac{1}{2}(u_{ji,k}^{[1]} + u_{ki,j}^{[1]}), \tag{24}$$

Next, the function $A^{[f]}$ that satisfies

$$\nabla^2 A^{[f+1]} = A^{[f]} \tag{25}$$

is considered. $A^{[f]}$ is generally given by

$$A^{[f]} = \frac{r^{2f}}{2\pi[(2f)!!]^2}[\ln(\frac{1}{r}) + C_q + \mathrm{sgn}(f-1)\sum_{e=2}^{f}\frac{1}{e}]. \tag{26}$$

Accordingly, from eqns (24), (26), (29) and (30), $\varepsilon_{ijk}^{[f]}$ is obtained as [4]

$$\begin{aligned}\varepsilon^{[f]}{}_{ijk} = \frac{2r^{2f-3}}{2\pi G\{(2f)!!\}^2}\Big\langle \frac{2}{(1-\nu)}\big[2f(f-1)(f-2)F_2 + (3f^2-6f+2)\big]r,_i r,_j r,_k \\ +2[2f^2(f-1)F_2 + f(3f-2)](\delta_{ik}r,_j + \delta_{ij}r,_k) \\ +\big[2f(f-1)F_2 + (2f-1)\big](\delta_{ij}r,_k + \delta_{ki}r,_j + \delta_{jk}r,_i)\Big\rangle,\end{aligned} \tag{27}$$

where

$$F_2 = \ln(r) + C_q - \sum_{e=2}^{f}\frac{1}{e}. \tag{28}$$

2.5 Internal stresses

The stress-strain relationship is given by

$$\sigma_{ijk}^{[1]} = \lambda\delta_{jk}u_{mi,m}^{[1]} + \mu(u_{ji,k}^{[1]} + u_{ki,j}^{[1]}), \tag{29}$$

where Lame's constant λ and μ are defined as

$$\lambda = \frac{2\nu G}{1-2\nu} \tag{30}$$

$$\mu = \mathrm{G}. \tag{31}$$

The internal stress is given by

$$\dot{\sigma}_{ij} = \int_\Gamma [-\sigma_{kij}^{[1]}\dot{p}_k - S_{kij}\dot{u}_k]d\Gamma + \int_\Omega \varepsilon_{ijks}^{[1]}\dot{\sigma}_{ks}^{[1]}d\Omega - \dot{\sigma}_{lij}^{[1]}, \tag{32}$$

where $\dot{\sigma}_{lj}^{[1]}$ is the initial stress obtained from the initial strain. The functions S_{kij} and $\sigma_{ijks}^{H[1]}$ in eqn (32) are given by

$$S_{kij} = \frac{G}{2\pi(1-\nu)r^2}[2\frac{\partial r}{\partial n}\{(1-2\nu)\delta_{ij}r_{,k} + \nu(\delta_{ik}r_{,j} + \delta_{jk}r_{,i}) - 4r_{,i}r_{,j}r_{,k}\} \\ +2\nu(n_i r_{,j} r_{,k} + n_j r_{,i} r_{,k}) + (1-2\nu)(2n_k r_{,i} r_{,j} + n_j \delta_{ik} + n_i \delta_{jk}) - (1-4\nu)n_k \delta_{ij}] \tag{33}$$

$$\varepsilon_{ijks}^{[1]} = \frac{1}{4\pi(1-\nu)r^2}[2(1-2\nu)(\delta_{ik}r,_j r,_s + \delta_{jk}r,_i r,_s - \delta_{ij}\delta_{ks} + 2\delta_{ij}r,_k r,_s) \\ +2\nu(\delta_{is}r,_j r,_k + \delta_{jk}r,_i r,_s + \delta_{ik}r,_j r,_s + \delta_{js}r,_i r,_k) + 2\delta_{ks}r,_i r,_j - 8r,_i r,_j r,_k r,_s]. \tag{34}$$

The function $\sigma_{ijks}^{H[f]}$ is defined as

$$\nabla^2 \varepsilon_{ijks}^{[f+1]} = \varepsilon_{ijks}^{[f]}. \tag{35}$$

Using eqn (35) and Green's theory, eqn (32) becomes

$$\dot{\sigma}_{ij} = \int_\Gamma [-\sigma_{kij}^{[1]} \dot{p}_k - S_{kij}\dot{u}_k] d\Gamma - \sum_{f=1}^{2}(-1)^f \int_\Gamma [\frac{\partial \varepsilon_{ijks}^{[f+1]}}{\partial n}\dot{\sigma}_{ks}^{[f]S} - \varepsilon_{ijks}^{[f+1]}\frac{\partial \dot{\sigma}_{ks}^{[f]S}}{\partial n}] d\Gamma \\ +\sum_{m=1}^{M} \varepsilon_{ijks}^{[3]}(p,q)\dot{\sigma}_{ks(m)}^{[3]P} - \dot{\sigma}_{lij}^{[1]}. \tag{36}$$

Using eqn (35) and the relationship between displacement and strain, $\varepsilon_{ijks}^{[f]}$ is obtained as

$$\varepsilon_{ijks}^{[f]}(p,q) = \frac{r^{2f-4}}{\pi(1-\nu)[(2f)!!]^2}\langle\{-2[8f^3\nu - 3f^2(1+6\nu) + 2f(3+4\nu) - 2] \\ -4f(f-1)(f-2)(2f\nu-1)F_2\}\delta_{is}r,_j r,_k \\ -[6f^2\nu - 2f(1+2\nu) + 1 + 2f(f-1)(2f\nu-1)F_2]\delta_{is}\delta_{jk} \\ +\{8(2f-3)[1+f(f-3)] + f(f-1)(f-2)(f-3)F_2\}r,_i r,_j r,_k r,_s \\ +2[3f^2 - 6f + 2 + 2f(f-1)(f-2)F_2]\delta_{jk}r,_i r,_s \\ -2\{4f^3(1-\nu) - 3f^2(4-3\nu) + 2f(5-2\nu) - 2 \\ -2f(f-1)(f-2)[1-f(1-\nu)]F_2\}(\delta_{ij}r,_k r,_s + \delta_{ik}r,_j r,_s + \delta_{sj}r,_k r,_i + \delta_{sk}r,_j r,_i) \\ -\{1 + 6f^2(1-\nu) - 2f(3-2\nu) + 2f(f-1)[2f(1-\nu)-1]F_2\}(\delta_{ij}\delta_{ks} + \delta_{ik}\delta_{js})\rangle. \tag{37}$$

2.6 Initial stress

An iterative process is used for elastoplastic analysis. The equivalent plastic strain increment is denoted as $d\varepsilon^P{}_e$. The von Mises yield criterion is considered. The deviatoric stress tensor S_{ij} is obtained from the stress rate in eqn (36). The plastic strain increments $d\varepsilon^P{}_{ij}$ are determined using the Prandtl-Reuss equation

$$d\varepsilon_{ij}^P = S_{ij}d\lambda, \tag{38}$$

where $d\lambda$ is a proportionality factor. Using this equation, the provisional plastic strain increments $d\varepsilon^P_{ij}$ are calculated. Using the initial stress rate on the boundary and at internal points, the initial stress rate is interpolated using eqn (18). The displacement and traction rates are obtained from eqn (20) as

$$HU=GP+D_1E^{[1]}(Q)-D_2V^{[1]}+D_3V^{[2]}+D_4E^{[3]P}(q). \tag{39}$$

The stress rate is obtained from eqn (36) as

$$\sigma=G^AP-H^AU+D^A{}_1E^{[1]}(Q)-D^A{}_2V^{[1]}+D^A{}_3V^{[2]}+D^A{}_4E^{[3]P}(q)-\sigma_I. \tag{40}$$

The details of this procedure are the same as those of the procedure given in Ref. [1], with the exception of the interpolation. In this theory, the surface stress rate must be calculated to obtain the initial strain on the boundary. This surface stress rate can be obtained by the differentiation of node displacements. The initial stress rate converges using the above iterative process. New coordinates of boundary elements and internal points are calculated. If a length of the new boundary element is larger than the standard length L_0, the boundary element is divided into two boundary elements. If the distance between a boundary element and an internal point is smaller than the standard length R_0, the internal point is deleted to prevent a numerical error in the internal stress. Subsequently, the next load increment (displacement increment) is added. Coulomb's friction coefficient is considered on the contact surfaces.

3 NUMERICAL EXAMPLE

The large plastic deformation problem of a plate (length 20 mm) subjected to tension, as shown in Fig. 2, is solved using the von Mises yield criterion. The number of discretized boundary elements is 400 and the number of internal points used for interpolation is 133, as shown in Fig. 2. Young's modulus E of 210 GPa, Poisson's ratio v of 0.3 and yield stress σ_0 of 200 MPa are assumed. For large deformation, Poisson's ratio v of 0.48 is assumed. It is assumed that the number of time steps for the tension (10 mm) and the substeps for convergence are 100 and 50, respectively. Displacement load curves are shown in Fig. 3 with theoretical values for work hardening values H=0 and 840 MPa.

The elastoplastic problem of a plane strain punch is next solved using the von Mises yield criterion. This example consists of a rigid flat punch indented into a solid plane strain

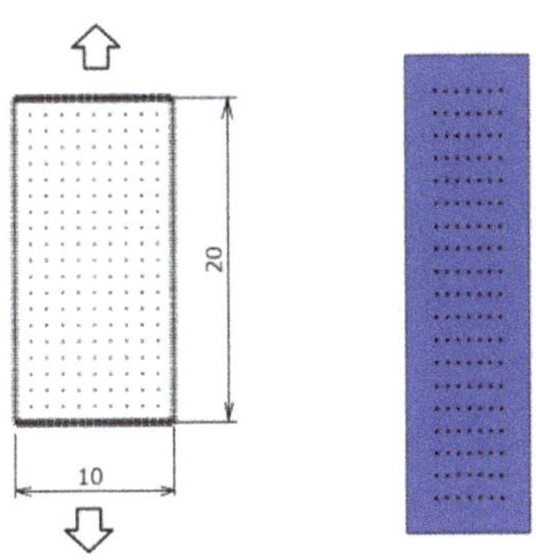

Figure 2: Tension of plate.

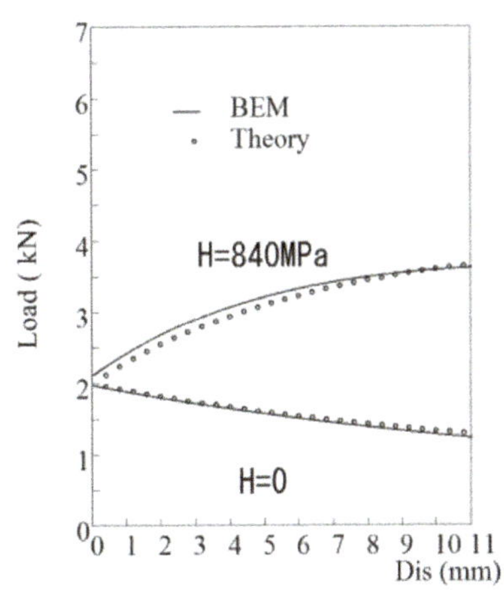

Figure 3: History of load.

specimen, as shown in Fig. 4. First, it is assumed that there is no friction between the specimen and the punch. One-quarter of the region is used for calculation. The number of discretized boundary elements is 400, and the number of internal points for interpolation is 153, as shown in Fig. 5. Young's modulus E of 210 GPa, Poisson's ratio ν of 0.3, and yield stress σ_0 of 200 MPa are assumed. It is assumed that the number of time steps for compression (5 mm) and substeps for convergence are 100 and 50, respectively. Displacement load curves without friction are shown in Fig. 6 with theoretical values for work hardening values H of 0 and 420 MPa. Second, the distributions of equivalent plastic strain and stress considering the friction coefficient ($\mu = 0.1$) are shown in Figs 7 and 8, respectively.

Finally, the constrained upsetting problem of a plane strain punch in Fig. 9 is solved using the von Mises yield criterion. It is assumed that there is no slip between the specimen and the punch. One-quarter of the region is used for calculation. The number of discretized boundary elements is 400 and the number of internal points used for interpolation is 153, as shown in Fig. 5. Young's modulus E of 210 GPa, Poisson's ratio ν of 0.3, and yield stress σ_0 of 200 MPa are assumed. It is assumed that the number of time steps for compression (5 mm) and substeps for convergence are 100 and 50, respectively. A work hardening value H of 420 MPa is assumed. Distributions of equivalent plastic strain and stress are shown in Figs 10 and 11, respectively.

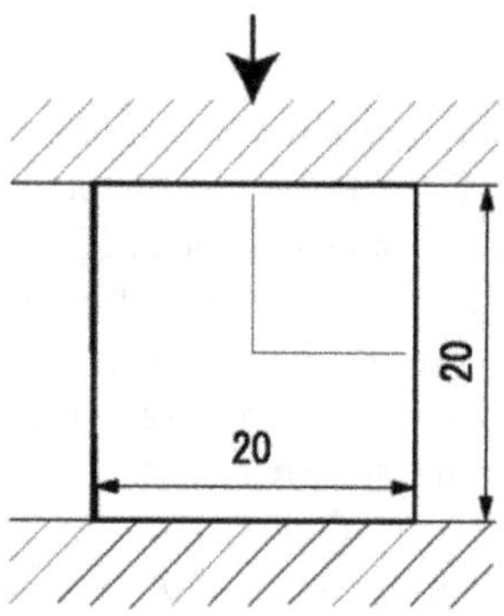

Figure 4: Plane strain punch problem.

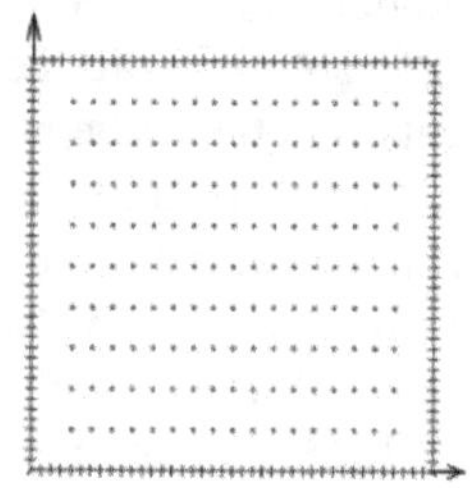

Figure. 5: Boundary elements and internal points in quarter region (Number of boundary elements: 100).

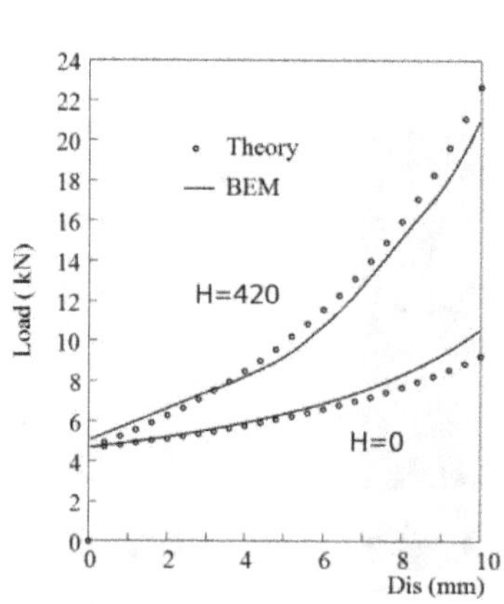

Figure 6: History of load ($\mu = 0$).

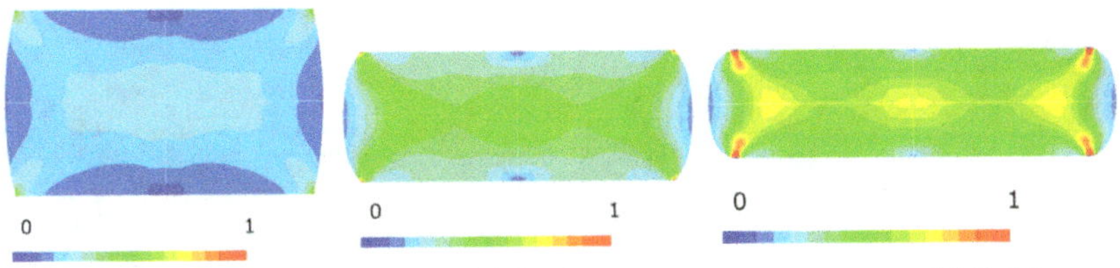

(a) Z= 5mm (b) Z= 8mm (c) Z=10mm

Figure 7: Distribution of plastic strain ($\mu = 0.1$).

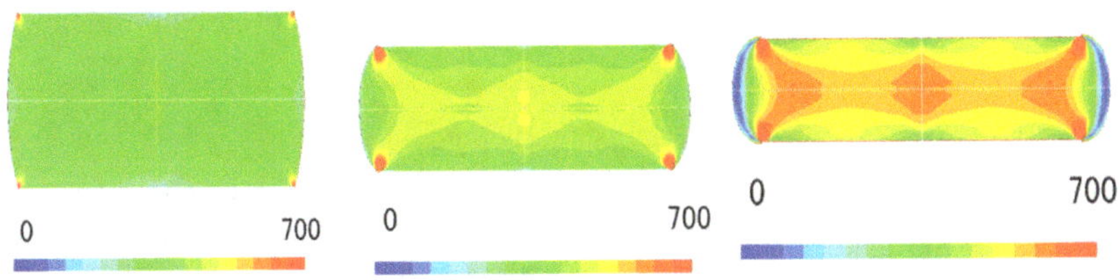

(a) Z= 5mm (b) Z= 8mm (c) Z=10mm

Figure 8: Equivalent stress distribution ($\mu = 0.1$, MPa).

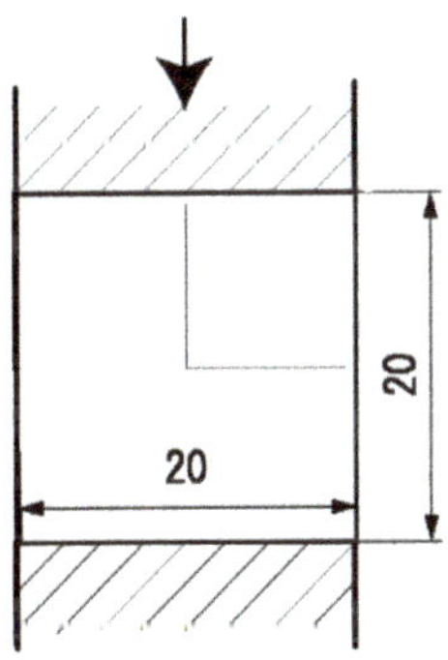

Figure 9: Constrained upsetting problem.

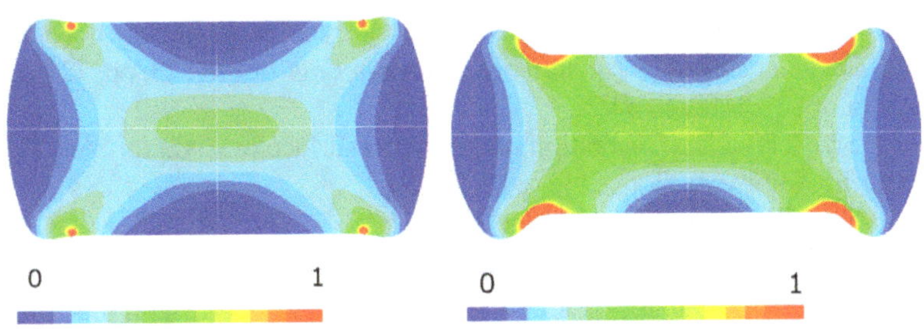

Z=5 mm Z=10mm

Figure 10: Distribution of equivalent strain in constrained upsetting problem.

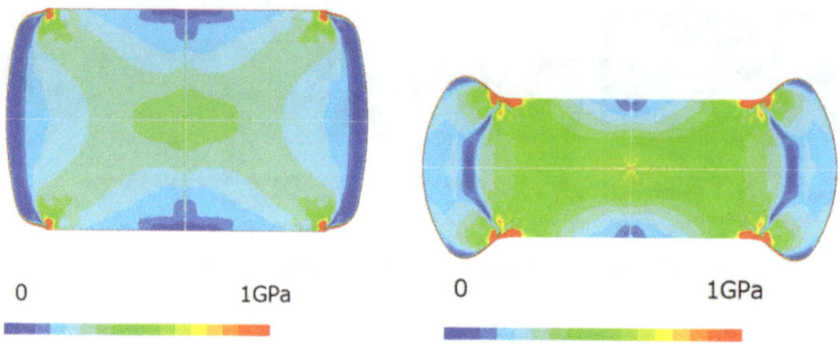

Figure 11: Equivalent stress distribution in constrained upsetting problem.

4 CONCLUSION

It was shown that two-dimensional large plastic deformation analysis can be carried out without the use of internal cells, by the triple-reciprocity BEM. The fundamental solutions for this analysis were shown. In this method, the strong singularity that appears in the calculation of internal stress by the ordinary BEM becomes weak. With the use of numerical examples, the effectiveness and accuracy of this method were demonstrated. In this method, the merit of BEM, which is the ease of data preparation, is not lost because internal cells are not necessary. If FEM is used, remeshing (rezoning) is necessary for large-plastic-deformation analysis. Only boundary elements are remeshed in this triple-reciprocity BEM.

REFERENCES

[1] Telles, J.C.F., *The boundary element method applied to inelastic problems*, Springer-Verlag, Berlin, 1983.

[2] Brebbia, C.A., Telles, J.C.F. & Wrobel, L.C., *Boundary element techniques-theory and applications in engineering*, Springer-Verlag, Berlin, pp. 252–266, 1984.

[3] Ochiai, Y. & Kobayashi, T., Initial stress formulation for elastoplastic analysis by improved multiple-reciprocity boundary element method. *Engineering Analysis with Boundary Elements*, **23**, pp. 167–173, 1999.
https://doi.org/10.1016/s0955-7997(98)00066-6

[4] Ochiai, Y. & Kobayashi, T., Initial strain formulation without internal cells for elastoplastic analysis by triple-reciprocity BEM. *International Journal for Numerical Method in Engineering*, **50**, pp. 1877–1892, 2001.
https://doi.org/10.1002/nme.100

[5] Ochiai, Y., Three-dimensional elastoplastic analysis by triple-reciprocity boundary element method. *Communications in Numerical Methods in Engineering*, **23**, pp. 721–732, 2007.
https://doi.org/10.1002/cnm.922

[6] Ochiai, Y. & Sekiya, T., Generation of free-form surface in CAD for dies. *Advances in Engineering Software*, **22**, pp. 113–118, 1995.
https://doi.org/10.1016/0965-9978(94)00030-m

[7] Ochiai, Y. & Yasutomi, Z., Improved method generating a free-form surface using integral equations. *Computer-Aided Geometric Design*, **17**(3), pp. 233–245, 2000.
https://doi.org/10.1016/s0167-8396(99)00047-3

[8] Ochiai, Y., Nishitani, H. & Sekiya, T., Stress analysis with arbitrary body force by boundary element method. *Engineering Analysis with Boundary Elements*, **17**, pp. 295–302, 1996.
https://doi.org/10.1016/s0955-7997(96)00031-8
[9] Ochiai, Y., Multidimensional numerical integration for meshless BEM. *Engineering Analysis with Boundary Elements*, **27**(3), pp. 241–249, 2003.
https://doi.org/10.1016/s0955-7997(02)00112-1
[10] Ochiai, Y. & Sladek, V., Numerical treatment of domain integrals without internal cells in three-dimensional BIEM formulations. *CMES* (Computer Modeling in Engineering & Sciences), **6**(6), pp. 525–536, 2004.
[11] Ochiai, Y., Meshless unsteady thermo-elastoplastic analysis by triple-reciprocity boundary element method. *CMES* (Computer Modeling in Engineering & Sciences), **79**(2), pp. 83–101, 2011.
https://doi.org/10.3970/cmes.2011.079.083
[12] Ochiai, Y., Three-dimensional thermo-elastoplastic analysis by triplereciprocity boundary element method. *Engineering Analysis with Boundary Elements*, **35**(3), pp. 478–488, 2011.
https://doi.org/10.1016/j.enganabound.2010.08.018
[13] Ochiai, Y., *Large plastic deformation analysis without use of internal cells by triple-reciprocity BEM*, Proceedings of 6th International Conference on Boundary Element Techniques, p. 293, 2005.

KANSA RBF METHOD FOR NONLINEAR PROBLEMS

M. A. JANKOWSKA[1], A. KARAGEORGHIS[2] & C. S. CHEN[3]
[1]Institute of Applied Mechanics, Poznan University of Technology, Poznan, Poland.
[2]Department of Mathematics and Statistics, University of Cyprus, Nicosia, Cyprus.
[3]Department of Mathematics, University of Southern Mississippi, Hattiesburg, USA.

ABSTRACT
We apply the Kansa–radial basis function (RBF) collocation method to two– dimensional nonlinear boundary value problems. The system of nonlinear equations resulting from the Kansa–RBF discretization is solved by directly applying a standard nonlinear solver. In a natural way, the value of the shape parameter in the RBFs employed in the approximation is included in the unknowns to be determined. The numerical results of some examples are presented and analysed.
Keywords: collocation, Kansa method, radial basis functions

1 INTRODUCTION

Radial basis function (RBF) methods have become popular in recent years in approximation theory as well as in the numerical solution of partial differential equations. The most widely used RBF method for the latter class of problems is the RBF collocation method due to Kansa [1], known as the *Kansa method.* The popularity of this method is due to its meshlessness which means that only a set of points is required in the discretization of the continuous problem. This renders the implementation of the method particularly easy, especially for problems in complex geometries and/or in three dimensions. A disadvantage of the method is the (unknown) optimal choice of the shape parameter which is found in most RBFs. Various techniques have been proposed for the determination of an appropriate value of the shape parameter, see e.g. [2–5].

We apply the Kansa RBF method to two-dimensional nonlinear boundary value problems. The discretization of such problems leads to systems of nonlinear equations which may be solved by using standard software. Moreover, we include the (unknown) value of the shape parameter in the set of unknowns of the nonlinear problem. The solution of the nonlinear problem thus yields not only the coefficients in the RBF approximation but, also, an appropriate value of the shape parameter. The implementation of the proposed technique is very simple and the results of several numerical experiments reveal it is satisfactorily accurate.

2 THE PROBLEM

We consider the boundary value problem in $\mathbb{R}^2$

$$\mathcal{L}u = f \quad \text{in} \quad \Omega, \tag{1a}$$

subject to the boundary condition

$$\mathcal{B}u = g \quad \text{on} \quad \partial\Omega, \tag{1b}$$

where $\mathcal{L}$ is a nonlinear elliptic operator and $\mathcal{B}$ is a linear (or nonlinear) operator describing the boundary condition.

3 THE KANSA METHOD

In the Kansa method [1] we approximate the solution u of boundary value problem (1) by a linear combination of RBFs [6–8]

DOI: 10.2495/CMEM-V6-N6-1000-1007

$$u_N(x,y)=\sum_{n=1}^{N} a_n\phi_n(x,y), \quad (x,y)\in\bar{\Omega}. \tag{2}$$

In this work we shall be using the *normalized multiquadric* basis function

$$\phi_n(x,y)=\Phi(r_n)=\sqrt{(cr_n)^2+1}, \text{ where } r_n^2=(x-x_n)^2+(y-y_n)^2, \tag{3}$$

and the points $\{(x_n,y_n)\}_{n=1}^{N}$ are usually referred to as *centers*. Moreover, c is the *shape parameter* and such parameters are often present in RBFs and the determination of their optimal value remains a major challenge.

The coefficients $\{a_n\}_{n=1}^{N}$ in equation (2) are determined from the collocation equations

$$\mathcal{L}u_N(x_m,y_m)=f(x_m,y_m), \quad m=1,\ldots,M_{int}, \tag{4a}$$

$$\mathcal{B}u_N(x_m,y_m)=g(x_m,y_m), \quad m=M_{int}+1,\ldots, M_{int}+M_{bry}, \tag{4b}$$

where $M_{int}+M_{bry}=M$ and the points $\{(x_m,y_m)\}_{m=1}^{M}\in\bar{\Omega}$ are the *collocation points*. Note that, in general, the collocation points are not the same as the centers and M≥N.

Since the operator $\mathcal{L}$ is nonlinear, the system of M equations resulting from equations (4a)–(4b) is nonlinear and will be solved in a (nonlinear) least squares sense. Moreover, we shall include the value of the shape parameter c of the RBF in the unknowns. We shall therefore be minimizing of the functional

$$S(a,c):=\sum_{m=1}^{M_{int}}[\mathcal{L}u_N(x_m,y_m)-f(x_m,y_m)]^2 + \sum_{m=M_{int}+1}^{M}[\mathcal{B}u_N(x_m,y_m)-g(x_m,y_m)]^2, \tag{5}$$

where the vector $\boldsymbol{a}=[a_1, a_2, \ldots, a_N]^T$. Thus the goal is to determine not only the vector of coefficients $\boldsymbol{a}$ but also the value of the shape parameter c. The minimization of functional (5) is carried out using the MATLAB© [9] optimization toolbox routine `lsqnonlin` which solves nonlinear least squares problems using, by default, a subspace trust region method. The routine `lsqnonlin` does not require the user to provide the gradient and, in addition, it offers the option of imposing lower and upper bounds on the elements of the vector of unknowns $x=[\boldsymbol{a}, c]^T$ through the vectors `lb` and `ub`. We can thus easily impose the constraints on the values of the coefficients and in particular on the values of the shape parameter. In all numerical experiments we chose the constraints $0<c<8$ and took the initial vector of coefficients $\boldsymbol{a}=0$.

4 NUMERICAL RESULTS

In the numerical examples considered in this section, we calculated the approximate solution u_N at L test points on a grid in $\bar{\Omega}$. In the case of Examples 1 and 2 the analytical solution is known. Hence, we calculated the maximum relative error E defined by

$$E=\frac{\| u-u_N \|_{\infty,\bar{\Omega}}}{\| u \|_{\infty,\bar{\Omega}}} \tag{6}$$

and the root mean square error (RMSE) *from*

$$\varepsilon = \sqrt{\frac{\sum_{l=1}^{L}[u(x_l, y_l) - u_{\mathrm{N}}(x_l, y_l)]^2}{L}}. \tag{7}$$

4.1 Example 1

We first consider an example from [10, 11] where

$$\mathcal{L}u = -\varepsilon^2 \Delta u - u + u^3 \text{ in } \Omega,$$

subject to the homogeneous Dirichlet boundary condition

$$\mathcal{B}u = u = 0 \text{ on } \partial\Omega,$$

where $\Omega = (0, 1) \times (0, 1)$, the exact solution is given by

$$u(x, y) = (1 + \mathrm{e}^{-1/\varepsilon} - \mathrm{e}^{-x/\varepsilon} - \mathrm{e}^{(x-1)/\varepsilon})(1 + \mathrm{e}^{-1/\varepsilon} - \mathrm{e}^{-y/\varepsilon} - \mathrm{e}^{(y-1)/\varepsilon})$$

and ε is a known constant. Clearly, the function f in (1a) is obtained by calculating $\mathcal{L}u$ for the u given above. We chose a uniform $M \times M$ grid in $\bar{\Omega}$ for the collocation points and a uniform $N \times N$ grid in $\bar{\Omega}$ for the centres. This arrangement means that we have a total of $\mathrm{M} = M^2$ collocation points and $\mathrm{N} = N^2$ centres. Moreover, we have $\mathrm{M}_{\mathrm{int}} = M^2 - 4M + 4$ interior collocation points and $\mathrm{M}_{\mathrm{bry}} = 4M–4$ boundary points.

The set of test points was taken to be a uniform 51 × 51 grid in $\bar{\Omega}$. Then, for the routine **lsqnonlin** we chose the initial value of the shape parameter $c_0 = 4$ and the maximum number of iterations **niter** equal to 100. We performed calculations for $\varepsilon = 0.1, 0.15, 0.25, 0.5$ and 1. It was observed that the accuracy of the results obtained was poor for $\varepsilon = 0.1$. However, when using continuation from $\varepsilon = 1$ to $\varepsilon = 0.5$ to $\varepsilon = 0.25$ to $\varepsilon = 0.15$ to $\varepsilon = 0.1$, satisfactory results were obtained as is shown in Table 1. By continuation we mean that the final values of all the unknown parameters in **lsqnonlin** for $\varepsilon = 1$ were saved and used as the initial values for the unknown parameters in **lsqnonlin** for $\varepsilon = 0.5$ and so on.

Table 1: Example 1: Results with continuation.

M	N	M	N	ε	c	E	RMSE
15	14	225	196	1.000	1.959	8.616(-3)	2.544(-5)
				0.500	2.078	5.492(-3)	1.911(-4)
				0.250	2.444	1.207(-2)	4.176(-3)
				0.150	2.784	1.548(-2)	6.299(-3)
				0.100	2.906	2.937(-2)	1.408(-2)
16	15	256	225	1.000	4.006	3.502(-3)	5.884(-5)
				0.500	4.049	4.258(-3)	5.331(-4)
				0.250	4.095	8.841(-3)	3.863(-3)
				0.150	4.175	1.331(-2)	7.524(-3)
				0.100	4.210	2.736(-2)	1.341(-2)

(Continued)

Table 1: (*Continued*)

M	N	M	N	ε	c	E	RMSE
22	21	484	441	1.000	4.016	7.360(-4)	1.271(-5)
				0.500	4.096	9.611(-4)	1.237(-4)
				0.250	4.280	2.643(-3)	1.153(-3)
				0.150	4.420	1.076(-2)	5.360(-3)
				0.100	4.658	1.434(-2)	6.366(-3)
24	23	576	529	1.000	4.014	6.191(-4)	7.844(-6)
				0.500	4.123	7.583(-4)	8.109(-5)
				0.250	4.311	4.237(-3)	1.041(-3)
				0.150	4.411	1.595(-2)	5.992(-3)
				0.100	4.686	1.821(-2)	5.606(-3)

4.2 Example 2

We next consider an example from [12, 13] where the governing equation is

$$\mathcal{L}u = \Delta u - 4u^3 = 0 \quad \text{in} \quad \Omega, \tag{8}$$

subject to Dirichlet boundary conditions corresponding to the exact solution

$$u(x,y) = \frac{1}{4+x+y}.$$

The computational domain Ω is peanut shaped and its boundary $\partial\Omega$ is defined parametrically by

$$x = r(\vartheta)\cos\vartheta,\ y = r(\vartheta)\sin\vartheta,\ \text{where } r(\vartheta) = 0.3\sqrt{\cos 2\vartheta + \sqrt{1.1 - \sin^2 2\vartheta}},$$

$0 \le \vartheta \le 2\pi$. The collocation points are distributed as follows:

$$(x_{ij}, y_{ij}) = r_j\sqrt{\cos 2\vartheta_i + \sqrt{1.1 - \sin^2 2\vartheta_i}}\,\left(\cos\vartheta_i, \sin\vartheta_i\right), \tag{9}$$

$$\vartheta_i = \frac{2\pi(i-1)}{M_\vartheta},\ i = 1,\ldots,M_\vartheta,\ \text{and } r_j = \frac{0.3j}{M_r},\ j = 1,\ldots,M_r.$$

A more uniform distribution of points is obtained as follows, see e.g [14]. We first calculate the length of the boundary curve from

$$S = \int_0^{2\pi} \sqrt{r(\vartheta)^2 + r'(\vartheta)^2 d\vartheta},$$

which may be evaluated using the MATLAB© routine **quadl** which evaluates an integral using adaptive Lobatto quadrature within a user–prescribed accuracy. Since there will be M_ϑ

collocation points on the boundary, we take the length of each subsegment to be S/M_ϑ. The angles ϑ_k which give such a distribution are obtained by choosing $\vartheta_1 = 0$ and then solving, serially for $k = 1,\dots, M_\vartheta-1$, the nonlinear equations

$$F(t)=\sqrt{(r(t)\cos t-r(\vartheta_k)\cos\vartheta_k)^2+(r(t)\sin t-r(\vartheta_k)\sin\vartheta_k)^2}-\frac{S}{M_\vartheta}=0,$$

to yield $t = \vartheta_k$, $k = 2,\dots, M_\vartheta$, respectively. The solution of the nonlinear equations may be carried out using the MATLAB© routine `fzero`. The angles ϑ_k, $k = 1,\dots, M_\vartheta$, define equally spaced points on the boundary curve.

The collocation points are now distributed as follows:

$$(x_{ij},y_{ij})=r_j\sqrt{\cos 2\vartheta_i+\sqrt{1.1-\sin^2 2\vartheta_i}}\,(\cos\vartheta_i,\sin\vartheta_i,$$

$$i=1,\dots,M_\vartheta,\ j=1,\dots,M_r,\ \text{where } r_j=0.3\left(\frac{j}{M_r}\right)^p,\ j=1,\dots,M_r.$$

The parameter $p \in (0,1)$ is chosen so that the collocation points are pushed away from the centre of the domain to yield a more uniform distribution of collocation points. An appropriate value of p was found to be 2/3. The centres are chosen in an identical way with M_ϑ and M_r replaced by N_ϑ and N_r, respectively. This arrangement means that we have a total of $\mathrm{M} = M_\vartheta M_r$ collocation points and $\mathrm{N} = N_\vartheta N_r$ centres. Moreover, we have $\mathrm{M}_{\mathrm{int}} - M_\vartheta(M_r-1)$ interior collocation points and $\mathrm{M}_{\mathrm{bry}} = M_\vartheta$ boundary points. An even more uniform distribution of collocation points and centres maybe obtained by using the *Halton* points [8, Appendix A.1]. These may be generated using the MATLAB© code `haltonseq.m`. In this approach we take M_b boundary collocation points and M_i interior points, and N_b boundary centres points and N_i interior centres. A typical distribution of collocation points using the first distribution, the more uniform distribution and the Halton points is presented in Fig. 1. The maximum relative error E and RMSE were calculated on a 21×21 grid in $\bar{\Omega}$ generated using a distribution similar to (9) and clearly different than the collocation points. The maximum number of iterations in `lsqnonlin` was set to 100 and we took $c_0 = 4$. In Table 2 we present some numerical results for various numbers of degrees of freedom. The corresponding results with the improved distribution of collocation points and centres are presented in Table 3. The results obtained with comparable numbers of degrees of freedom using Halton points are presented in Table 4.

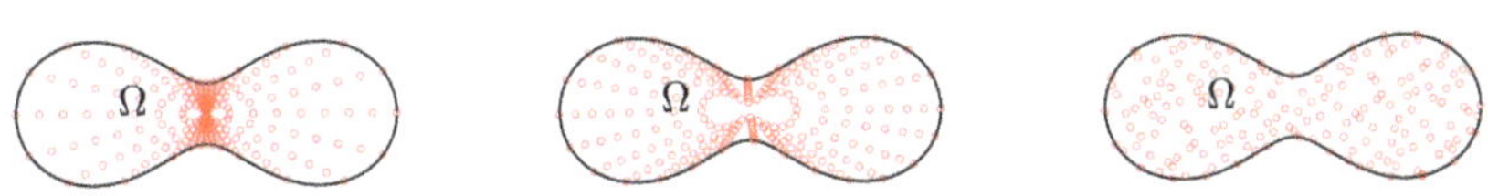

(a) First distribution (b) Improved distribution (c) Halton points

Figure 1: Example 2. Computational domain and typical distributions of collocation points.

Table 2: Example 2: Results for various numbers of degrees of freedom.

M_ϑ	M_r	M	N_ϑ	N_r	N	c	E	RMSE
12	10	120	10	8	80	2.095	3.574(-3)	2.395(-4)
14	12	168	12	10	120	2.134	4.099(-3)	2.550(-4)
16	14	224	14	12	168	2.730	7.854(-3)	6.251(-4)
18	16	288	16	14	224	2.862	6.227(-3)	4.396(-4)
20	18	360	18	16	288	2.854	8.104(-3)	6.215(-4)
22	20	440	20	18	360	3.086	5.796(-3)	2.717(-4)
24	22	528	22	20	440	3.102	5.296(-3)	2.931(-4)
26	24	624	24	22	528	3.172	7.541(-3)	5.154(-4)
28	26	728	26	24	624	2.770	4.691(-3)	3.456(-4)
30	28	840	28	26	728	2.668	3.088(-3)	2.255(-4)

Table 3: Example 2: Results for various numbers of degrees of freedom with improved distribution of points.

M_ϑ	M_r	M	N_ϑ	N_r	N	c	E	RMSE
12	10	120	10	8	80	2.416	9.236(-3)	1.192(-3)
14	12	168	12	10	120	2.830	9.333(-3)	1.162(-3)
16	14	224	14	12	168	2.934	7.185(-3)	6.291(-4)
18	16	288	16	14	224	2.367	2.620(-3)	1.979(-4)
20	18	360	18	16	288	2.380	1.820(-3)	1.406(-4)
22	20	440	20	18	360	2.762	3.773(-3)	3.266(-4)
24	22	528	22	20	440	3.064	2.369(-3)	1.059(-4)
26	24	624	24	22	528	3.168	2.274(-3)	1.327(-4)
28	26	728	26	24	624	3.133	2.054(-3)	1.571(-4)
30	28	840	28	26	728	3.197	2.046(-3)	1.648(-4)

Table 4: Example 2: Results for various numbers of degrees of freedom with Halton points.

M_b	M_i	M	N_b	N_i	N	c	E	RMSE
12	67	79	10	42	52	2.426	4.365(-3)	2.459(-4)
14	107	121	12	67	79	2.358	3.143(-3)	2.234(-4)
16	158	174	14	107	121	2.485	2.244(-3)	1.240(-4)
18	206	224	16	158	174	3.117	6.190(-3)	2.983(-4)
20	247	267	18	206	224	3.162	2.011(-3)	1.597(-4)
22	300	322	20	247	267	3.192	3.187(-3)	2.046(-4)
24	370	394	22	300	322	3.171	3.519(-3)	1.976(-4)
26	452	478	24	370	394	3.173	4.830(-3)	2.480(-4)
28	545	573	26	452	478	3.109	3.735(-3)	1.908(-4)
30	616	646	28	545	573	3.095	1.760(-3)	1.446(-4)
32	692	724	30	616	646	3.164	1.825(-3)	1.557(-4)
34	786	820	32	692	724	3.091	2.929(-3)	1.960(-4)

5 CONCLUSIONS

A Kansa–RBF collocation method has been applied for the solution of nonlinear boundary value problems. The discretization leads to a system of nonlinear equations which is solved using the nonlinear least–squares minimization MATLAB© routine `lsqnonlin`. The determination of the shape parameter, which remains a challenging problem in the application of Kansa–RBF collocation methods, is resolved by taking the unknown value of the shape parameter to be part of the unknowns in `lsqnonlin`. The main advantage of the proposed technique is the ease with which it can be implemented while the results of several numerical experiments reveal that it leads to accurate numerical results with a relatively small number of iterations.

REFERENCES

[1] Kansa, E.J., Multiquadrics—a scattered data approximation scheme with applications to computational fluid-dynamics. II. Solutions to parabolic, hyperbolic and elliptic partial differential equations. *Computers & Mathematics with Applications,* **19**, pp. 147–161, 1990.
https://doi.org/10.1016/0898-1221(90)90271-k

[2] Kansa, E.J. & Hon, Y.C., Circumventing the ill-conditioning problem with multiquadric radial basis functions: applications to elliptic partial differential equations. *Computers & Mathematics with Applications,* **39**(7–8), pp. 123–137, 2000.
https://doi.org/10.1016/s0898-1221(00)00071-7

[3] Karageorghis, A., Chen, C.S. & Liu, X-Y., Kansa-RBF algorithms for elliptic problems in axisymmetric domains. *SIAM Journal on Scientific Computing*, **38**, pp. A435–A470, 2016.
https://doi.org/10.1137/15m1037974

[4] Larsson, E. & Fornberg, B., A numerical study of some radial basis function based solution methods for elliptic PDEs. *Computers & Mathematics with Applications,* **46**, pp. 891–902, 2003.
https://doi.org/10.1016/s0898-1221(03)90151-9

[5] Rippa, S., An algorithm for selecting a good value for the parameter c in radial basis function interpolation. *Advances in Computational Mathematics*, **11**, pp. 193–210, 1999.
https://doi.org/10.1023/a:1018975909870

[6] Buhmann, M.D., *Radial Basis Functions: Theory and Implementations*, Cambridge Monographs on Applied and Computational Mathematics, vol. **12**, Cambridge University Press, Cambridge, 2003.

[7] Chen, W., Fu, Z.J. & Chen, C.S., *Recent Advances in Radial Basis Function Collocation Methods*, Springer Briefs in Applied Sciences and Technology, Springer, Heidelberg, 2014.

[8] Fasshauer, G.E., *Meshfree Approximation Methods with MATLAB*, Interdisciplinary Mathematical Sciences, vol. **6**, World Scientific Publishing Co. Pte. Ltd., Hackensack, NJ, 2007.

[9] The MathWorks, Inc., 3 Apple Hill Dr., Natick, MA, *Matlab.*

[10] Fasshauer, G.E., Newton iteration with multiquadrics for the solution of nonlinear PDEs. *Computers & Mathematics with Applications,* **43**, pp. 423–438, 2002.
https://doi.org/10.1016/s0898-1221(01)00296-6

[11] Fasshauer, G.E., Gartland, E.C. & Jerome, J.W., Newton iteration for partial differential equations and the approximation of the identity. *Numerical Algorithms*, **25**, pp. 181–195, 2000.
https://doi.org/10.1023/a:1016609007255

[12] Tsai, C.C., Homotopy method of fundamental solutions for solving certain nonlinear partial differential equations. *Engineering Analysis with Boundary Elements*, **36**, pp. 1226–1234, 2012.
https://doi.org/10.1016/j.enganabound.2012.02.012

[13] Tsai, C.C., Liu, C.S. & Yeih, W.C., Fictitious time integration method of fundamental solutions with Chebyshev polynomials for solving Poisson-type nonlinear PDEs. *Computer Modeling in Engineering and Sciences*, **56**, pp. 131–151, 2010.

[14] Li, M., Chen, C.S. & Karageorghis, A., The MFS for the solution of harmonic boundary value problems with non-harmonic boundary conditions. *Computers & Mathematics with Applications*, **66**, pp. 2400–2424, 2013.
https://doi.org/10.1016/j.camwa.2013.09.004

A CONCEPT OF SEPARATED NUMERICAL FORMULATIONS FOR THE SOLUTION AND EVALUATION OF COMPLEX FIELD PROBLEMS

ANDRÉ BUCHAU & MATTHIAS JÜTTNER
University of Stuttgart, Institute for Theory of Electrical Engineering, Stuttgart, Germany.

ABSTRACT

Nowadays, a variety of numerical methods and numerical formulations exits to solve complex or coupled field problems in three dimensions. Most of them are generally applicable to nearly arbitrary kind of field problems. On the other hand, some highly optimized methods are available, which are predestined for the solution of a specific kind of problem. Especially in the case of weakly coupled multiphysics problems, a mixture of several numerical methods is very advantages to benefit from different properties of numerical methods for diverse physical sub-problems. A very promising approach for a flexible coordination of the related solution process is the application of software agents. Then, the results of one sub-problem are converted into boundary values or volume source distributions for another sub-problem and software agents choose solution methods independently for each sub-problem. Furthermore, two main aspects have to be considered in applications of numerical methods. First, the solution of a boundary value problem should be computed efficiently and second, the solution is evaluated for visualization and interpretation of obtained results. In practice, it is difficult to choose a single appropriate method, which is well suited both for the solution of a problem and its evaluation, since the demands differ in both cases. Here, a concept is presented to apply various numerical methods successfully to the solution and evaluation of complex field problems. Attention is mainly turned on the integration of boundary element methods into the concept of mixed numerical formulations.

Keywords: boundary element methods, coupled problems, finite element methods, post-processing, software agent systems, visualization.

1 INTRODUCTION

In the last years, the interest on the numerical solution of complex and coupled field problems has grown noticeably. Modern computers are equipped with relatively large main memory and modern central processing units (CPU) support the parallel computation of more than eight threads even for floating point arithmetic. Hence, large problems can be solved in acceptable time. Furthermore, very good implementations of general applicable numerical methods like the finite element method (FEM) or boundary element method (BEM) exist. Along with the increasing computation capability, the demand on numerical field simulations grows, too. Instead of stationary simulations, time-dependent problems including non-linear material properties and different physical aspects are examined. Afterwards, advanced post-processing techniques enable expressive evaluations and visualizations of computed results.

A classical approach for the solution of the aforementioned problems is to treat them monolithically as a single complex problem. All physical aspects are summarized into one equation system and one numerical method and one formulation is chosen, which solves the total problem. The post-processing is executed using the same model along with the same numerical method as for the solution of the equation system. An advantage of such an approach is that the numerical model is very precise and clear. On the other hand, the properties of the chosen numerical method must fit to all steps of the simulation process. Hence, the choice of an appropriate numerical method is a compromise between general

DOI: 10.2495/CMEM-V6-N6-1008-1018

applicability and efficiency. Furthermore, this choice is a challenging task, especially for users of numerical methods.

A decomposition of a single field problem into several sub-problems is a very promising approach to tackle complex problems. Numerical methods can be chosen separately for the solution of each sub-problem and for its post-processing. That means, properties of numerical methods fit much better to the demands of partial simulation steps than in the case of only one method for the total problem. On the other hand, data must be prepared for an exchange between different sub-problems.

Here, a very flexible and powerful concept is presented, which combines advantages of different numerical methods and formulations for the solution of complex or coupled field problems. The focus is on boundary element methods including various formulations, compression techniques, and post-processing approaches. Next, several mixtures of boundary element methods with other numerical methods are discussed. Classical coupling of numerical methods is considered as well as modern techniques like software agents.

2 BOUNDARY ELEMENT METHODS

The main idea of the boundary element method (BEM) is to describe a field problem only based on values on domain boundaries (Brebbia *et al.* [1]). Hence, a modelling and discretization of surfaces suffices to completely define the considered problem. Therefore, BEM is a very attractive method for science and engineering applications (Buchau [2]). Especially open boundary problems or problems with large distances between objects are solved very efficiently and with high accuracy.

2.1 BEM formulations

A concrete BEM formulation depends on the specific physical problem, which is solved. Furthermore, several formulations are possible for the same problem. For instance, a static electric field $\boldsymbol{E}$ can be described completely using a scalar electric potential u with

$$\boldsymbol{E} = -\nabla u. \tag{1}$$

Based on Green's theorem the potential u in a point $\boldsymbol{r}$ within a closed domain Ω is obtained from the value of u at the domain boundary $\partial\Omega$ and its normal derivative $\frac{\partial u}{\partial n'}$ by

$$u(\boldsymbol{r}) = \int_{\partial\Omega} G(\boldsymbol{r},\boldsymbol{r}') \frac{\partial u(\boldsymbol{r}')}{\partial n'} \mathrm{d}A' - \int_{\partial\Omega} \frac{\partial G(\boldsymbol{r},\boldsymbol{r}')}{\partial n'} u(\boldsymbol{r}') \mathrm{d}A' \tag{2}$$

with Green's function of three-dimensional Laplace equation

$$G(\boldsymbol{r},\boldsymbol{r}') = \frac{1}{4\pi} \frac{1}{|\boldsymbol{r}-\boldsymbol{r}'|}. \tag{3}$$

Equation (2) is only applicable, if objects with a closed surface and a defined volume are considered. In the case of very thin objects, it would be necessary to add artificial boundaries to close all surfaces. Then, another BEM formulation

$$u(\boldsymbol{r}) = \frac{1}{\epsilon_0} \int_A \sigma(\boldsymbol{r}') G(\boldsymbol{r},\boldsymbol{r}') \mathrm{d}A' \tag{4}$$

based on surface charge densities σ is more suited. The free space permittivity is ϵ_0 and A are the surfaces of closed or thin objects.

An important distinction between eqn (2) and eqn (4) is the scope where the equations can be applied. Equation (2) is valid only inside a closed domain. In the case of multiple domains, several sets of boundary values are required to describe the field for each domain. In contrast, eqn (4) is valid in total space and no distinction between domains is necessary even in the case of multiple closed domains.

In practice, both BEM formulations are applied for the solution of static electric field problems. The direct formulation using eqn (2) is for instance very advantages for problems with layered objects, since then the matrix of the corresponding linear system of equations is relatively sparse. The indirect formulation based on eqn (4) is well suited for thin structures. In total, there are pros and cons for both BEM formulations and it depends on the concrete example in which formulation is more precise and more efficient.

However, the computation of the solution of a problem is an important step but only a first step in numerical methods. Evaluation and interpretation of results is mandatory in engineering applications.

A common approach is to use the same BEM formulation for both, the computation of the solution of problem and the post-processing. Then, computed values on boundaries can be used directly for post-processing. Furthermore, it is necessary to store only formerly unknown values and to integrate known boundary values.

That means in the example of a static electric field that the potential is given at the electrodes and it is not necessary to store potential values in the element nodes of electrodes. Only its normal derivative is computed and stored for each element node. During the post-processing, eqn (2) is evaluated using the given boundary values and using the computed missing boundary values.

An advantage of such a self-evident approach is that the amount of required main memory is minimized and no unnecessary computations are executed. A significant drawback is that the approach is very restricted. All details of the formulation, which are necessary for the solution of the problem, must be considered and implemented for the post-processing, too.

Nowadays, enough main memory is available for most applications. Hence, a more flexible approach, which consumes a little more memory, is recommended. Here, the solution of the problem and the post-processing are consequently separated. That means the solution of the problem and its post processing are computed independently of each other.

The solution of the problem is calculated using a BEM formulation, which is predestined for a stable and efficient solution of the problem. All computational steps are optimized for the solution process. Then, the solution is converted into boundary values for the post-processing. These values contain both original boundary values and computed values on the boundaries.

The post-processing step is significantly simplified, if all given and computed boundary values are treated in the same manner. In case of a direct formulation like eqn (2) is applied, potential and its normal derivative are stored for each element node. Then, no distinction is necessary for the evaluation of eqn (2) during the post-processing. Furthermore, the BEM formulation can be changed for the post-processing. For instance, the solution of the problem can be obtained using a direct BEM formulation and the post-processing can be done based on an indirect formulation and vice versa.

Another benefit of that separation is that different BEM implementations can be used. The solution is computed and stored in a general applicable format independently of the formulation and problem description. Another process imports the boundary element mesh along with all nodal values. That enables flexible field computations and impressive visualization techniques for arbitrary BEM problems (Buchau & Rucker [3]).

2.2 Matrix compression techniques

Since the matrix of the linear system of equations is dense or block-wise dense, a matrix compression technique is required to solve complex field problems. In the last decades, several matrix compression techniques have been developed. They exploit the dependency of Green's function on the distance between source point and evaluation point and approximate Green's function in the case of far-field interactions.

One class of matrix approximation techniques is based on series expansions of Green's function along with hierarchical grouping schemes for the boundary elements. The most popular representative of this class is the fast multipole method (FMM) (Greengard & Rokhlin [4]). The series expansion depends on the used Green's function. Hence, each kind of problem or basic equation, respectively, requires that series expansion in spherical coordinates for three-dimensional problems. Furthermore, transformations of these series expansions must be available, which convert the coefficients of the series expansion from one local coordinate system into coefficients of the series expansion of another local coordinate system. Unfortunately, the computation of these transformations is relatively expansive. On the other hand, the FMM is a very flexible tool both for the solution of the problem and for the post-processing, since it depends only on geometrical position of boundary elements and evaluation points (Buchau *et al.* [5]). In both cases, very high compression rates are obtained and very large problems can be solved efficiently.

Another class of compression techniques is based on direct matrix operations. This is used for instance for the adaptive cross approximation technique (ACA) (Bebendorf [6]). The ACA applies a hierarchical matrix clustering based on the distance between source point and evaluation point. Thus, an excellent numbering of element nodes is required to achieve high compression rates. Each matrix block is then approximated by a few rows and columns. Since the computed rows and columns correspond to those of the original dense matrix, the computation of matrix elements is the same for the ACA as for the classical BEM. Therefore, the ACA is applicable to general BEM formulations. Some modifications of matrix assembly are necessary to avoid multiple expensive computations of integrals over the same boundary element. The ACA is very advantageous for the solution of the linear system of equations with a large number of iterations, for instance, in combination with a time-dependent or non-linear solver, because matrix by vector products are evaluated very fast. On the other hand, post-processing based on ACA is limited to regular grids of evaluation points but a flexible post-processing is not possible.

In total, matrix compression techniques are very attractive for BEM computations, since both the computational costs and memory requirements are significantly reduced (Buchau *et al.* [7]). The choice of an appropriate matrix compression technique depends on the concrete problem and BEM formulation as well as on the desired post-processing. If the same matrix compression technique is applied both to the solution of the problem and the post-processing, an optimal method is difficult to find. Here, a mixture of matrix compression techniques is recommended. A separated treatment of solution step and post-processing step as discussed in the previous section enables an independent choice of matrix compression technique, too. Furthermore, it is possible to choose different algorithms or implementations for the same matrix compression method.

The structure of the matrix of the linear system of equations is predefined by the given boundary element mesh. Hence, a matrix compression method can be highly optimized and parallelized for that case. An important property of a fixed boundary element mesh is that clustering and grouping for ACA or FMM is fixed, too. Both matrix block of ACA and octree structure along with transformation coefficients of the FMM can be precomputed and reused in each iteration step of an iterative solver of the linear system of equations (Axelsson [8]).

If not a fixed grid is used for the post-processing but a flexible meshfree post-processing is applied (Buchau & Rucker [3]), adapted versions of matrix compression techniques are more efficient (Buchau & Rucker [9]). There, the position of evaluation points is determined during the calculation of visualization objects like field lines. Hence, evaluation points are added to the hierarchical octree based grouping scheme of FMM during the post-processing. This requires a flexible implementation of both the octree and the FMM algorithm. Furthermore, a reversed FMM algorithm reduces the number of FMM operations significantly. A drawback is that optimization and parallelization is more difficult.

As demonstrated in that example, even the same matrix compression technique should be used in very different manners for the solution of the problem and for the post-processing. Hence, different implementations are often advisable. Then, all aspects of performance and flexibility can be considered appropriately.

2.3 Post-processing

Post-processing BEM operations differ significantly from BEM operations for the solution of the problem, not only for the matrix compression technique as shown in the previous section. A second important topic is BEM integrals, especially singular or nearly singular integrals.

In many cases, the number of considered boundary element types is limited. Hence, a small number of singular or nearly singular integral types have to be implemented. Then, these integrals are normally highly optimized to shorten CPU time.

In the case of post-processing, different integral kernels are used. The reason is that not only the field value, which is the solution of the problem, but also derivatives of that field value are evaluated. Of course, an optimized implementation with low computational costs is favoured but flexibility has higher priority in that case. Singular, nearly singular, nearly strong singular or nearly hyper-singular integrals can be implemented in a relatively general way. For instance, sub-division of boundary elements for an application of local polar coordinates can be generalized. Then, special treatment of these integrals is shifted into weights of integration points in the implementation. The overhead for some additional multiplications, which can be summarized in an optimized implementation, is relatively small but a huge number of element types, formulations, and field values can be taken into account.

Another advantage of a separated post-processing is that different techniques can be applied. A classical approach is to create a grid or auxiliary volume mesh and to compute field values in that mesh nodes. Then, post-processing tools for FEM are applicable. The main drawback of such an approach is that field values are computed in points, which are not necessary for visualization. On the other hand, implementation is relatively easy and already existing code can be reused.

If post-processing is separated from the solution of the problem, only boundary values of for instance the potential are transferred to the post-processing tool. The evaluation of these values can be done with an arbitrary BEM formulation, an arbitrary compression technique and an arbitrary implementation of BEM integrals. The post-processing computations can be integrated into the visualization tool or visualization objects are computed separately and then transferred to the visualization tool (Buchau & Rucker [3]).

For instance, basic concepts of FMM and isovalue search are common (Wilhelms & Gelder [10], Buchau & Rucker [11]). An integration of BEM with FMM into an algorithm for computation of isosurfaces offers the opportunity to interweave both. The same octree is then used to group boundary elements and evaluation points for a reversed FMM algorithm and FMM series expansions are applied to interpolate field values for an isovalue search method with high accuracy.

Furthermore, an integration of BEM operations directly into the post-processing computations enable a much better control of accuracy of computed visualization objects. Adaptive integration methods like a Runge–Kutta–Fehlberg method adjust step size automatically for precise field line computations (Fehlberg [12]). However, a fixed grid of evaluation points would limit the total accuracy by rough interpolations between the evaluation points but a direct evaluation of BEM integrals improves accuracy significantly although the computational costs are kept low.

3 COUPLED NUMERICAL METHODS

Boundary element methods are applicable to a wide range of problems. However, non-linear or inhomogeneous media cannot be integrated into BEM easily. In that case, the finite element method (FEM) is very attractive. A classical approach is to couple BEM and FEM by integrating both methods into one monolithic equation system. Another approach is to extend a coupling of BEM with other methods within the meaning of the concept presented in the previous section.

Although FEM is well suited to consider nearly arbitrary material properties, modelling of large surrounding air domains is very expensive. Boundary layers modelled with infinite elements represent alternatives, but unfortunately field values must also be computed precisely in air domains in many engineering applications. A very interesting approach is to apply FEM for the solution of the problem and BEM for the post-processing. Then, the FEM model is acceptable small and the surrounding domain is discretized as small as possible. Of course, field values in the surrounding domain are too inaccurate but field values at the surfaces of examined objects are accurate enough. These values are used for a BEM post-processing.

A typical situation for that approach is the computation of the radiated electromagnetic field of a smartphone antenna. The structure of a modern smartphone is very complex especially the antenna system. Hence, it is advisable to model the device using FEM. Then, a very small air domain along with absorbing boundary conditions suffices to compute the electromagnetic field of the device with high accuracy. Unfortunately, the radiated electromagnetic field in the air domain is of large interest for the design of a smartphone. However, a small air domain avoids visualization of the radiation pattern in the far-field. Here, a BEM post-processing is a promising alternative method. The complex structure of the device does not matter, since only boundary values on the device surface are needed for BEM. The infinite surrounding air domain is very precisely considered by Green's function. Especially a mesh-free BEM post-processing results in accurate and expressive visualizations of the radiated electromagnetic field.

A combination of BEM with other methods like FEM for general problems or physical optics or universal theory of diffraction for electromagnetic wave problems is very attractive for the computation of the solution of the problem. Instead of a classical coupling of these methods, an application of software agent systems offers completely new possibilities (Jüttner *et al.* [13]). A complex problem is subdivided into smaller problems similar to domain decomposition methods. Each sub-problem is solved by a software agent, whose aim is to solve its local sub-problem most efficiently. The advantage of that approach is that arbitrary numerical methods along with their implementation can be used without modification of the corresponding code. Only an interface for data exchange and its coordinate based evaluation is required, which enables the independent selection of an appropriate numerical method for each sub-problem. Thus, boundary or domain values are computed and exchanged with software agents of all other sub-problems. Software agents not only coordinate the solution process on their own, but they also have the ability of choosing a numerical method. Hence, this approach is very attractive for a combination of several well-developed software packages with different numerical methods. Same holds for the visualisation of simulation results. Since software agents solve only sub-problems, the post-processing can benefit from the separated results by processing data only within the region interest.

The concept of coupling of different numerical methods for the solution of a complex problem can be extended to the solution of multiphysics field problems (Jüttner *et al.* [14]). There, different physical aspects of a complex problem are assigned to software agents, which are on the top layer of the solution process. Each physical sub-problem is handled by a software agent, which chooses the numerical method for the solution of the sub-problem. Based on performance criteria rated by a user this results in a mixture of different numerical methods within a solution sequence. Furthermore, also different numerical libraries and solver configurations can compete within a calculation process to provide the result as fast as possible (Jüttner *et al.* [15]). Alternatively, also the calculation of parameter sweeps and optimisation tasks benefit from this naturally parallel setup.

A significant advantage of solving a multiphysics field problem with the help of software agents is that not only numerical methods are chosen for each physical problem but also different meshes can be used. Examples are overlapping domains within a multiphysics problem where the physics suggest different meshes based on the considered different material properties. The values, which are needed for another physical sub-problem, are computed in the same manner as the aforementioned post-processing approach. Furthermore, a software agent is able to fetch values from other agents, which compute these values on demand.

For some multiphysics problems the coupling is relatively weak. For instance, if the temperature distribution inside an electric conductor is approximately homogeneous, electric conductivity is homogeneous, too. Then, a BEM can be used for the solution of the sub-problem. If the temperature distribution is inhomogeneous along with inhomogeneous electric conductivity, a FEM is preferred. This change of the numerical method easily realized by the software agent system. This also includes iterative domain decomposition approaches based on the Schwarz formulation. Sub-problems with a strong coupling benefit from the redundancy and the fault tolerance during the calculation when using a segregated and distributed approach.

4 NUMERICAL RESULTS

Two examples are considered to demonstrate that the presented concepts work reliable and advantageous in practice.

4.1 High voltage insulator

The electric field of a high voltage insulator is studied as a first example. Since the insulator has linear material properties, an indirect BEM formulation has been chosen to compute the electric field strength between the electrodes inside the insulator and in the surrounding air domain.

The surfaces of the device have been discretized using second order quadrilateral boundary elements. A Galerkin method is applied to obtain a linear system of equations, which is solved iteratively using GMRES with a Jacobi preconditioner. The matrix of the equation system is compressed with the help of a classical FMM. Singular integrals are computed using highly optimized code. All methods for the solution of the problem are implemented in C++ and parallelized with OpenMP standard.

The results are extracted for a post-processing based on another BEM implementation. There, the focus was on flexibility and not on maximum performance. The post-processing software is implemented in C# and has interfaces to the above-mentioned BEM solver and to commercial FEM code. Meshes are imported too. Since the post-processing is independent of the solution, order of elements or type of elements can be changed. Especially linear triangles are often used in visualization because of hardware supported rendering capabilities. Singular integrals are implemented in a very flexible manner to enable the treatment of all kind of boundary elements. Visualization algorithms are integrated to apply a powerful meshfree computation of field lines (Fig. 1) or isosurfaces (Fig. 2).

4.2 Wet casing of an antenna including a lens

Now, the propagated wave of an antenna within its wet casing with an inserted lens is evaluated. Therefore, a FEM model of the electric field as dependent variable is created and

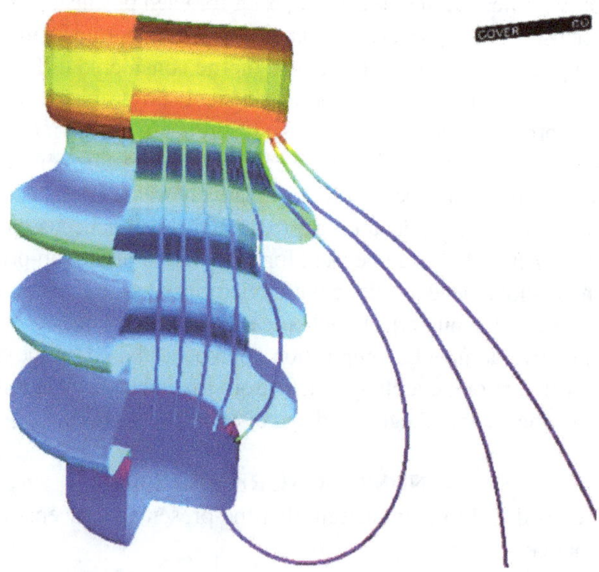

Figure 1: Field lines of the electric field strength of a high voltage insulator.

Figure 2: Isosurface of the electric potential of a high voltage insulator.

updated in a next engineering step by an additional description of the temperature. This leads to a bi-directionally coupling between the dependent variables and enables the regard of electric losses within the materials and their temperature dependent behaviour within the simulation.

The applied software agent system performs a physics based decomposition of the problem. Since the initial model only contains a description of the electric field, redundant resources are used automatically to evaluate a best performing solver. The proposed solver is used after updating the model. The physics based decomposition now generated sub-problems that use the already calculated results as initial values for the consequently following iterative calculation process. This decomposition further allows the separated treatment of the mixed complex and real degrees of freedom by using optimised numerical libraries. A sub-problem dependent application of the methods BEM and FEM for different domains is also possible (Jüttner *et al.* [15]).

The overall result of the calculation process is shown in Fig. 3. For its calculation, different discretisations for the sub-problems were used. Different shape functions were used by the software agents due to the considered physics. Distinct discretisation were chosen since the material properties within the physics differ.

5 CONCLUSIONS

A concept for an application of different numerical methods for the solution of complex field problems has been shown. Especially boundary element methods are predestined for such a coupling of numerical methods. A separation of solution of the problem and post-processing

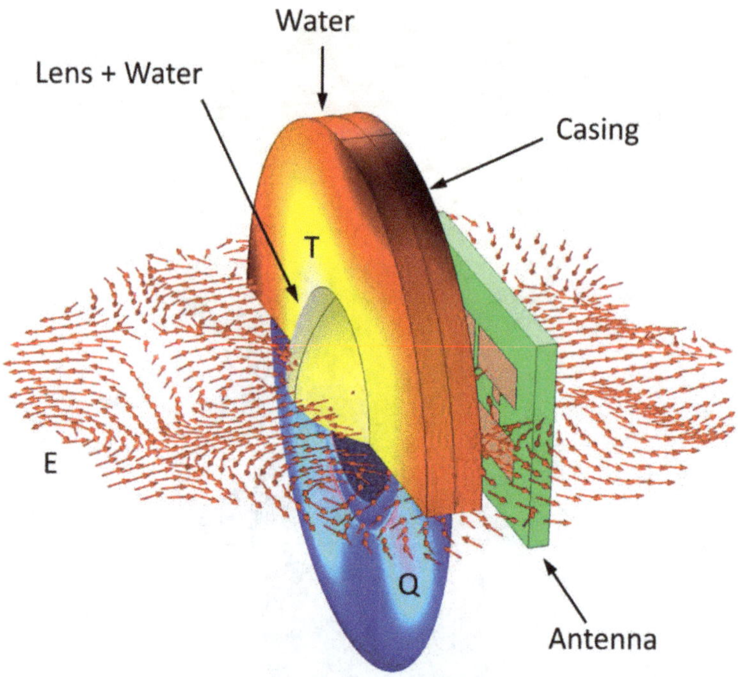

Figure 3: Multiphysics result for a heated lens within its casing by a patch antenna array.

enables rich possibilities for efficient and flexible computations. Different algorithms and implementations of the same method or different methods can be combined in nearly arbitrary manner. Modern software techniques like software agents support this concept and extend it to the solution of complex multiphysics field problems.

REFERENCES

[1] Brebbia, C.A., Telles, J.C.F. & Wrobel, L.C., (eds.), *Boundary Element Techniques*, Springer-Verlag: Berlin and New York, 1984.

[2] Buchau, A., *Application of Fast Boundary Element Methods in Engineering and Science*. Proceedings of ICCES 2017 International Conference on Computational & Experimental Engineering and Science, Funchal, 2017.

[3] Buchau, A. & Rucker, W.M., Feasibility of a meshfree post-processing for boundary element methods. *Boundary Elements and Other Mesh Reduction Methods XXXVIII*, pp. 327–338, 2015.
https://doi.org/10.2495/bem380261

[4] Greengard, L. & Rokhlin, V., *A new version of the Fast Multipole Method for the Laplace equation in three dimensions.* Acta Numerica, pp. 229–269, 1997.

[5] Buchau, A., Rieger, W. & Rucker, W.M., Fast field computations with the fast multipole method. *COMPEL*, **20**(2), pp. 547–561, 2001.
https://doi.org/10.1108/03321640110383861

[6] Bebendorf, M., Approximation of boundary element matrices. *Numerische Mathematik*, **86**(4), pp. 565–589, 2000.
https://doi.org/10.1007/pl00005410

[7] Buchau, A., Rucker, W.M., Rain, O., Rischmüller, V., Kurz, S. & Rjasanow, S., Comparison between different approaches for fast and efficient 3D BEM computations. *IEEE Transactions on Magnetics*, **39**(3), pp. 1107–1110, 2003. https://doi.org/10.1109/tmag.2003.810167

[8] Axelsson, O., *Iterative Solution Methods*, Cambridge University Press, Cambridge, UK, 1996

[9] Buchau, A. & Rucker, W.M., Meshfree computation of field lines across multiple domains using fast boundary element methods. *IEEE Transactions on Magnetics*, **51**(3), pp. 1–4, 2015. https://doi.org/10.1109/tmag.2014.2359520

[10] Wilhelms, J. & van Gelder, A., Octrees for faster isosurface generation. *ACM Transactions on Graphics*, **11**(3), pp. 201–227, 1992. https://doi.org/10.1145/130881.130882

[11] Buchau, A. & Rucker, W.M., A meshfree isosurface computation method for boundary element methods. *International Journal of Computational Methods and Experimental Measurements*, **5**(5), pp. 647–658, 2017.

[12] Fehlberg, E., Klassische Runge-Kutta-Formeln fünfter und siebenter Ordnung mit Schrittweiten-Kontrolle. *Computing*, **4**(2), pp. 93–106, 1969. https://doi.org/10.1007/bf02234758

[13] Jüttner, M., Buchau, A., Rauscher, M., Rucker, W.M. & Göhner, P., *Software Agent Based Domain Decomposition Method*, 15th International IGTE Symposium, Graz, Austria, pp. 89–94, 2012.

[14] Jüttner, M., Buchau, A., Vögeli, D., Rucker, W.M. & Göhner, P., *Iterative Software Agent Based Solution of Multiphysics Problems,* Scientific Computing in Electrical Engineering, Springer, pp. 123–131, 2016. https://doi.org/10.1007/978-3-319-30399-4_13

[15] Jüttner, M., Grabmaier, S., Vögeli, D., Rucker, W.M. & Göhner, P., Coupled multiphysics problems as market place for competing autonomous software agents. *IEEE Transactions on Magnetics*, **53**(6), pp. 1–4, 2017. https://doi.org/10.1109/tmag.2017.2659228

FUNDAMENTAL SOLUTIONS FOR THE GENERAL LAMINATE PROBLEM WITH THE STRESS FUNCTION FORMALISM

STAVROS SYNGELLAKIS
Wessex Institute, UK.

ABSTRACT

The linear coupled stretching-bending problem for general laminates is here formulated with the mid-plane stress function and the lateral deflection as independent field variables. A mathematical similarity between the two problems is achieved by introducing a re-arranged mid-plane strain tensor as one of the dependent variables. As a step towards a genuine boundary element solution for this problem, its fundamental solutions are derived using a Fourier transform approach. First, the transforms of the solutions are obtained in terms of the transform space variables and their inverses are deduced using complex integral calculus. Through the use of these fundamental solutions, boundary integral equations of the linear coupled stretching-bending problem are formulated without the presence of any irreducible domain integrals. Issues regarding the numerical implementation of this formulation are raised and discussed.

Keywords: boundary integral equations, coupled stretching-bending, fundamental solutions, general laminates, stress function formalism.

1 INTRODUCTION

Composites have become attractive alternatives to traditional engineering materials because of their high specific stiffness and strength. They can be engineered to meet specific structural requirements through the choice of reinforcement and matrix materials, the volume fractions of fibre and matrix and the fabrication method. Laminated plates, in particular, are designed to have desired properties through the choice of layer orientation, number of layers in a given direction, thickness of individual layers, type of layer and the layer stacking sequence.

Laminated plates are modelled and analysed as elastic anisotropic solids. The anisotropy of a laminate depends on the anisotropy of its individual layers and the layer stacking sequence. The most complex form of such anisotropy, exhibited by general, non-symmetrically laminated plates, causes coupled stretching-bending response to applied in-plane and lateral forces.

Exact analyses of general laminates have been performed in cases of special material symmetry that eliminates a number of elastic constants without removing the coupling effect. Static solutions under simple loadings [1] were extended to free flexural vibration and buckling problems [2, 3] under a variety of boundary conditions [4]. A post-buckling analysis under simple in-plane loading was also performed [5]. Numerical post-buckling solutions, based on finite element or finite difference methods and accounting for the stretching-bending coupling can also be found in the technical literature [6].

Through the boundary element method (BEM), complex plate geometries as well as loading and boundary conditions can be accommodated with the dimensionality of the problem being reduced by one. Numerical analysis schemes for general laminates, based on boundary element methodology, have been proposed although not fully implemented. Previous such analyses relied on the fundamental solutions of the two uncoupled, extensional and flexural problems [7]. The coupling of the two problems resulted in irreducible domain integrals and the proposed solution relied on the adoption of an iterative scheme.

DOI: 10.2495/CMEM-V6-N6-1019-1032

For a genuine BEM formulation, the fundamental solutions for the general laminate problem are essential. Explicit forms of such solutions have been derived [8] through an elaborate approach leading to expressions of considerable complexity. The mathematical basis of a BEM scheme relying on these solutions has also been developed [9].

The in-plane extensional problem can be formulated in two ways, that is, either in terms of the stress function [1, 3, 5, 7] or the in-plane displacements [2, 4] as field variables. Although the latter approach is more versatile in accounting for in-plane boundary conditions, the former is more mathematically elegant and possibly more suited to the case of stability problems when the applied contour traction is usually known. Fundamental solutions for the general laminate problem with the displacement formalism have been obtained using Fourier transforms [10]. In this paper, the same approach is applied to the problem formulated in terms of the stress function and the lateral deflection as field variables. The mathematical similarity between the extensional and flexural problems facilitates the derivation of fundamental solutions in very similar compact forms. Starting from a reciprocity relation, integral equations are derived by repeated applications of the Green's theorem. Introducing the derived fundamental solutions into these integral equations leads to a boundary element formulation without any irreducible domain integrals.

2 LINEAR LAMINATE THEORY

According to the classical lamination theory, the plate is assumed to be perfectly laminated consisting of an arbitrary number of discrete layers, each individual layer being homogeneous through its thickness and in a state of plane stress. The laminate is also assumed to deform according to Kirchhoff's assumptions for the bending of thin plates. According to this theory, the membrane forces $N_{\alpha\beta}$ and bending moments $M_{\alpha\beta}$ are related to the curvatures $\kappa_{\alpha\beta}$ and mid-plane strains $\varepsilon_{\alpha\beta}$ by

$$N_{\alpha\beta} = A_{\alpha\beta\gamma\delta}\varepsilon_{\gamma\delta} + B_{\alpha\beta\gamma\delta}\kappa_{\gamma\delta} \tag{1}$$

$$M_{\alpha\beta} = B_{\alpha\beta\gamma\delta}\varepsilon_{\gamma\delta} + D_{\alpha\beta\gamma\delta}\kappa_{\gamma\delta} \tag{2}$$

where $A_{\alpha\beta\gamma\delta}$ and $D_{\alpha\beta\gamma\delta}$ are, respectively, the extensional and flexural rigidities, $B_{\alpha\beta\gamma\delta}$ are extensional-flexural coupling coefficients and repeated Greek indices mean summation over their range, which is from 1 to 2. It is evident from their derivation that all three material tensors are symmetric with respect to their pair of indices $\alpha\beta$ and $\gamma\delta$. Strains and curvatures are, respectively, related to the mid-plane displacements u_α and deflection w by

$$\varepsilon_{\alpha\beta} = \frac{1}{2}(u_{\alpha,\beta} + u_{\beta,\alpha}) \tag{3}$$

$$\kappa_{\alpha\beta} = -w_{,\alpha\beta} \tag{4}$$

where a comma followed by a lower index indicates differentiation with respect to the corresponding co-ordinate. Finally, forces and moments should satisfy the equations of equilibrium

$$N_{\alpha\beta,\beta} + f_\alpha = 0 \tag{5}$$

$$M_{\alpha\beta,\alpha\beta} + q = 0 \tag{6}$$

where f_α is the body force assumed to be derivable from a potential function Φ according to

$$f_\alpha = -\Phi_{,\alpha} \tag{7}$$

and q is the lateral pressure.

The field equations are complemented by the boundary conditions. Along the smooth portions of a boundary contour Γ with normal and tangent unit vectors **n** and **s**, respectively, the field variables should satisfy the conditions

$$p_{\alpha} = n_{\beta} N_{\alpha\beta} \tag{8}$$

$$V_n = n_{\alpha} M_{\alpha\beta},_{\beta} + \frac{\partial M_{ns}}{\partial s} \tag{9}$$

$$M_n = n_{\alpha} n_{\beta} M_{\alpha\beta} \tag{10}$$

where p_{α}, V_n and M_n are, respectively, the in-plane traction, shear force and bending moment, while, at any corner point j, the discontinuity jump of the twisting moment M_{ns}, given by

$$M_{ns} = s_{\alpha} n_{\beta} M_{\alpha\beta}, \tag{11}$$

should be equal to a concentrated force C_j.

If forces and moments are pre-scribed along the whole or part of Γ, p_{α}, V_n, M_n and C_j have known values there; alternatively, mid-plane displacements, deflection and deflection gradient θ_n may be pre-scribed along the whole or part of Γ, in which case p_{α}, V_n, M_n and C_j become the boundary unknowns of the problem.

3 STRESS FUNCTION FORMALISM

The problem can be re-formulated in terms of a stress function F such that

$$N_{\alpha\beta} = L_{\alpha\beta} F + \Phi \delta_{\alpha\beta} \tag{12}$$

where $\delta_{\alpha\beta}$ is the Kronecker delta and the operator $L_{\alpha\beta}$ defined by

$$L_{\alpha\beta} = \frac{\partial^2}{\partial x_{\kappa} \partial x_{\kappa}} \delta_{\alpha\beta} - \frac{\partial^2}{\partial x_{\alpha} \partial x_{\beta}} \tag{13}$$

In-plane equilibrium, that is, eqn (5) with the body force given by eqn (7), is identically satisfied by the distributed internal forces given by the expressions of eqn (12). The stress function also needs to satisfy the compatibility condition imposed on the mid-plane strains, namely

$$L_{\alpha\beta} \varepsilon_{\alpha\beta} = \bar{\varepsilon}_{\alpha\beta},_{\alpha\beta} = 0 \tag{14}$$

where

$$\bar{\varepsilon}_{\alpha\beta} = \varepsilon_{\kappa\kappa} \delta_{\alpha\beta} - \varepsilon_{\alpha\beta} \tag{15}$$

is a re-arranged strain tensor introduced so that the compatibility condition, eqn (14), is mathematically similar to the second equilibrium eqn (6). Compatibility is imposed on the field variables by re-arranging the constitutive eqn (1). For this purpose, it is necessary to introduce the inverse of the extensional rigidity tensor $A_{\alpha\beta\gamma\delta}$ through

$$A^{-1}_{\alpha\beta\gamma\delta} A_{\gamma\delta\lambda\mu} = \delta_{\alpha\lambda} \delta_{\beta\mu} \tag{16}$$

Next, $N_{\alpha\beta}$ and $\kappa_{\alpha\beta}$ are substituted from eqns (12) and (4), respectively, into constitutive eqn (1), which is solved for $\varepsilon_{\alpha\beta}$ and the result substituted into eqn (15) to give

$$\bar{\varepsilon}_{\alpha\beta} = \hat{A}_{\alpha\beta\gamma\delta} F,_{\gamma\delta} + \hat{B}_{\alpha\beta\gamma\delta} w,_{\gamma\delta} + \bar{A}_{\alpha\beta\kappa\kappa} \Phi \tag{17}$$

where

$$\hat{A}_{\alpha\beta\gamma\delta} = A^{-1}_{\kappa\kappa\lambda\lambda}\delta_{\alpha\beta}\delta_{\gamma\delta} - A^{-1}_{\alpha\beta\kappa\kappa}\delta_{\gamma\delta} - A^{-1}_{\kappa\kappa\gamma\delta}\delta_{\alpha\beta} + A^{-1}_{\alpha\beta\gamma\delta} \tag{18}$$

$$\hat{B}_{\alpha\beta\gamma\delta} = \bar{A}_{\alpha\beta\lambda\mu}B_{\lambda\mu\gamma\delta} \tag{19}$$

and

$$\bar{A}_{\alpha\beta\gamma\delta} = A^{-1}_{\kappa\kappa\gamma\delta}\delta_{\alpha\beta} - A^{-1}_{\alpha\beta\gamma\delta} \tag{20}$$

It should be noted that, among the tensors defined above, only that given by eqn (18) is symmetric with respect to the pair of indices $\alpha\beta$ and $\gamma\delta$.

The second constitutive equation also needs to be expressed in terms of the same field variables. Thus the expression for $\varepsilon_{\alpha\beta}$ obtained by substituting $N_{\alpha\beta}$ and $\kappa_{\alpha\beta}$ from eqns (12) and (4), respectively, into eqn (1) and $\kappa_{\alpha\beta}$ from eqn (4) are substituted into eqn (2) to give

$$M_{\alpha\beta} = \hat{B}_{\gamma\delta\alpha\beta}F,_{\gamma\delta} - \hat{D}_{\alpha\beta\gamma\delta}w,_{\gamma\delta} + \bar{B}_{\alpha\beta\kappa\kappa}\Phi \tag{21}$$

where

$$\hat{D}_{\alpha\beta\gamma\delta} = D_{\alpha\beta\gamma\delta} - B_{\alpha\beta\lambda\mu}A^{-1}_{\lambda\mu\nu\xi}B_{\nu\xi\gamma\delta} \tag{22}$$

is the reduced flexural stiffness tensor and

$$\bar{B}_{\alpha\beta\gamma\delta} = B_{\alpha\beta\lambda\mu}\, A^{-1}_{\lambda\mu\gamma\delta}. \tag{23}$$

It is evident from eqn (22) that the reduced flexural stiffness tensor is symmetric with respect to the pair of indices $\alpha\beta$ and $\gamma\delta$ while the tensor defined by eqn (23) is not. However, it is evident from the new constitutive eqns (17) and (21) that the full elasticity tensor comprising the 4th-order tensors defined by eqns (18), (19), (22) and relating mid-plane strains and bending moments to the second derivatives of the field variables is symmetric.

The requirement that the mid-plane strains satisfy compatibility, eqn (14), and the bending moments satisfy equilibrium, eqn (6), leads to the 4th order differential equations

$$\hat{A}_{\alpha\beta\gamma\delta}F,_{\alpha\beta\gamma\delta} + \hat{B}_{\alpha\beta\gamma\delta}w,_{\alpha\beta\gamma\delta} = -\bar{A}_{\alpha\beta\kappa\kappa}\Phi,_{\alpha\beta} \tag{24}$$

$$\hat{B}_{\gamma\delta\alpha\beta}F,_{\alpha\beta\gamma\delta} - \hat{D}_{\alpha\beta\gamma\delta}\, w,_{\alpha\beta\gamma\delta} = -q - \bar{B}_{\alpha\beta\kappa\kappa}\,\Phi,_{\alpha\beta} \tag{25}$$

The formulation is completed with the determination of the boundary conditions that replace eqn (8), that is, conditions satisfied by F along the boundary contour. Using eqns (12), it has been shown that, at any point $Q(\bar{x}_1, \bar{x}_2)$ along the boundary,

$$F = \int_O^Q [(x_1 - \bar{x}_1)p_2 - (x_2 - \bar{x}_2)p_1]\mathrm{d}\Gamma \tag{26}$$

$$\frac{\partial F}{\partial n} = -s_1(Q)\int_O^Q p_1 \mathrm{d}\Gamma - s_2(Q)\int_O^Q p_2 \mathrm{d}\Gamma \tag{27}$$

where O is an arbitrarily located origin. According to eqn (26), F can be physically interpreted as the resultant moment about Q of the traction over OQ. Similarly, eqn (27) describes the normal derivative of F as the component of the resultant traction over OQ in the direction $-\mathbf{s}$ at Q. It is evident from eqns (26) and (27) that F and its normal derivative can be identified along Γ only if traction has been specified along the whole boundary. If kinematic mid-plane conditions are imposed on parts of or the whole boundary, additional relations need to be adopted for a consistent formulation of the problem.

4 FUNDAMENTAL SOLUTIONS

A fundamental solution is defined as the response of an infinite domain to unit concentrated action, that is, force or moment. The force is applied at the source point P(ξ_a) while the response is determined at the field point Q(x_a) as illustrated in Fig. 1. Solutions to eqns (24) and (25) are therefore sought for mid-plane body forces

$$f_a = \delta_{\lambda a}\delta\,(\mathbf{x}-\xi) \tag{28}$$

that is, unit forces acting in directions parallel to the co-ordinate axes, a transverse unit point force

$$q = \delta\,(\mathbf{x}-\xi) \tag{29}$$

and a unit moment about the direction normal to unit in-plane vector **m** given by,

$$q = \frac{\partial\delta(\mathbf{x}-\xi)}{\partial m(\xi)} = m_a\delta_{,a}\,(\mathbf{x}-\xi) \tag{30}$$

These four solutions are represented by pairs of functions $(F_\lambda^*, w_\lambda^*)$ satisfying eqns (24) and (25) with right-hand sides given by eqn (28) for λ = 1,2, by eqn (29) for λ = 3 and by eqn (30) for λ = 4.

5 FOURIER TRANSFORMS

In a two-dimensional space, the Fourier transform of a function $f(\mathbf{x})$ is given by [11]

$$\hat{f}\,(\zeta) = \frac{1}{2\pi}\int_\Omega f(\mathbf{x})\,\mathrm{e}^{\mathrm{i}\zeta\cdot\mathbf{x}}\,\mathrm{d}\Omega$$

where ζ is the two-dimensional independent variable in the Fourier space. For the purposes of the present analysis, it is important to note that the Fourier transform of the derivative $f_{,a}$ is $(-\mathrm{i}z_a)\hat{f}\,(\zeta)$ where i is the unit imaginary number. Thus, taking the Fourier transforms of both sides of eqns (24) and (25) with the right-hand sides given by eqns (28), (29) and (30) converts the partial differential into algebraic operators comprising polynomials in ζ_a.

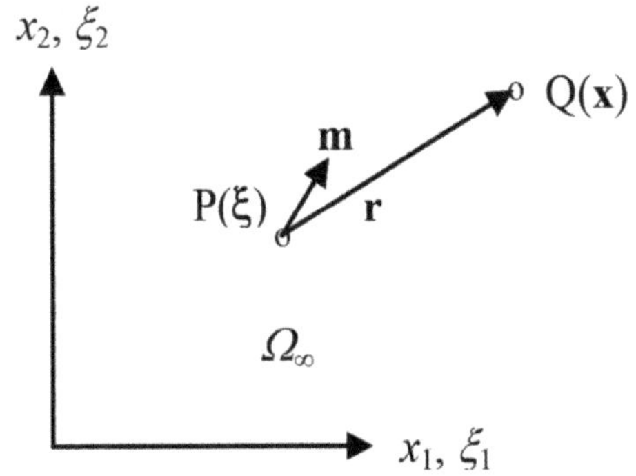

Figure 1: Infinite laminated plate under unit forces/moment at source point P(ξ).

Thus, the Fourier transforms of the fundamental solutions for the field variables satisfy the equations

$$\hat{A}_{\alpha\beta\gamma\delta}\,\zeta_\alpha\zeta_\beta\zeta_\gamma\zeta_\delta\hat{F}^*_\lambda + \hat{B}_{\alpha\beta\gamma\delta}\,\zeta_\alpha\zeta_\beta\zeta_\gamma\zeta_\delta\,\hat{w}^*_\lambda = \frac{b_{1\lambda}}{2\pi}\mathrm{e}^{\mathrm{i}\zeta\cdot\xi} \tag{31}$$

$$\hat{B}_{\gamma\delta\alpha\beta}\,\zeta_\alpha\zeta_\beta\zeta_\gamma\zeta_\delta\,\hat{F}^*_\lambda - D_{\alpha\beta\gamma\delta}\,\zeta_\alpha\zeta_\beta\zeta_\gamma\zeta_\delta\,\hat{w}^*_\lambda = \frac{b_{2\lambda}}{2\pi}\mathrm{e}^{\mathrm{i}\zeta\cdot\xi} \tag{32}$$

where

$$b_{1\lambda} = -\mathrm{i}\hat{A}_{\lambda\beta\kappa\kappa}\zeta_\beta,\ b_{2\lambda} = -\mathrm{i}\hat{B}_{\lambda\beta\kappa\kappa}\zeta_\beta,\ \lambda = 1,2$$

$$b_{13} = 0,\ b_{23} = -1$$

$$b_{14} = 0,\ b_{24} = \mathrm{i}m_\kappa\zeta_\kappa$$

The system of equations (31) and (32) can now be easily solved for the Fourier transforms of the fundamental solutions. The results take the form

$$\hat{F}^*_\lambda = \frac{g_\lambda(\zeta_1,\zeta_2)}{\Delta(\zeta_1,\zeta_2)}\frac{\mathrm{e}^{\mathrm{i}\zeta\cdot\xi}}{2\pi} \tag{33}$$

$$\hat{w}^*_\lambda = \frac{h_\lambda(\zeta_1,\zeta_2)}{\Delta(\zeta_1,\zeta_2)}\frac{\mathrm{e}^{\mathrm{i}\zeta\cdot\xi}}{2\pi} \tag{34}$$

where

$$\Delta(\zeta_1,\zeta_2) = (\hat{A}_{\alpha\beta\gamma\delta}\,\hat{D}_{\kappa\mu\nu\xi} + \hat{B}_{\kappa\mu\nu\xi}\,\hat{B}_{\gamma\delta\alpha\beta})\zeta_\alpha\zeta_\beta\zeta_\gamma\zeta_\delta\zeta_\kappa\zeta_\mu\zeta_\nu\zeta_\zeta, \tag{35}$$

$$g_\lambda(\zeta_1,\zeta_2) = -\mathrm{i}(\hat{A}_{\lambda\beta\kappa\kappa}\,\hat{D}_{\alpha\mu\gamma\delta} + \hat{B}_{\lambda\beta\kappa\kappa}\,\hat{B}_{\alpha\mu\gamma\delta})\zeta_\alpha\zeta_\beta\zeta_\gamma\zeta_\delta\zeta_\mu;\ \lambda = 1,2 \tag{36}$$

$$h_\lambda(\zeta_1,\zeta_2) = \mathrm{i}(\hat{B}_{\lambda\beta\kappa\kappa}\,\hat{A}_{\alpha\mu\gamma\delta} - \hat{A}_{\lambda\beta\kappa\kappa}\,\hat{B}_{\gamma\delta\alpha\mu})\zeta_\alpha\zeta_\beta\zeta_\gamma\zeta_\delta\zeta_\mu;\ \lambda = 1,2 \tag{37}$$

$$g_3(\zeta_1,\zeta_2) = \hat{B}_{\alpha\beta\gamma\delta}\zeta_\alpha\zeta_\beta\zeta_\gamma\zeta_\delta, \tag{38}$$

$$h_3(\zeta_1,\zeta_2) = -\hat{A}_{\alpha\beta\gamma\delta}\,\zeta_\alpha\zeta_\beta\zeta_\gamma\zeta_\delta, \tag{39}$$

$$g_4(\zeta_1,\zeta_2) = -\mathrm{i}m_\kappa\,\hat{B}_{\alpha\beta\gamma\delta}\zeta_\alpha\zeta_\beta\zeta_\gamma\zeta_\delta\zeta_\kappa = -\mathrm{i}m_\kappa\zeta_\kappa g_3(\zeta_1,\zeta_2) \tag{40}$$

$$h_4(\zeta_1,\zeta_2) = \mathrm{i}m_\kappa\,\hat{A}_{\alpha\beta\gamma\delta}\,\zeta_\alpha\zeta_\beta\zeta_\gamma\zeta_\delta\zeta_\kappa = -\mathrm{i}m_\kappa\zeta_\kappa h_3(\zeta_1,\zeta_2) \tag{41}$$

Thus $\Delta(\zeta_1, \zeta_2)$ is an 8th order homogeneous polynomial in either ζ_1 or ζ_2 whose coefficients are expressed in terms of the material constants. Similarly, the functions $g_\lambda(\zeta_1, \zeta_2)$ and $h_\lambda(\zeta_1, \zeta_2)$ are homogeneous polynomials of 5th order for $\lambda = 1,2,4$ and of 4th order for $\lambda = 3$.

6 INVERSE FOURIER TRANSFORMS

The similarity of the expressions for the Fourier transforms of the stress function and the deflection suggests that the derivation of only one inverse transform needs to be described in detail. In this derivation, only the distinction between 4th and 5th order polynomials in the numerator needs to be made.

6.1 Stress function

The inverse of the Fourier transform given by eqn (33) is written:

$$F_{\lambda}^{*}(\mathbf{x}-\boldsymbol{\xi})=\frac{1}{2\pi}\int_{\Omega_\infty}\hat{F}_{\lambda}^{*}(\boldsymbol{\zeta})\mathrm{e}^{-\mathrm{i}\boldsymbol{\zeta}\cdot\mathbf{x}}\mathrm{d}\Omega_{\zeta}=\frac{1}{4\pi^{2}}\int_{\Omega_\infty}\frac{g_{\lambda}(\zeta_1,\zeta_2)}{\Delta(\zeta_1,\zeta_2)}\mathrm{e}^{-\mathrm{i}\boldsymbol{\zeta}\cdot(\mathbf{x}-\boldsymbol{\xi})}\mathrm{d}\Omega_{\zeta} \tag{42}$$

The integration on the right-hand sides of eqn (42) is performed by separation of variables:

$$F_{\lambda}^{*}=\frac{1}{4\pi^{2}}\int_{-\infty}^{\infty}\mathrm{e}^{-\mathrm{i}\zeta_1 r_1}\mathrm{d}\zeta_1\int_{-\infty}^{\infty}\frac{g_{\lambda}(\zeta_1,\zeta_2)}{\Delta(\zeta_1,\zeta_2)}\mathrm{e}^{-\mathrm{i}\zeta_2 r_2}\mathrm{d}\zeta_2 \tag{43}$$

where $r_a = x_a - \xi_a$. Integration with respect to ζ_2 is performed using complex variable calculus [12]. For this purpose, a complex variable $z = \zeta_2 + \mathrm{i}\zeta'$ is introduced and a complex function $H(z)$ defined by

$$H(z)=\frac{g_{\lambda}(\zeta_1,z)}{\Delta(\zeta_1,z)}\mathrm{e}^{-\mathrm{i}r_2 z} \tag{44}$$

$H(z)$ is integrated around the contour shown in Fig. 2 comprising the segment of the real axis from $-R$ to R and the semicircle C_R in the upper half of the complex plane.

Thus, for any real value of R,

$$\int_{-R}^{R}\frac{g_{\lambda}(\zeta_1,\zeta_2)}{\Delta(\zeta_1,\zeta_2)}\mathrm{e}^{-\mathrm{i}\zeta_2 r_2}\mathrm{d}\zeta_2+\int_{C_R}H(z)\mathrm{d}z=2\pi\mathrm{i}\sum_{k}\mathrm{Res}(a_k) \tag{45}$$

where a_k represent the poles of $H(z)$ within the contour. It has been shown [12] that the contour integral on the left-hand side tends to zero as R goes to infinity provided that $r_2 < 0$. As R increases, the semicircle expands and eventually encompasses all the poles of $H(z)$. To determine the residues associated with these poles, it is noted that the 8th order polynomial $\Delta(\zeta_1, z)$ has four pairs of distinct complex conjugate roots [13] which can be easily expressed in terms of ζ_1 by introducing the parameter ρ such that $z = \zeta_2 = \rho\zeta_1$; this transforms eqn (35) into

$$\Delta(\zeta_1, z) = \zeta_1^8\Delta(1, \rho)$$

The roots of $\Delta(1, \rho)$ are represented by

$$\rho_k=\rho_k'+\mathrm{i}\rho_k'',\bar{\rho}_k=\rho_k'-\mathrm{i}\rho_k'',\rho_k''>0;k=1,2,3,4$$

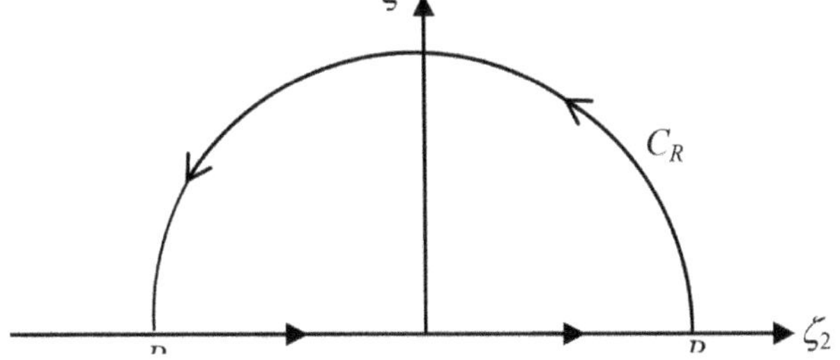

Figure 2: Contour for complex function integration.

where a bar above a symbol represents the complex conjugate of the corresponding parameter. The denominator in eqn (44) can thus be written

$$\Delta(\zeta_1,\zeta_2)=\Delta_8\prod_{i=1}^{4}(\zeta_2-\rho_i\zeta_1)(\zeta_2-\bar{\rho}_i\zeta_1)$$

where Δ_8 is the coefficient of ρ^8 which can be expressed explicitly in terms of material constants according to eqn (35). Before proceeding with the application of eqn (45), it is, at this stage, necessary to make the distinction between ζ_1 being positive or negative.

(a) $\zeta_1>0$

In this case, $H(z)$ has four poles $\rho_k\zeta_1$ of order 1, that is, simple poles, in the upper half-plane and the respective residues are

$$\mathrm{Res}(\rho_K\zeta_1)=[(z-\rho_k\zeta_1)H(z)]_{z=\rho_k\zeta_1}=\frac{g_\lambda(1,\rho_k)e^{-\mathrm{i}r_2\rho_k\zeta_1}}{2\mathrm{i}\Delta_8\beta_k\zeta_1^a}$$

where exponent a in the denominator of the right-hand side is equal to either 2 or 3, depending on whether g_λ is a 5th order or 4th order polynomial, respectively, and

$$\beta_k=\rho_k''\prod_{i=1,i\neq k}^{4}(\rho_k-\rho_i)(\rho_k-\bar{\rho}_i)$$

Thus, from eqn (45)

$$\int_{-\infty}^{\infty}\frac{g_\lambda(\zeta_1,\zeta_2)}{\Delta(\zeta_1,\zeta_2)}\mathrm{e}^{-\mathrm{i}\zeta_2 r_2}\mathrm{d}\zeta_2=\frac{\pi}{\Delta_8}\sum_{k=1}^{4}\frac{g_\lambda(1,\rho_k)\mathrm{e}^{-\mathrm{i}r_2\rho_k\zeta_1}}{\beta_k\zeta_1^a}$$

(b) $\zeta_1<0$

In this case, the complex conjugates of ρ_k, multiplied by ζ_1, are the four simple poles of $H(z)$ in the upper half-plane; hence,

$$\mathrm{Res}(\bar{\rho}_k\zeta_1)=[(z-\bar{\rho}_k\zeta_1)H(z)]_{z=\bar{\rho}_k\zeta_1}=\frac{g_\lambda(1,\bar{\rho}_k)e^{-\mathrm{i}\zeta_1 r_2\bar{\rho}_k}}{-2\mathrm{i}\Delta_8\bar{\beta}_k\zeta_1^a}$$

and

$$\int_{-\infty}^{\infty}\frac{g_\lambda(\zeta_1,\zeta_2)}{\Delta(\zeta_1,\zeta_2)}\mathrm{e}^{-\mathrm{i}\zeta_2 r_2}\mathrm{d}\zeta_2=-\frac{\pi}{\Delta_8}\sum_{k=1}^{4}\frac{g_\lambda(1,\bar{\rho}_k)\mathrm{e}^{-\mathrm{i}\zeta_1 r_2\bar{\rho}_k}}{\bar{\beta}_k\zeta_1^a}$$

It is thus possible to obtain the inverse transform given by eqn (43) over the whole range of ζ_1:

$$F_\lambda^*=\frac{1}{4\pi\Delta_8}\sum_{k=1}^{4}\frac{g_\lambda(1,\rho_k)}{\beta_k}\int_0^{\infty}\frac{\mathrm{e}^{-\mathrm{i}\zeta_1 v_k}}{\zeta_1^a}\mathrm{d}\zeta_1-\frac{1}{4\pi\Delta_8}\sum_{k=1}^{4}\frac{g_\lambda(1,\bar{\rho}_k)}{\bar{\beta}_k}\int_{-\infty}^{0}\frac{\mathrm{e}^{-\mathrm{i}\zeta_1\bar{v}_k}}{\zeta_1^a}\mathrm{d}\zeta_1 \tag{46}$$

where $v_k=r_1+\rho_k r_2$ while $a=2$ for $\lambda=1,2,4$ and $a=3$ for $\lambda=3$. By substituting $\zeta_1=s$ in the first integral of the right hand side of eqn (46) and $\zeta_1=-s$ in the second, it is possible to show that the second term is simply minus the complex conjugate of the first. This leads to

$$F_\lambda^*=\frac{1}{2\pi\Delta_8}\mathrm{Re}\left[\sum_{k=1}^{4}\frac{g_\lambda(1,\rho_k)}{\beta_k}\int_0^{\infty}\frac{\mathrm{e}^{-\mathrm{i}sv_k}}{s^a}\mathrm{d}s\right]$$

Since $r_2 < 0$, $\mathrm{Re}(-\mathrm{i}v_k) < 0$; thus $\exp(-\mathrm{i}v_k s)$ is bounded as s goes to infinity. Under this condition, integration by parts leads to

$$\int_0^\infty \frac{\mathrm{e}^{-\mathrm{i}sv_k}}{s^2}\mathrm{d}s = -\mathrm{i}v_k \int_0^\infty \frac{\mathrm{e}^{-\mathrm{i}sv_k}}{s}\mathrm{d}s \tag{47}$$

and

$$\int_0^\infty \frac{\mathrm{e}^{-\mathrm{i}sv_k}}{s^3}\mathrm{d}s = \frac{v_k^2}{2}\int_0^\infty \frac{\mathrm{e}^{-\mathrm{i}sv_k}}{s}\mathrm{d}s \tag{48}$$

Then the Cauchy principal value of the divergent integral on the right-hand side of eqns (47) and (48) can be obtained by making use of a property of the Laplace transforms [12], which gives

$$\mathrm{P}\int_0^\infty \frac{\mathrm{e}^{-\mathrm{i}sv_k}}{s}\mathrm{d}s = -\ln v_k \tag{49}$$

Thus, finally, the fundamental solutions for the in-plane stress function is provided in the form of the real functions

$$F_\lambda^* = \frac{1}{2\pi A_8}\mathrm{Re}\left[\sum_{k=1}^{4}\frac{\mathrm{i}g_\lambda(1,\rho_k)v_k \ln v_k}{\beta_k}\right];\ \lambda = 1,2,4 \tag{50}$$

$$F_3^* = -\frac{1}{4\pi A_8}\mathrm{Re}\left[\sum_{k=1}^{4}\frac{g_3(1,\rho_k)v_k^2 \ln v_k}{\beta_k}\right] \tag{51}$$

If $r_2 > 0$, for the contour integral over C_R to tend to zero as R goes to infinity, the semicircle should lie in the lower half-plane. The process described above is repeated leading to the same final result.

6.2 Deflection

The derivation of the fundamental solutions for the deflection follows exactly the same steps as that for the stress function. Thus, the final expressions are the same as those given by eqns (50) and (51) but with function h_λ replacing g_λ, namely,

$$w_\lambda^* = \frac{1}{2\pi A_8}\mathrm{Re}\left[\sum_{k=1}^{4}\frac{ih_\lambda(1,\rho_k)v_k \ln v_k}{\beta_k}\right];\ \lambda = 1, 2, 4 \tag{52}$$

$$w_3^* = -\frac{1}{4\pi A_8}\mathrm{Re}\left[\sum_{k=1}^{4}\frac{h_3(1,\rho_k)v_k^2 \ln v_k}{\beta_k}\right] \tag{53}$$

7 BOUNDARY INTEGRAL EQUATIONS

BEM can now be applied to the analysis of a finite laminated plate, whose domain Ω is bounded by the contour Γ as shown in Fig. 3. Apart for the in-plane body forces and the lateral pressure, the plate may also be subjected to various edge loads such as traction **p**. A BEM solution scheme can be developed by formulating first the reciprocity relation

$$\int_{\Omega}(\bar{\varepsilon}_{\alpha\beta}-\bar{A}_{\alpha\beta\kappa\kappa}\Phi)F'_{,\alpha\beta}+(M_{\alpha\beta}-\bar{B}_{\alpha\beta\kappa\kappa}\Phi)w'_{,\alpha\beta})\mathrm{d}\Omega$$
$$=\int_{\Omega}[(\bar{\varepsilon}'_{\alpha\beta}-\bar{A}_{\alpha\beta\kappa\kappa}\Phi')F_{,\alpha\beta}+(M'_{\alpha\beta}-\bar{B}_{\alpha\beta\kappa\kappa}\Phi')w_{,\alpha\beta}]\mathrm{d}\Omega \tag{54}$$

where

$$(\bar{\varepsilon}_{\alpha\beta},F,M_{\alpha\beta},w) \text{ and } (\bar{\varepsilon}'_{\alpha\beta},F',M'_{\alpha\beta},w')$$

are two equilibrium states due, respectively, to body force-lateral pressure combinations (**f**, q) and (**f'** q'). As such they satisfy compatibility, eqn (14), and equilibrium, eqn (6). This reciprocity relation is proven by taking into account the constitutive eqns (17) and (21) as well as the symmetries of material tensors.

Repeated integrations by parts and applications of Green's theorem transform eqn (54) into

$$\int_{\Omega}(\bar{\varepsilon}_{\alpha\beta,\alpha\beta}F'-\bar{\varepsilon}'_{\alpha\beta,\alpha\beta}F)\mathrm{d}\Omega+\int_{\Omega}(M_{\alpha\beta,\alpha\beta}w'-M'_{\alpha\beta,\alpha\beta}w)\mathrm{d}\Omega$$
$$+\int_{\Omega}\bar{A}_{\alpha\beta\kappa\kappa}(\Phi_{,\beta}F'_{,\alpha}-\Phi'_{,\beta}F_{,\alpha})\mathrm{d}\Omega+\int_{\Omega}\bar{B}_{\alpha\beta\kappa\kappa}(\Phi_{,\beta}w'_{,\alpha}-\Phi'_{,\beta}w_{,\alpha})\mathrm{d}\Omega$$
$$=I_F(F,F')+J_F(F,F')+I_w(w,w')+J_w(w,w')+I_{\Phi}(F,F',w,w') \tag{55}$$

It is noted that the first domain integral on the left-hand side of eqn (55) always vanishes on account of strains satisfying compatibility eqn (14). The right-hand side of eqn (55) comprises contour integrals and corner jump terms defined by the equations

$$I_F(F,F')=\int_{\Gamma}\left(\kappa_n F'-\varepsilon_{ss}\frac{\partial F'}{\partial n}+\varepsilon'_{ss}\frac{\partial F}{\partial n}-\kappa'_n F\right)\mathrm{d}\Gamma \tag{56}$$

$$I_w(w,w')=\int_{\Gamma}\left(V_n w'-M_n\frac{\partial w'}{\partial n}+M'_n\frac{\partial w}{\partial n}-V'_n w\right)\mathrm{d}\Gamma \tag{57}$$

$$I_{\Phi}(F,F',w,w')=\int_{\Gamma}n_{\beta}\left[\Phi(\bar{A}_{\alpha\beta\kappa\kappa}F'_{,\alpha}+\bar{B}_{\alpha\beta\kappa\kappa}w'_{,\alpha})-\Phi'(\bar{A}_{\alpha\beta\kappa\kappa}F_{,\alpha}+\bar{B}_{\alpha\beta\kappa\kappa}w_{,\alpha})\right]\mathrm{d}\Gamma \tag{58}$$

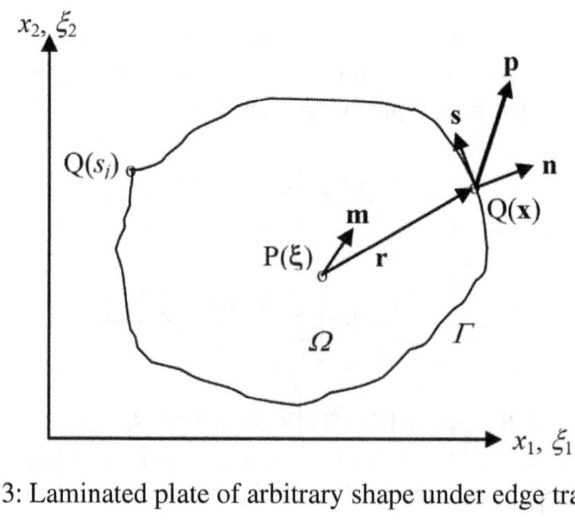

Figure 3: Laminated plate of arbitrary shape under edge traction **p**.

$$J_F(F, F') = \sum_{j=1}^{K} \left[\bar{\varepsilon}_{ns} F' - \bar{\varepsilon}'_{ns} F \right]_j \tag{59}$$

$$J_w(w, w') = \sum_{j=1}^{K} \left[M_{ns} w' - M'_{ns} w \right]_j \tag{60}$$

where K is the number of corner points along Γ. In eqn (56),

$$\varepsilon_{ss} = \bar{\varepsilon}_{nn} = \frac{\partial u_s}{\partial s}$$

is the normal strain in the direction of **s**, while

$$\kappa_n = n_\alpha \bar{\varepsilon}_{\alpha\beta},_\beta + \frac{\partial}{\partial s}(s_\alpha n_\beta \bar{\varepsilon}_{\alpha\beta}) = \frac{\partial \bar{\varepsilon}_{nn}}{\partial n} + 2\frac{\partial \bar{\varepsilon}_{ns}}{\partial s} = -\frac{\partial^2 u_n}{\partial s^2}$$

represents the change of curvature of the boundary Γ due to the normal in-plane displacement u_n. It is worth noting the ε_{ss} and κ_n are multiplied, respectively, by a force acting in the **s**-direction and a moment about an axis perpendicular to the plane of the plate; it is thus evident that these products represent work terms.

It is next assumed that the loading system (**f**, q) is the one actually applied while (**f'** q') comprises the unit actions given by eqns (28), (29) and (30) causing the fundamental states derived in the previous section. Then, the non-vanishing domain integrals on the left-hand side of eqn (55) are reduced to the left-hand side expressions given below:

$$\int_\Omega (M_{\alpha\beta},_{\alpha\beta} w^*_\lambda - M^*_{\lambda\alpha\beta},_{\alpha\beta} w)\mathrm{d}\Omega = \begin{cases} -\int_\Omega q w^*_\lambda \mathrm{d}\Omega & \text{for } \lambda=1,2 \\ -\int_\Omega q w^*_3 \mathrm{d}\Omega + w(\xi) \\ -\int_\Omega q w^*_4 \mathrm{d}\Omega + \dfrac{\partial w(\xi)}{\partial m} \end{cases}$$

$$-\int_\Omega \bar{A}_{\alpha\beta\kappa\kappa}(f_\beta F^*_\lambda,_\alpha - f^*_{\lambda\beta} F,_\alpha)\mathrm{d}\Omega = \begin{cases} -\int_\Omega \bar{A}_{\alpha\beta\kappa\kappa} f_\beta F^*_\lambda,_\alpha \mathrm{d}\Omega + \bar{A}_{\alpha\lambda\kappa\kappa} F,_\alpha(\xi) & \text{for } \lambda=1,2 \\ \int_\Omega \bar{A}_{\alpha\beta\kappa\kappa} f_\beta F^*_\lambda,_\alpha \mathrm{d}\Omega & \text{for } \lambda=3,4 \end{cases}$$

$$-\int_\Omega \bar{B}_{\alpha\beta\kappa\kappa}(f_\beta w^*_\lambda,_\alpha - f^*_{\lambda\beta} w,_\alpha)\mathrm{d}\Omega = \begin{cases} -\int_\Omega \bar{B}_{\alpha\beta\kappa\kappa} f_\beta w^*_\lambda,_\alpha \mathrm{d}\Omega + \bar{B}_{\alpha\lambda\kappa\kappa} w,_\alpha(\xi) & \text{for } \lambda=1,2 \\ -\int_\Omega \bar{B}_{\alpha\beta\kappa\kappa} f_\beta w^*_\lambda,_\alpha \mathrm{d}\Omega & \text{for } \lambda=3,4 \end{cases}$$

The domain integrals in the above expressions depend on known quantities, they can therefore be evaluated. The variable of each free term is transferred to the boundary where unit vector **m** becomes identical to unit normal **n**; this generates a factor k, which equals ½ if P(ξ) lies on a smooth segment of Γ. The boundary variable $F,_\alpha$ can be evaluated since

$$F,_\alpha = n_\alpha \frac{\partial F}{\partial n} + s_\alpha \frac{\partial F}{\partial s}$$

and both $F(s)$ and its normal derivative are given, respectively, by eqns (26) and (27). The boundary deflection and its normal derivative can be known or unknowns depending on the specified boundary conditions. Thus, the deflection gradient $w,_{\alpha}$ can either be evaluated or modelled along Γ using

$$w,_{\alpha} = n_{\alpha}\frac{\partial w}{\partial n} + s_{\alpha}\frac{\partial w}{\partial s}$$

The contour integrals on the left-hand side of eqn (55) depend on boundary variables multiplied by kernels derivable from the fundamental solutions using differentiation and the constitutive equations. Four of the boundary variables are unknown and need to be modelled using boundary elements. Special modelling schemes are required for the corner variables if any of them are not known.

Thus, the final form of the boundary integral equations on which a BEM scheme could be based is

$$\begin{aligned} & I_F(F,F_\lambda^*)+J_F(F,F_\lambda^*)+I_w(w,w_\lambda^*)+J_w(w,w_\lambda^*)+I_\Phi(F,F_\lambda^*,w,w_\lambda^*) \\ & +\int_\Omega (\bar{A}_{\alpha\beta\kappa\kappa} f_\beta F^*_{\lambda},_{\alpha} + \bar{B}_{\alpha\beta\kappa\kappa} f_\beta w^*_{\lambda},_{\alpha} + q w_\lambda^*)\mathrm{d}\Omega \\ & = \begin{cases} k[\bar{A}_{\alpha\lambda\kappa\kappa} F,_{\alpha}(\xi) + \bar{B}_{\alpha\lambda\kappa\kappa} w,_{\alpha}(\xi)] & \text{for } \lambda=1,2 \\ k w_\lambda(\xi) & \text{for } \lambda=3,4 \end{cases} \end{aligned} \tag{61}$$

where

$$w_3(\xi) = w(\xi) \text{ and } w_4(\xi) = \frac{\partial w}{\partial n}(\xi) \tag{62}$$

8 CONCLUDING REMARKS

Fundamental solutions were derived in closed, compact and elegant form; to be practically useful, their parameters depending on the material constants need to be initially evaluated. For this purpose, tensor products such as those appearing in eqn (35)-(41) should be expanded to identify the coefficients of the respective polynomials and the roots of the 8th degree polynomial $\Delta(1, \rho)$ determined. With all their constant parameters known, their dependence on space co-ordinates is quite simple and their variation can be evaluated without high computational cost. From the nature of the roots of $\Delta(1, \rho)$ it is evident that complex number coding would be necessary in any computer implementation.

Expressions for the various components of the fundamental states appearing as weighting functions in domain and contour integrals should also be derived. The derivatives of the fundamental solutions also have simple forms although entering them into constitutive equations would generate longer expressions.

The potential of using the obtained fundamental solutions in BEM schemes for general laminates was demonstrated through the derivation of a set of four boundary integral equations involving four boundary unknowns. In contrast to the formulation based on just the three displacement components as field variables, the present one using the stress function and deflection generates integral equations in identical form and thus susceptible to a more efficient and flexible numerical manipulation.

A computer-implemented BEM formulation solving the general laminate problem under any loading and boundary conditions is expected to present many challenges. Practically relevant simplifications are possible, such as the elimination of applied in-plane body forces from the formulation. Two issues that require rigorous mathematical treatment is the determination of the in-plane force potential for the fundamental states and the derivation of expressions for the free term coefficient k when the source point is a corner point. The latter issue is obviated if discontinuous boundary elements are used adjacent to a contour corner.

As already pointed out, it is expedient to specify only traction boundary conditions along the boundary for this formulation to work without further analysis and modelling. This is usually the case with stability analyses but the geometric nonlinearity of such problems introduce additional analytical and numerical complexities. Similar challenges would be faced in attempts to extend the BEM scheme to dynamic problems.

REFERENCES

[1] Reissner, E. & Stavsky, Y., Bending and stretching of certain types of heterogeneous aeolotropic elastic plates. *Transactions of the ASME, Journal of Applied Mechanics*, **28**(3), pp. 402–408, 1961.
https://doi.org/10.1115/1.3641719

[2] Whitney, J.M. & Leissa, A.W., Analysis of heterogeneous, anisotropic plates. *Transactions of the ASME, Journal of Applied Mechanics*, **36**, pp. 261–266, 1969.
https://doi.org/10.1115/1.3564618

[3] Whitney, J.M. & Leissa, A.W., Analysis of a simply supported laminated anisotropic rectangular plate. *AIAA Journal*, **8**(1), pp. 28–33, 1970.
https://doi.org/10.2514/3.5601

[4] Whitney, J.M., The effect of boundary conditions on the response of laminated composites. *Journal of Composite Materials*, **4**(2), pp. 192–203, 1970.
https://doi.org/10.1177/002199837000400205

[5] Harris, G.Z., The buckling and post-buckling behaviour of composite plates under biaxial loading. *International Journal of Mechanical Sciences*, **17**, pp. 187–202, 1975.
https://doi.org/10.1016/0020-7403(75)90052-1

[6] Noor, A.K., Mathers, M.D. & Anderson, M.S., Exploiting symmetries for efficient post-buckling analysis of composite plates. *AIAA Journal*, **15**(1), pp. 24–32, 1977.
https://doi.org/10.2514/3.60601

[7] Syngellakis, S., A boundary element approach to buckling of general laminates. *WIT Transactions on Modelling and Simulation*, **53**, pp. 145–155, 2012.
https://doi.org/10.2495/be120131

[8] Hwu, C., Green's function for the composite laminates with bending extension coupling. *Composite Structures*, **63**, pp. 283–292, 2004.
https://doi.org/10.1016/s0263-8223(03)00175-2

[9] Hwu, C., Boundary integral equations for general laminated plates with coupled stretching-bending deformation. *The Royal Society, Proceedings: Mathematical, Physical and Engineering Sciences*, **466**(2116), pp. 1027–1054, 2010.
https://doi.org/10.1098/rspa.2009.0432

[10] Syngellakis, S., Fundamental solutions for the coupled extension-flexure laminate problem. *WIT Transactions on Modelling and Simulation*, **61**, pp. 235–246, 2015.
https://doi.org/10.2495/bem380191

[11] Courant, R. & Hilbert, D., *Methods of Mathematical Physics vol. II: Partial Differential Equations*, Interscience Publishers: New York, 1953.

[12] Hildebrand, F.B., *Advanced Calculus for Applications*, Prentice-Hall, Inc.: Englewood Cliffs, N.J., 1962.

[13] Cheng, Z.-Q. & Reddy, J.N., Octet formalism for Kirchhoff anisotropic plates. *The Royal Society, Proceedings: Mathematical, Physical and Engineering Sciences*, **458**(2022), pp. 1499–1517, 2002.
https://doi.org/10.1098/rspa.2001.0934

AN APPLICATION OF TOPOLOGY OPTIMISATION TO DEFECT IDENTIFICATION IN TWO-DIMENSIONAL ELASTODYNAMICS WITH THE BEM AND $\mathcal{H}$-MATRIX METHOD

KEI MATSUSHIMA, HIROSHI ISAKARI, TORU TAKAHASHI & TOSHIRO MATSUMOTO
Nagoya University, Japan.

ABSTRACT
This paper presents a numerical method for topology optimisation for two-dimensional elastodynamics based on the level set method and the boundary element method (BEM) accelerated by the $\mathcal{H}$-matrix method and its application to identifications of defects in an infinite elastic medium. Gradient-based topology optimisation methods require design sensitivity, which is obtained by solving some boundary value problems. The BEM is employed for this sensitivity analysis because the BEM can deal with infinite domains rigorously without any approximation. However, the computational cost in the BEM is expensive, and this is a serious drawback since we need to repeat sensitivity analysis even for a single optimisation process. In this study, the $\mathcal{H}$-matrix method is used as an acceleration method of the BEM for the reduction of the computational cost of the sensitivity analysis. Also proposed is a method to improve the efficiency of the $\mathcal{H}$-matrix method by exploiting a property of the kernel function of the elastodynamic fundamental solution. Some numerical examples are demonstrated, and the effectiveness of the proposed method is confirmed.
Keywords: boundary element method, defect identification, elastic wave, level set method, topological derivative, topology optimisation, $\mathcal{H}$-matrix method

1 INTRODUCTION

Recently, non-destructive testing has attracted attention in engineering, and a number of works have been made to develop an effective method to image defects in an elastic medium from some measured data [1]. To obtain such data, elastic waves are often used. For example, an elastic wave propagates in an elastic medium and is scattered by a defect. The scattered wave can then be measured as the displacement on some part of the boundary of the elastic medium. This is a so-called inverse scattering problem and can be treated as a topology optimisation problem with an elastic wave propagation phenomenon detecting the unknown defect shape and location. The objective functional of the corresponding optimisation problem is often defined to minimise a discrepancy between the measured data and solutions of the wave problem in the elastic domain with estimated defects. Topological derivative, which is defined as a rate of change of the objective functional (in this case, the discrepancy) when an infinitesimal defect is allocated in the medium, is useful for this approach. Some topological derivatives in three-dimensional elastodynamics are derived in [2, 3]. The adjoint variable method is often used for the computation of topological derivatives, where we have to solve two kinds of boundary value problems: the direct and adjoint problems. For the analyses of these problems, the boundary element method (BEM) is preferred to the finite element method for a certain class of the problems because the BEM can deal with the infinite domain rigorously.

The computational cost of the BEM is highly expensive, therefore, the BEM is usually implemented with an acceleration method such as the fast multipole method (FMM) [4, 5] and $\mathcal{H}$-matrix method [6]. In the present optimisation scheme, a direct solver based on

DOI: 10.2495/CMEM-V6-N6-1033-1042

LU-decomposition is more effective than iterative solvers such as GMRES because both the coefficient matrices corresponding to the direct and adjoint problems are the same, thus LU-decomposition is more efficient than iterative solvers. In fact, we employed the $\mathcal{H}$-matrix method for a topology optimisation problem for two-dimensional electromagnetic field, and we demonstrated its effectiveness in [7].

In this study, we employ the $\mathcal{H}$-matrix method as an acceleration method of the BEM and apply it to the defect identification based on the topology optimisation with the level set method.

2 TOPOLOGY OPTIMISATION WITH THE LEVEL SET METHOD

2.1 Inverse scattering problem and corresponding topology optimisation problem

In the present study, we assume that all defects are assumed to be cavities having traction-free surfaces. Let $\Omega \subset \mathrm{R}^2$ be an infinite domain of a linear isotropic elastic medium under plane strain condition. As shown in Fig. 1, let us assume that an incident elastic wave u^{in} propagates in Ω and is scattered by the defects $\mathrm{R}^2 \backslash \overline{\Omega}$. Assuming time harmonic oscillations with the time dependence of $\mathrm{e}^{-\mathrm{i}\omega t}$, where ω is the angular frequency, the displacement u and stress σ in Ω are governed by the following boundary value problem:

$$\sigma_{ji,j}(x) + \rho\omega^2 u_i(x) = 0 \quad x \in \Omega, \tag{1}$$

$$t_i(x) = \sigma_{ji}(x) n_j(x) = 0 \qquad x \in \Gamma, \tag{2}$$

$$\text{Radiation condition for } u^{\mathrm{sc}} \text{ as } |x| \to \infty, \tag{3}$$

where ρ is the mass density, $\Gamma = \partial\Omega$ is the boundary of Ω, $u^{\mathrm{sc}} = u - u^{\mathrm{in}}$ is the scattered field, t is the traction, and n is the unit outward normal vector to Ω.

Suppose that we measure the displacements on some observation points x_m^{obs} $(m = 1, \ldots, M)$ and identify the distribution of the defects from the measured displacements $\hat{u}(x_m^{\mathrm{obs}})$. This inverse scattering problem can be treated as a topology optimisation problem minimising the objective functional:

$$J = \frac{1}{2} \sum_{m=1}^{M} \left| u\left(x_m^{\mathrm{obs}}\right) - \hat{u}\left(x_m^{\mathrm{obs}}\right) \right|^2. \tag{4}$$

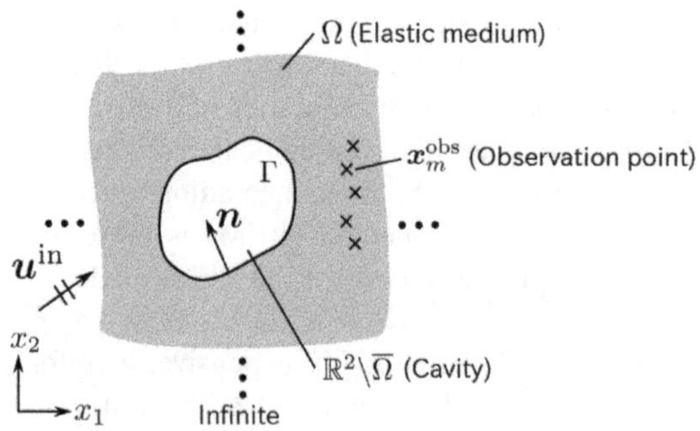

Figure 1: Problem statement.

2.2 Level set method

For simplicity, we assume $R^2\backslash\bar{\Omega} = D\backslash\bar{\Omega}$, i.e., there exists no defect outside of a certain bounded fixed design domain D. To express the distribution of the domains Ω and $D\backslash\bar{\Omega}$, we employ the level set method in which the defects $D\backslash\bar{\Omega}$, elastic medium Ω, and its boundary Γ are expressed by a level set function ϕ, as follows:

$$\Omega = \{x \mid 0 < \phi(x) \leq 1\}, \tag{5}$$

$$\Gamma = \{x \mid \phi(x) = 0\}, \tag{6}$$

$$D\backslash\bar{\Omega} = \{x \mid -1 \leq \phi(x) < 0\}, \tag{7}$$

The level set method converts the topology optimisation problem into a problem in which we find ϕ that minimises the objective functional J. We assume that ϕ is the solution of the following initial-boundary value problem [8, 9]:

$$\frac{\partial\phi(x,t)}{\partial t} = -KT(x,t) + \tau\nabla^2\phi(x,t) \quad x \in D, \tag{8}$$

$$\phi(x,0) = \phi_0 \quad x \in D, \tag{9}$$

$$\phi(x,t) = c \quad x \in \partial D, \tag{10}$$

where K is a positive constant, t is a fictitious time, τ is a positive constant that works as a regularisation parameter for the distribution of ϕ, and ϕ_0 is the initial distribution of ϕ corresponding to the initial geometry at $t = 0$. The Dirichlet condition (10) with a constant $c \in (0,1]$ restricts that the defects can only exist inside D. T denotes the topological derivative, defined by

$$\delta J(x) = T(x)\nu(\varepsilon) + o(\nu(\varepsilon)), \tag{11}$$

where $\delta J(x)$ is the variation of J when an cylindrical defect of radius ε centred at x appears and ν is a scalar function which vanishes as $\varepsilon \to 0$. For the objective function J defined by (4), T is obtained as

$$T(x) = \mathrm{Re}\left[\frac{\lambda + 2\mu}{4\mu(\lambda + \mu)}\{4\sigma_{ij}(x)\tilde{\sigma}_{ij}(x) - \sigma_{ii}(x)\tilde{\sigma}_{jj}(x)\} - \rho\omega^2 u_i(x)\tilde{u}_i(x)\right], \tag{12}$$

where $\tilde{u}$ and $\tilde{\sigma}$ are the displacement and stress, respectively, of the adjoint field problem governed by

$$\tilde{\sigma}_{ji,j}(x) + \rho\omega^2\tilde{u}_i(x) + \sum_{m=1}^{M}\overline{\left(u_i\left(x_m^{\text{obs}}\right) - \hat{u}_i\left(x_m^{\text{obs}}\right)\right)}\,\delta\left(x - x_m^{\text{obs}}\right) = 0 \quad x \in \Omega, \tag{13}$$

$$\tilde{t}_i(x) = \tilde{\sigma}_{ji}(x)n_j(x) = 0 \quad x \in \Gamma, \tag{14}$$

$$\text{Radiation condition for } \tilde{u}_i(x) \text{ as } |x| \to \infty. \tag{15}$$

2.3 Algorithm of the level set-based topology optimisation

As described in the previous section, we solve the initial-boundary value problem (8)–(10) to obtain the optimal configuration. In this study, we perform the topology optimisation by the following procedure:

1. Determine an initial configuration and initialise a level set function ϕ. The values of both ϕ and the topological derivative T are computed and stored at all the lattice points arranged in the fixed design domain D.
2. Interpolate ϕ linearly between each adjacent lattice point and calculate the point of intersection between the segment of two neighboring lattice points and the zero-valued isoline of ϕ. Then, connect these intersection points of ϕ and interpolate them with the Hermite spline curve. Finally, divide the generated isoline curve into linear boundary elements of approximately equal lengths [10].
3. Solve the direct problem (1)–(3) by the BEM.
4. Compute the value of the objective function J and the gradient of J. If the absolute value of the gradient is less than a preset parameter, terminate the procedure; otherwise, go to 5.
5. Solve the adjoint problem by the BEM and compute the topological derivative T from the numerical solutions of the direct and adjoint problems.
6. Update ϕ according to (8)–(10) and go to 2.

The entire step consisting of the steps 2 to 6 is called as an *optimisation step* in this paper.

3 SENSITIVITY ANALYSIS WITH THE BEM AND $\mathcal{H}$-MATRIX METHOD

3.1 Formulation of the BEM in two-dimensional elastodynamics

In order to compute the topological derivative T, we need to obtain both the displacement and stress in the direct and adjoint fields governed by (1)–(3) and (13)–(15), respectively.

We obtain the boundary integral equations equivalent to the direct problem (1)–(3) by the Burton-Miller formulation [11] as follows:

$$\begin{bmatrix} \frac{1}{2}I + D_{11} + aN_{11} & D_{12} + aN_{12} \\ D_{21} + aN_{21} & \frac{1}{2}I + D_{22} + aN_{22} \end{bmatrix} \begin{pmatrix} u_1 \\ u_2 \end{pmatrix} = \begin{pmatrix} u_1^{in} + at_1^{in} \\ u_2^{in} + at_2^{in} \end{pmatrix}, \tag{16}$$

where $\alpha \in \mathrm{C}$ is a coupling parameter, I is the identity operator, D_{ij} and N_{ij} are the integral operators defined for an arbitrary scalar function w as

$$(D_{ij}w)(x) = -\text{v.p.}\int_{\Gamma} C_{kljm} G_{ki,l}(x,y) n_m(y) w(y) d\Gamma_y, \tag{17}$$

$$(N_{ij}w)(x) = -\text{p.f.}\int_{\Gamma} C_{impq} C_{kljn} G_{kp,lq}(x,y) n_m(x) n_n(y) w(y) d\Gamma_y, \tag{18}$$

where v.p. and p.f. indicate Cauchy's principal value and the finite part of divergent integral, respectively. C is the elasticity tensor and is written using Kronecker's delta δ_{ij} and Lamé's constants λ and μ as

$$C_{ijkl} = \lambda\delta_{ij}\delta_{kl} + \mu(\delta_{ik}\delta_{jl} + \delta_{il}\delta_{jk}), \tag{19}$$

and G_{ij} is the fundamental solution of two-dimensional elastodynamic problems given by

$$G_{ij}(x, y) = \frac{i}{4\mu}\left\{H_0^{(1)}(k_{\mathrm{T}}\,|\,x-y\,|)\delta_{ij} + \frac{1}{k_{\mathrm{T}}^2}\frac{\partial^2}{\partial y_i \partial y_j}\left(H_0^{(1)}(k_{\mathrm{T}}\,|\,x-y\,|) - H_0^{(1)}(k_{\mathrm{L}}\,|\,x-y\,|)\right)\right\}, \tag{20}$$

where $H_0^{(1)}$ is the Hankel function of first kind with order 0 and k_{L} and k_{T} are the wavenumbers of the longitudinal and transverse wave, respectively.

3.2 $\mathcal{H}$-matrix method

As described in Section 2.3, iterative sensitivity analyses are required for our optimisation method, causing significantly large computation time. In this study, we employ the $\mathcal{H}$-matrix method to the BEM and accelerate the sensitivity analyses.

The algorithm of the $\mathcal{H}$-matrix method is composed of two processes: hierarchical blocking of a coefficient matrix into submatrices and low rank approximation of the submatrices.

3.2.1 Hierarchical blocking of a coefficient matrix

Let us consider a system of linear algebraic equations which is obtained by discretising the BIEs (16) by the collocation method with constant elements. In the $\mathcal{H}$-matrix method, a set of boundary elements is called a cluster. Let I be the set of all the boundary elements and obtain a binary tree T whose root cluster is I by the following recursive procedure:

1. Set $C = I$.
2. If the number of elements which are included in C is larger than a preset parameter $n_{\min}$, then go to 3; otherwise set C to be a leaf cluster, which has no son cluster.
3. Make a rectangle which encloses all elements in C.
4. Divide the longer side of the rectangle into two parts of equal size. Correspondingly, C is also divided into two subsets C_1 and C_2 so that each subset consists of elements whose collocation point is included in one of the two rectangles.
5. Renumber the elements so that the elements in the same cluster have the element numbers close to each other.
6. Set C_1 and C_2 to be son clusters of C.
7. For $i = 1; 2$, set $C = C_i$ and go to 2.

Next, we consider the blocking procedure of the coefficient matrix in (16). For illustrative purpose, we pick one of the four submatrices in (16). We divide the matrix into some submatrices according to the cluster and classify them into admissible and inadmissible blocks, which will be low lank-approximated or not, respectively, by the following procedure:

1. Set $C_r = I$ and $C_c = I$.
2. If C_r and C_c satisfy the admissible condition defined as

$$\min\{\mathrm{diam}\ C_r,\ \mathrm{diam}\ C_c\} \leq \eta\,\mathrm{dist}(C_r, C_c), \tag{21}$$

where $\eta \in \mathrm{R}$ is a preset parameter and 'diam' and 'dist' denote a diameter and distance of a cluster, respectively, then set the submatrices whose row indices and column indices correspond to C_r and C_c to be admissible blocks, respectively; otherwise, go to 3

3. If both C_r and C_c have sons C_{r1}, C_{r2} and C_{c1}, C_{c2} in T, respectively, then for all $(\tilde{C}_r, \tilde{C}_c)$ in combinations $(C_{r1}, C_{r2}) \times (C_{c1}, C_{c2})$, set $C_r = \tilde{C}_r$ and $C_c = \tilde{C}_c$ and go to 2; otherwise set the submatrices whose row indices and column indices correspond to C_r and C_c to be inadmissible blocks, respectively.

By the above two procedures, we obtain a coefficient matrix which is hierarchically divided into some submatrices and each submatrix is classified into an admissible or inadmissible block.

3.2.2 Low rank approximation with the ACA

An arbitrary matrix $A \in C^{m \times n}$ can be approximated by two matrices $U \in C^{m \times k}$ and $V \in C^{n \times k}$ $(k < \text{rank } A)$ as

$$A \simeq UV^H, \tag{22}$$

and this is called a low rank approximation of a matrix. In this study, the adaptive cross approximation (ACA) is employed for the low rank approximation. In the ACA, we can obtain U, V and their rank k such that they satisfy the following condition with a given tolerance ε_{ACA}:

$$\| UV^H - A \|_F \leq \varepsilon_{\text{ACA}} \| A \|_F, \tag{23}$$

where $\| \cdot \|_F$ denotes the Frobenius norm. The details of the algorithm of the ACA and its application to the $\mathcal{H}$-matrix method can be found in [6].

3.2.3 Efficiency improvement of the $\mathcal{H}$-matrix method

In the process of the BEM, it is time-consuming not only to solve a system of linear algebraic equations but also to compute entries of its coefficient matrix. Each of the entries is calculated as an integral of the second-order tensors (17) and (18). The most time-consuming part of the calculation is the computation of the Hankel function $H_n^{(1)}$ which depends on positions of a collocation point and integration points of a numerical integration but is independent of indices of the tensors. Hence we can reduce the computational time by storing the values of $H_n^{(1)}$ and reuse them for all indices when we compute the four entries which indicate influence from the same boundary element to the same collocation point.

In the $\mathcal{H}$-matrix method, however, we do not need all of the four entries to construct a coefficient matrix, because not all the entries are required for low rank approximation in the ACA. In this study, we employ this store-and-reuse strategy only for inadmissible blocks and reduce the computational time to construct the matrix.

4 NUMERICAL EXAMPLES

In this section, we demonstrate two numerical examples and confirm the effectiveness of the proposed method. The elastic medium is assumed to be steel (mass density $\rho = 7.8 \times 10^3$ [kg], Young's modulus $E = 205$ [GPa], Poisson's ratio $\upsilon = 0.3$). The parameters of the $\mathcal{H}$-matrix method are set as $\varepsilon_{\text{ACA}} = 10^{-5}$, $n_{\text{min}} = 32$, $\eta = 0.7$, and the LU decomposition is used as a direct solver in the boundary element analysis. All of the linear algebraic operations in the LU decomposition are accelerated by the $\mathcal{H}$-matrix method. The incident wave u^{in} is set as a plane P-wave propagating in x_1 direction with the frequency 3 [kHz]. We set the number of the observation points $M = 400$ and distribute them around the fixed design domain D at

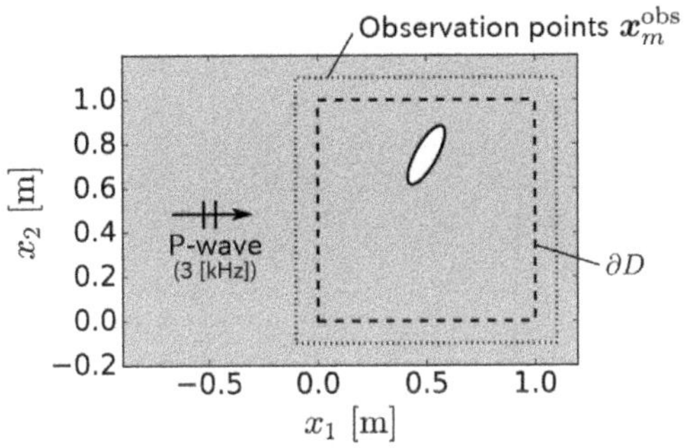

Figure 2: Identification of an elliptical defect.

regular intervals as shown in Fig. 2. The initial level set function ϕ_0 is uniformly set to be 1 in D. We first assume that an elliptical defect is allocated in D and use a numerical solution of the BEM, in which each defect is divided into 5,000 boundary elements, as the measured displacement $\hat{u}$. The example numerical calculations are performed on a computer with Intel Xeon CPU E5-1660 with 8 cores, and the code is parallelised using Open MP.

Figures 3 and 4 show results of the optimisations and histories of the objective function J in each case whether the $\mathcal{H}$-matrix method is utilised or not, respectively. ZGETRF routine by LAPACK is used for the LU decomposition in the case that the $\mathcal{H}$-matrix method is not used. Note that J is normalised by the amplitude of the plane incident wave. We see that the results are almost the same and give us a good estimation of the position and shape of the defect. Also, Fig. 5 shows computational time for one optimisation step versus the number of boundary elements in the step. From this comparison, we confirm that the $\mathcal{H}$-matrix method can reduce the computational time for the sensitivity analysis without losing its accuracy.

Next, we add a cylindrical defect to the previous configuration and perform the optimisation as shown in Fig. 6. The result of the optimisation is shown in Fig. 7 and indicates that we

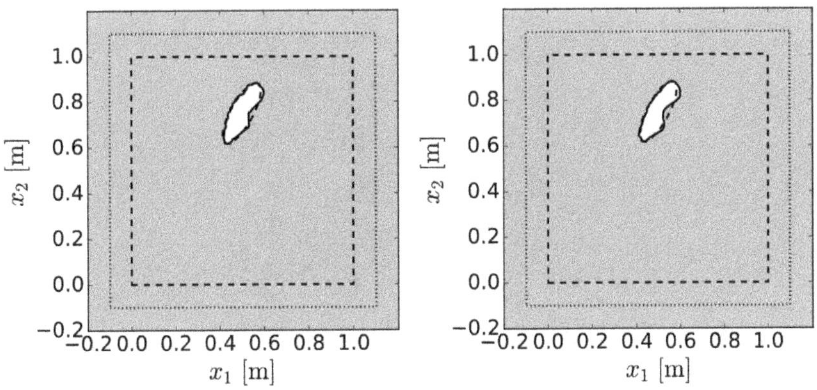

(a) BEM without $\mathcal{H}$-matrix method (b) BEM with $\mathcal{H}$-matrix method

Figure 3: Optimal configurations in the case of one elliptical defect.

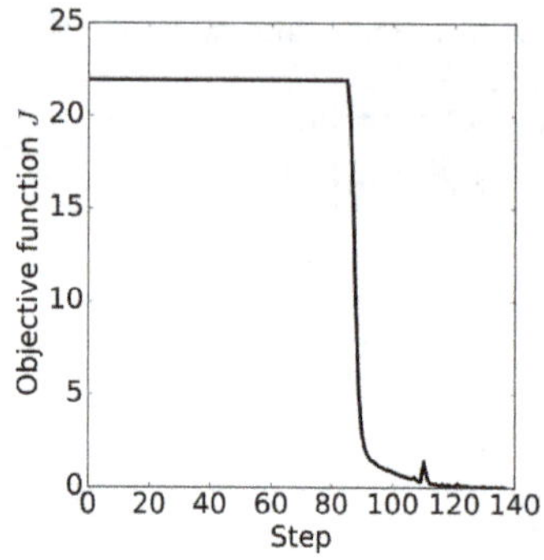

(a) BEM without $\mathcal{H}$-matrix method

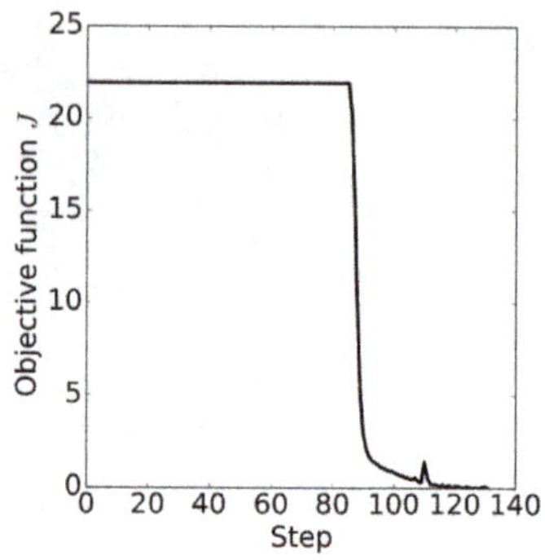

(b) BEM with $\mathcal{H}$-matrix method

Figure 4: Histories of the objective function J in the case of one elliptical defect.

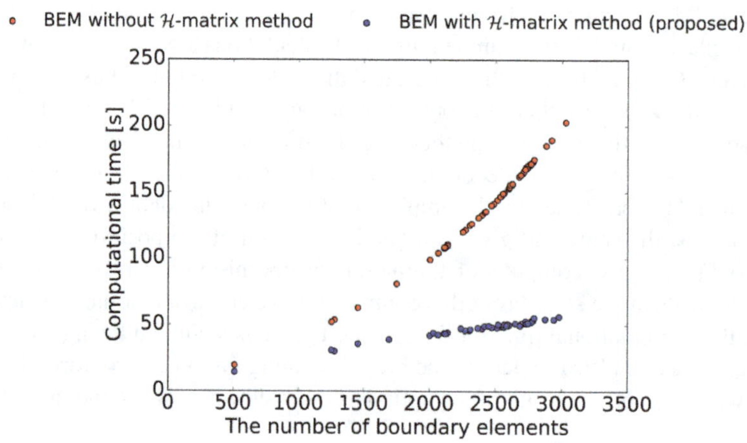

Figure 5: Computational time for one optimisation step.

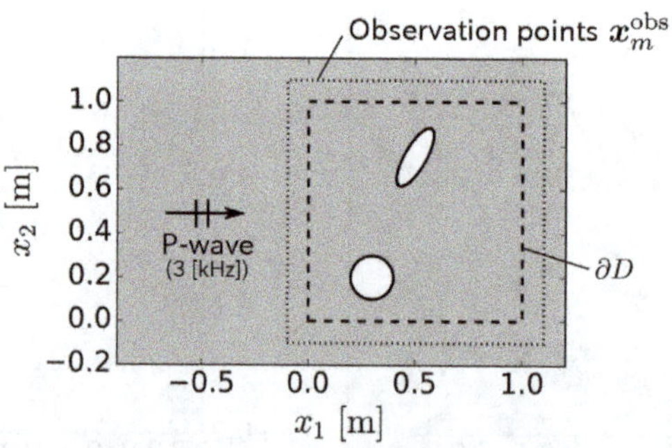

Figure 6: Identification of elliptical and cylindrical defects.

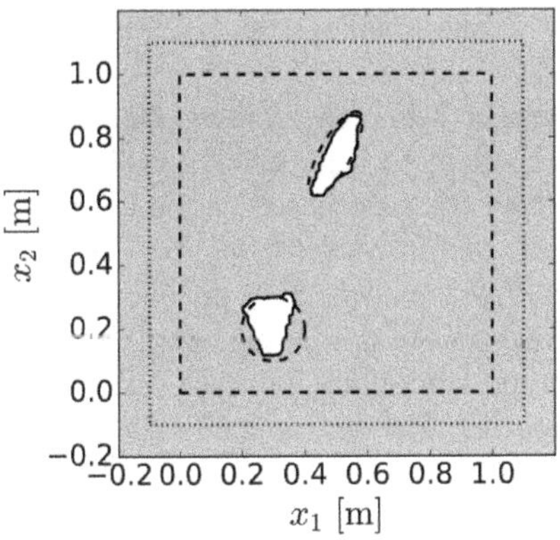

(a) Optimal configuration

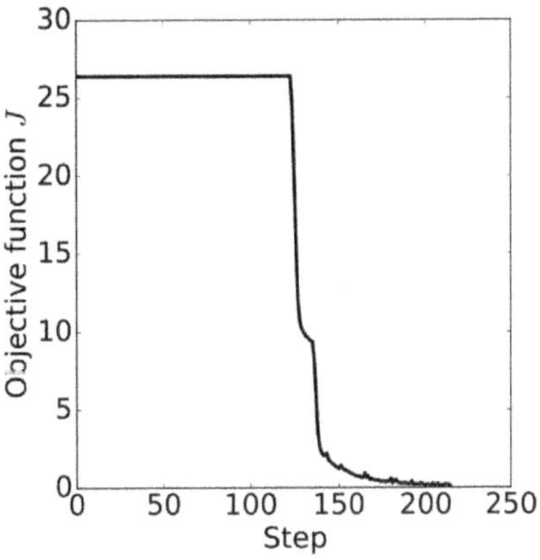

(b) History of the objective function J

Figure 7: Optimisation result in the case of elliptical and cylindrical defects.

can identify not only one but two defects though it seems to be less accurate than the case of one defect.

5 CONCLUSION

We have developed a numerical method for defect identification in an elastic medium based on the level set method and the BEM. We have employed the $\mathcal{H}$-matrix method to the BEM and reduce the computational cost. Finally, we have demonstrated numerical examples and confirmed that the proposed method is effective for the inverse scattering problem.

ACKNOWLEDGEMENT

This work was supported by JSPS KAKENHI Grant Number JP16H04255 and Number JP17K14146.

REFERENCES

[1] Bonnet, M. & Constantinescu, A., Inverse problems in elasticity. *Inverse Problems*, **21**(2), pp. R1–R50, 2005.
https://doi.org/10.1088/0266-5611/21/2/r01

[2] Guzina, B.B. & Bonnet, M., Topological derivative for the inverse scattering of elastic waves. *The Quarterly Journal of Mechanics and Applied Mathematics*, **57**(2), pp. 161–179, 2004.
https://doi.org/10.1093/qjmam/57.2.161

[3] Bonnet, M. & Guzina, B.B., Sounding of finite solid bodies by way of topological derivative. *International Journal for Numerical Methods in Engineering*, **61**(13), pp. 2344–2373, 2004.
https://doi.org/10.1002/nme.1153

[4] Greengard, L. & Rokhlin, V., A fast algorithm for particle simulations. *Journal of Computational Physics*, **73**(2), pp. 325–348, 1987.
https://doi.org/10.1016/0021-9991(87)90140-9

[5] Martinsson, P.G. & Rokhlin, V., A fast direct solver for boundary integral equations in two dimensions. *Journal of Computational Physics*, **205**(1), pp. 1–23, 2005. https://doi.org/10.1016/j.jcp.2004.10.033

[6] Bebendorf, M., *Hierarchical matrices*. Springer Berlin: Heidelberg, 2008.

[7] Nakamoto, K., Isakari, H., Takahashi, T. & Matsumoto, T., Acceleration of a multi-objective topology optimisation in 2D electro-magnetic field based on the level-set method and the boundary element method by the-$\mathcal{H}$ Matrix Method. *International Journal of Computational Methods and Experimental Measurements*, **5**(5), pp. 686–695, 2017.

[8] Yamada, T., Izui, K., Nishiwaki, S. & Takezawa, A., A topology optimization method based on the level set method incorporating a fictitious interface energy. *Computer Methods in Applied Mechanics and Engineering*, **199**(45), pp. 2876–2891, 2010. https://doi.org/10.1016/j.cma.2010.05.013

[9] Isakari, H., Kuriyama, K., Harada, S., Yamada, T., Takahashi, T. & Matsumoto, T., A topology optimisation for three-dimensional acoustics with the level set method and the fast multipole boundary element method. *Mechanical Engineering Journal*, **1**(4), pp. CM0039–CM0039, 2014.
https://doi.org/10.1016/j.jcp.2004.10.033

[10] Isakari, H., Nakamoto, K., Kitabayashi, T., Takahashi, T. & Matsumoto, T., A multi-objective topology optimisation for 2D electro-magnetic wave problems with the level set method and BEM. *European Journal of Computational Mechanics*, **25**(1–2), pp. 165–193, 2016.

[11] Burton, A.J. & Miller, G.F., The application of integral equation methods to the numerical solution of some exterior boundary-value problems. *Proceedings of the Royal Society of London A: Mathematical, Physical and Engineering Sciences, The Royal Society,* **323**, pp. 201–210, 1971.
https://doi.org/10.1098/rspa.1971.0097

A QUASI-STATIC INTERFACE DAMAGE MODEL WITH FRICTIONAL CONTACT – APPLICATIONS TO STEEL REINFORCED CONCRETE STRUCTURES

ROMAN VODIČKA* & FILIP KŠIŇAN
Technical University of Košice, Faculty of Civil Engineering, Slovakia.

ABSTRACT

A model for numerical analysis of compound structures made of various materials is presented. The mathematical concept of solution is based on quasi-static evolution of debonding processes occurring along the interface. It is formulated in terms of energies considering the stored energy represented by the elastic energy of the structures and dissipation due to damage processes, plastic slip at the interface or friction. The numerical solution includes a semi-implicit time stepping procedure, relying on splitting of the whole problem at a current time step into two problems of variational nature solved recursively. The space discretisation includes Symmetric Galerkin Boundary Element Method used to obtain the stored energies, and, in combination with the variational character of the recursive problems, also to calculate its gradients to be utilized in non-linear programming algorithms for finding the time-evolving solution. Numerical results are demonstrated for a steel-concrete interface frequently met in civil engineering applications to assess the model applicability in engineering practice.

Keywords: cohesive interface, contact with friction, interface damage, interface plastic slip, quadratic programming, quasi-static delamination, SGBEM.

1 INTRODUCTION

Various engineering applications consider analysis of compound structures made of various materials. Numerical analysis of such structures which additionally includes, at least at the material interface, also damage mechanisms, plastic deformations or frictional contact is a challenging problem, which to be solved requires advanced computational techniques.

Computational simulations of damage evolution also with the other non-linear phenomena eventually lead to problems of crack propagation at such material interfaces. All of them are currently under intensive development. There exist several fracture models which, unlike classical fracture mechanics, enable also prediction of crack initiation: cohesive zone models (CZM) [1–3], finite fracture mechanics models [4–6], or variationally based interface damage and plasticity models [7–9].

Damage and subsequent cracking of an interface are usually modelled by an internal variable introduced on the interface. The concept of such damage variable was introduced by Frémond [10]. In the present model, interface stress-strain relations providing the structural response similar to known CZMs [11, 12] are considered and also other features that are typical for interface cracks are taken into account.

First, it is known that an interface crack propagating in the opening fracture mode usually dissipates much less energy than in the shearing mode [13, 14]. Therefore, it is necessary to consider additional dissipated energy in shearing mode. This energy may come from developed plastic deformation in the vicinity of the evolving interface [15, 16], or the fracture energy may be dependent on the mode of arising crack throught the fracture mode-mixity angle [8, 17, 18]. Both of these possibilities are considered in the present model.

**ORCID: http://orcid.org/0000-0002-7702-3126*

DOI: 10.2495/CMEM-V6-N6-1043-1056

Second, when modeling cohesive behaviour of an interface by CZMs, the normal contact conditions are often represented by additional non-damageable stiffness in compression. Such an approach assumes the contact surface exhibiting certain elastic response also under compression, e.g. due to its roughness, or due to the presence of a thin layer of an adhesive. This approach is also suitable for mathematical analysis, even if friction is present [19, 20].

Third, viscosity can be considered in the domains to expand applicability of the model to more general states than elastic deformation [21, 22], and also to ensure uniqueness of the solution if a sufficiently dissipative model is used, e.g. Kelvin-Voigt, see [21], even if the elastic state is approximated by vanishing-viscosity solution concept [23].

The proposed mathematical approach is based on an energetic formulation looking for a kind of a weak solution, approximated by a time stepping procedure and by the Symmetric Galerkin Boundary Element Method (SGBEM) as in [11]. In the solution process, it utilizes nonlinear programming algorithms based on quadratic ones of [24].

In what follows, the model incorporating all mentioned phenomena is described briefly in Section 2, some aspects of the numerical solution concept are provided in Section 3. Finally, Section 4 shows some results obtained by the numerical solution. In this section some problems of civil engineering discussing the steel-concrete interfaces are presented: from the point of view of convergence of the used model in numerical solution and from practical aspect to asses the applicability of the model in practical simulations.

2 DESCRIPTION OF THE MODEL

We consider a bounded domain Ω split into several bodies by internal interfaces Γ_C. For the sake of simplicity only two 2D bodies will be considered. These bodies are supposed to occupy subdomains $\Omega^\eta \subset \mathbb{R}^2 (\eta = \mathrm{A},\mathrm{B})$ with bounded Lipschitz boundaries Γ^η, Fig. 1. However, it should be noted that the theory can be generalized to allow also 3D bodies. Let $\vec{n}$ denote the unit outward normal vector defined at smooth parts of Γ, let $\vec{s}$ denote the unit tangential vector such that it defines anti-clockwise orientation of Γ. A potential superscript index η refers to the pertinent subdomain.

The contact zone Γ_C is defined as the common part of Γ^A and Γ^B, i.e. $\Gamma_C = \Gamma^A \cap \Gamma^B$. The Dirichlet and Neumann boundary conditions are defined on the outer boundary disjoint parts Γ_D and Γ_N, respectively, at a current time t, cf. eqn (8) below. Additionally, Γ_D is considered far from the contact boundary, i.e.$\overline{\Gamma}_D \cap \overline{\Gamma}_C = \emptyset$.

The difference (the gap for displacements) on Γ_C of the functions defined on Ω^A and Ω^B will be denoted by $[\![\cdot]\!]$. In particular, the gap of displacements on the contact boundary Γ_C means $[\![u]\!] := u^A|_{\Gamma_C} - u^B|_{\Gamma_C}$. Further, we will use the gap of the normal and tangential displacements

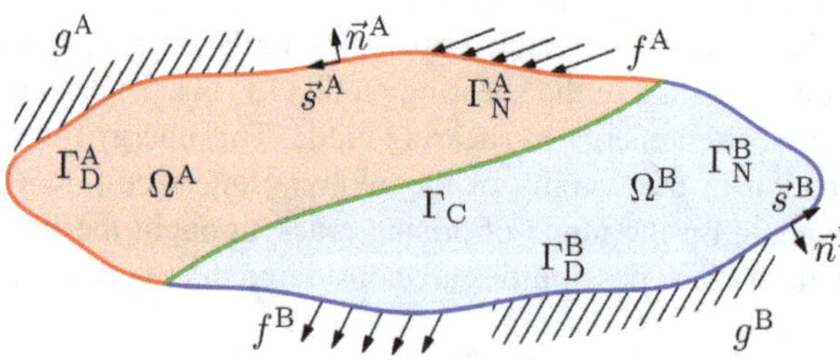

Figure 1: A sketch of two debonding domains.

$$[\![u]\!]_{\rm n} := [\![u]\!]\cdot\vec{n}^{\rm B} = -[\![u]\!]\cdot\vec{n}^{\rm A} = -u^{\rm B}|_{\Gamma_{\rm C}}\cdot\vec{n}^{\rm B} - u^{\rm A}|_{\Gamma_{\rm C}}\cdot\vec{n}^{\rm A}, \quad [\![u]\!]_{\rm s} := [\![u]\!] - [\![u]\!]\vec{n}^{\rm B}. \tag{1}$$

We will also use the convention that te 'dot' will stand for the partial time derivative $\frac{\partial}{\partial t}$. The model includes the force equilibrium on the contact interface $\Gamma_{\rm C}$:

$$[\![\sigma]\!]\vec{n} = 0 \quad \sigma := \mathbb{D}e(\dot{u}) + \mathbb{C}e(u) \tag{2a}$$

$$\sigma_{\rm n} - \kappa_{\rm n}(\zeta)[\![u]\!]_{\rm n} - \sigma_{\rm c} = 0, \text{ with } \sigma_{\rm c} := -\kappa_{\rm c}[\![u]\!]_{\rm n}, \tag{2b}$$

$$\begin{gathered}|\sigma_{\rm f}| < -\mu(\zeta)\sigma_{\rm c} \Rightarrow [\![\dot{u}]\!]_{\rm s} = 0, \quad |\sigma_{\rm f}| = -\mu(\zeta)\sigma_{\rm c} \Rightarrow \exists\lambda \geq 0 : \sigma_{\rm f} = \lambda[\![\dot{u}]\!]_{\rm s}, \\ \text{with } \sigma_{\rm f} := \sigma_{\rm s} - \kappa_{\rm s}(\zeta)([\![u]\!]_{\rm s} - \pi), \text{ and } \sigma_{\rm s} := \sigma\vec{n} - \sigma_{\rm n}\vec{n} \text{ with } \sigma_{\rm n} := \vec{n}^{\rm T}\sigma\vec{n},\end{gathered} \tag{2c}$$

where we consider a damageable interface represented by the normal and the tangential stiffnesses $\kappa_{\rm n}$ and $\kappa_{\rm s}$, respectively, depending on the scalar interface damage variable ζ limited to the range [0; 1], with zero values pertinent to the total interface damage supposed to produce an interface crack. Various functions for this dependence were presented in [11] and [25], e.g. having $\kappa_{\rm n}(\zeta)= \kappa_{\rm n0}\,\Phi(\zeta)$ and $\kappa_{\rm s}(\zeta)= \kappa_{\rm s0}\,\Phi(\zeta)$ with

$$\Phi(\zeta) = \frac{\beta\zeta}{1+\beta-\zeta}, \text{ with } \beta > 0 \tag{3}$$

leads to stress-strain relation of the bilinear CZM, [12].

There is also present the interface plastic slip π on $\Gamma_{\rm C}$ and, simultaneously, we suppose frictional character of the contact between the subdomains with Coulomb friction law and damage dependent friction coefficient μ for which we prefer to use the relation.

$$\mu(\zeta) = \mu_0(1-\zeta)^p, \text{ with } \mu_0, p > 0. \tag{4}$$

Furthermore, the flow rules for damage and plastic slip evolutions on $\Gamma_{\rm C}$ are considered:

$$\partial_{\dot{\zeta}} a_1([\![u]\!], \dot{\zeta}) + \frac{1}{2}\kappa_s'(\zeta)\,|[\![u]\!]_s - \pi|^2 + \frac{1}{2}\kappa_n'(\zeta, [\![u]\!]_n)^2 + N_{[0,1]}(\zeta) \ni \operatorname{div}_{\rm s}(\kappa_2\nabla_{\rm s}\zeta), \tag{5a}$$

$$\dot{\pi} \in N_{|\cdot|\leq\sigma_y}(\operatorname{div}_{\rm s}(\kappa_1\nabla_{\rm s}\pi) - \kappa_s(\zeta)([\![u]\!]_{\rm s} - \pi) + \kappa_{\rm H}\pi), \tag{5b}$$

$$\text{With boundary conditions on } \partial\Gamma_{\rm C}\ \nabla_{\rm s}\pi\cdot\vec{n}_{\rm s} = 0, \quad \nabla_{\rm s}\zeta\cdot\vec{n}_{\rm s} = 0, \tag{5c}$$

where N_K denotes the normal cone of the convex set K, see [8]. The a_1-term is supposed in the form

$$a_1([\![u]\!]; \dot{\zeta}) := \begin{cases} -G_{\rm c}([\![u]\!])\dot{\zeta}, & \text{if } \dot{\zeta} \leq 0 \text{ on } \Gamma_{\rm C}, \\ +\infty & \text{else,} \end{cases} \tag{6}$$

including the interface fracture energy $G_{\rm c}$, required to break a unit area of the interface, being considered to depend on the current state of the displacement jump $[\![u]\!]$. Any phenomenological law describing the dependence of $G_{\rm c}$ on current displacement gap can be used, say that of Benzeggagh [26], or Hutchinson [27].

$$G_c([\![u]\!]) = G_c^I\left(1+\tan^2\left(\frac{2}{\pi}\ \arctan\ \sqrt{\frac{G_c^{II}-G_c^I}{G_c^I}}\ \arctan\ \left|\frac{[\![u]\!]_s}{[\![u]\!]_n}\right|\right)\right). \tag{7}$$

The parameters G_c^I and G_c^{II} express the fracture energies in the pure Mode I (opening) and in the pure Mode II (shear), respectively.

The plastic slip in the interface, represented by the variable π, is initiated if the shearing stress reaches the the yield level σ_y and continues evolving with kinematic hardening whose pertinent stiffness parameter is κ_H.

In the domains, we consider the standard visco-elasticity (no inertia) using the Kelvin-Voigt rheology supplemented with standard boundary conditions:

$$-\mathrm{div}\sigma = h \quad \text{in } \Omega\setminus\Gamma_C, \quad u = g(t) \quad \text{on } \Gamma_D, \quad \sigma\vec{n} = f(t) \quad \Gamma_N. \tag{8}$$

In addition, we consider an initial-value problem by prescribing the following initial conditions

$$u\,|_{t=0} = u_0, \quad \pi\,|_{t=0} = \pi_0, \quad \zeta\,|_{t=0} = \zeta_0. \tag{9}$$

In the numerical solution, a variational structure of the described model is intended to be exploited. Therefore, let us briefly present energetics behind the boundary-value problem (BVP) eqns (2)–(9). The underlying overall free energy $\varepsilon = \varepsilon(u,\pi,\zeta)$ expressed here by mechanical internal energy can be split into its volumetric and interface parts as follows

$$\varepsilon(u,\pi,\zeta) := \int_{\Omega\setminus\Gamma_C} \frac{1}{2}\mathbb{C}e(u):e(u)\ \mathrm{d}x + \int_{\Gamma_C} \psi([\![u]\!],\pi,\zeta)\,\mathrm{d}S, \tag{10a}$$

Where the specific interface energy $([\![u]\!],\pi,\zeta)$ is expressed as

$$\psi([\![u]\!],\pi,\zeta) := \begin{cases} \dfrac{\kappa_s(\zeta)}{2}\left([\![u]\!]_s-\pi\right)^2 + \dfrac{\kappa_n(\zeta)}{2}\left([\![u]\!]_n\right)^2 + \dfrac{\kappa_c}{2}\left([\![u]\!]_n^-\right)^2 & \\ +\dfrac{\kappa_H}{2}|\pi|^2 + \dfrac{\kappa_1}{2}|\nabla_s\pi|^2 + \dfrac{\kappa_2}{2}|\nabla_s\zeta|^2 & \text{if } 0\le\zeta\le 1 \text{ a.e. on } \Gamma_C, \\ +\infty & \text{otherwise.} \end{cases} \tag{10b}$$

The other ingredient of the model is the overall pseudo-potential of dissipative forces $\mathscr{R}$, which also can be split into a volumetric and a surface contributions as follows:

$$\begin{aligned} \mathscr{R}(u,\zeta;\dot{u},\dot{\pi},\dot{\zeta}) := &\int_{\Omega\setminus\Gamma_C} \frac{1}{2}\mathbb{D}e(\vec{\dot{u}}):e(\dot{u})\,\mathrm{d}x \\ &+\int_{\Gamma_C} \mu(\zeta)\,|\,\kappa_c[\![u]\!]_n^-\,|\,|[\![\dot{u}]\!]_s| + \sigma_y|\dot{\pi}| + a_1([\![u]\!],\dot{\zeta})\mathrm{d}S \end{aligned} \tag{11}$$

where the a_1-term, cf. eqn (6), represents the potential (and also the specific dissipation rate) of the rate-independent delamination process on the contact boundary Γ_C.

Eventually, we need the (linear) functional of external mechanical loading $\mathscr{F}(t)$ which, after the standard transformation to time-constant (homogeneous) Dirichlet condition by the shift $u \to u + \bar{g}$ with $\bar{g}$ being an extension of g from eqn (8) on Ω is given by

$$\langle \mathscr{F}(t),u\rangle := \int_\Omega h\cdot u-(\mathbb{D}e(\dot{\bar{g}})+\mathbb{C}e(\bar{g})):e(u)\,\mathrm{d}x+\int_{\Gamma_N} f\cdot u\,\mathrm{d}S. \tag{12}$$

For simplicity and without eliminating interesting applications, we assume that Γ_D and Γ_C are far from each other so that one can assume $\bar{g}|_{\Gamma_C}=0$ and thus this shift transformation does not influence the flow rules on Γ_C.

In terms of these functionals, the evolution BVP can be expressed by the system of nonlinear variational inclusions as:

$$\partial_{\dot{u}}\mathscr{R}(u,\zeta;\dot{u})+\partial_u\mathscr{E}(u,\pi,\zeta)\ni\mathscr{F}(t), \tag{13a}$$

$$\partial_{\dot{\pi}}\mathscr{R}(\dot{\pi})+\partial_\pi\mathscr{E}(u,\pi,\zeta)\ni 0, \tag{13b}$$

$$\partial_{\dot{\zeta}}\mathscr{R}(u;\dot{\zeta})+\partial_\zeta\mathscr{E}(u,\pi,\zeta)\ni 0, \tag{13c}$$

where we reflected special form of $\mathscr{R}$ in relation to the used subdifferentials, e.g. $\partial_{\dot{\zeta}}\mathscr{R}$ does not depend on $\dot{u}$ and $\dot{\pi}$.

The system provides also energy balance relation. It can be obtained by testing eqns (13a), (13b), and (13c) by $\dot{u}$, $\dot{\pi}$, and $\dot{\zeta}$, respectively, which gives:

$$\frac{\mathrm{d}}{\mathrm{d}t}\mathscr{E}(u,\pi,\zeta)+\mathscr{D}(u,\zeta;\dot{u},\dot{\pi},\dot{\zeta})=\langle F(t),\dot{u}\rangle \tag{14}$$

Where $\mathscr{D}(u,\zeta;\dot{u},\dot{\pi},\dot{\zeta}):=\langle\partial_{\dot{u}}\mathscr{R}(u,\zeta;\dot{u}),\dot{u}\rangle+\langle\partial_{\dot{\pi}}\mathscr{R}(\dot{\pi}),\dot{\pi}\rangle+\langle\partial_{\dot{\zeta}}\mathscr{R}(u;\dot{\zeta}),\dot{\zeta}\rangle$ denotes the overall dissipation rate.

3 DISCRETISATION

Numerical solution requires both time discretization and spatial discretization. In what follows, we separately describe some aspects of both.

3.1 A semi-implicit fractional-step time discretisation

The time discretisation which ensures numerical convergence and stability of the obtained solution frequently includes decoupling of the system by a suitable fractional-step method. The splitting of the variables then should be done so that it guarantees separate convexity of the free energy $\mathscr{E}$ and additive splitting of the dissipation potential $\mathscr{R}$. Here, we apply splitting of the variables to *(u, π)* and ζ, assuming that (u, π) → $\mathscr{E}$ (u, π, ζ) is convex for all ζ, and ζ → $\mathscr{E}$ (u, π, ζ) is convex for all (u, π). This assumption leads to a two-step decoupled scheme, written in the form of eqn (13) as:

$$\tau\partial_{u_\tau^k}\mathscr{R}\left(u_\tau^{k-1},\zeta_\tau^{k-1};\frac{u_\tau^k-u_\tau^{k-1}}{\tau}\right)+\partial_{u_\tau^k}\mathscr{E}\left(u_\tau^k,\pi_\tau^k,\zeta_\tau^{k-1}\right)\ni\mathscr{F}(k\tau), \tag{15a}$$

$$\tau\partial_{\pi_\tau^k}\mathscr{R}\left(\frac{\pi_\tau^k-\pi_\tau^{k-1}}{\tau}\right)+\partial_{\pi_\tau^k}\mathscr{E}\left(u_\tau^k,\pi_\tau^k,\zeta_\tau^{k-1}\right)\ni 0, \tag{15b}$$

$$\tau\partial_{\zeta_\tau^k}\mathscr{R}\left(u_\tau^k,\zeta_\tau^{k-1};\frac{\zeta_\tau^k-\zeta_\tau^{k-1}}{\tau}\right)+\partial_{\zeta_\tau^k}\mathscr{E}\left(u_\tau^k,\pi_\tau^k,\zeta_\tau^k\right)\ni 0, \tag{15c}$$

Where the rates have been approximated by the finite differences, e.g. $\dot{\zeta} \approx \frac{\zeta_\tau^k - \zeta_\tau^{k-1}}{\tau}$, where ζ_τ^k denotes the solution at the instant $k\tau$ introduced by the time step τ.

The decoupled scheme provides a variational structure to the solved problem, with two recursive minimisations in each time step. First, the minimisation of

$$\mathscr{H}_1^k(u,\pi) := \tau\mathscr{R}\left(u_\tau^{k-1},\zeta_\tau^{k-1};\frac{u-u_\tau^{k-1}}{\tau},\frac{\pi-\pi_\tau^{k-1}}{\tau}\right)+\mathscr{E}\left(u,\pi,\zeta_\tau^{k-1}\right)-\left\langle\mathscr{F}(\tau^k),u\right\rangle \quad (16a)$$

provides $\left(u_\tau^k,\pi_\tau^k\right)$, and then the minimisation of

$$\mathscr{H}_2^k(\zeta) := \tau\mathscr{R}\left(u_\tau^k,\zeta_\tau^{k-1};\frac{\zeta-\zeta_\tau^{k-1}}{\tau}\right)+\mathscr{E}\left(u_\tau^k,\pi_\tau^k\zeta\right) \quad (16b)$$

provides ζ_τ^k. The minimisation system eqn (16) is to be solved recursively for k=1, …, T/τ, starting from k=1 with

$$u_\tau^0 = u_0, \quad \pi_\tau^0 = \pi_0, \quad \zeta_\tau^0 = \zeta_0. \quad (17)$$

It is important to see that the scheme eqn (15) has a variational character of both sub-problems for $\left(u_\tau^k,\pi_\tau^k\right)$ and for ζ_τ^k with convex functionals which ensures existence of the numerical solution and which makes the numerical solution relatively simple.

In fact, the functional $\mathscr{E}$ is quadratic with respect to $(u;\ \pi)$ so that various efficient numerical QP algorithms can be used in the solution, see [24] for a survey. Here, we have used Conjugate Gradient (CG) based algorithms with bound constraints [24] and implemented in a way similar to [16]. Considering $\mathscr{E}$ with respect to ζ , it is quadratic if the functions $\kappa(\zeta)$ in eqn (10b) are quadratic. Nevertheless, generally, e.g. also for the function in eqn (3), the function can be more general. In such a case, we apply the QP algorithm sequentially as in [11].

Additionally, energy balance from eqn (14) can be adopted for the used discretization when replacing the time derivatives by the time differences. Therefore, the tests of the pertinent inclusions in eqns (15a), (15b), and (15c) by $u_\tau^k - u_\tau^{k-1}, \pi_\tau^k - \pi_\tau^{k-1}$, and $\zeta_\tau^k - \zeta_\tau^{k-1}$, respectively, result in the following estimate:

$$\begin{aligned}\mathscr{E}\left(u_\tau^k,\pi_\tau^k,\zeta_\tau^k\right)+\tau\mathscr{D}\left(u_\tau^{k-1},\zeta_\tau^{k-1};\frac{u_\tau^k-u_\tau^{k-1}}{\tau},\frac{\pi_\tau^k-\pi_\tau^{k-1}}{\tau},\frac{\zeta_\tau^k-\zeta_\tau^{k-1}}{\tau}\right)\\ \le \mathscr{E}\left(u_\tau^{k-1},\pi_\tau^{k-1},\zeta_\tau^{k-1}\right)+\left\langle\mathscr{F}(k\tau),u_\tau^k-u_\tau^{k-1}\right\rangle\end{aligned} \quad (18)$$

which (im)balances the energies within the preformed time step and shows that the numerical solution does not have to conserve energy, even though the difference between the right-hand and the left-hand sides diminishes for refined discretisations. This relation can be verified by the numerical solution, cf. Fig. 3.

3.2 Notes on the spatial discretisation by SGBEM

The viscosity as a part of the model is intentionally used especially for better mathematical treatment of the model, as it provides a convergence property of the discretized solution, even in the case of vanishing viscosity, though the convergence towards the inviscid solution can hardly be proved, see [23]. Additionally, it can be a natural part of the physical model, though the numerical approach requires a specific physical situation with a slight restriction by assuming that $\mathbb{D}'' = \tau_r\mathbb{C}''$, where $\tau_r>0$ is given relaxation-time parameter. We also suppose that the relaxation time τ_r is the same for both subdomains and it is implemented by a simple trick from [22, 23]. It enables to reformulate the solution within the bulk domains in terms of an elastostatic problem solved by the conventional elastostatic SGBEM. The trick introduces a new fictitious displacement variable v at the k-th time step by the relation

$$v^k = u^k + \frac{\tau_r}{\tau}\left(u^k - u^{k-1}\right). \tag{19}$$

The functionals in eqn (16) are then reformulated in terms of the new variable v, see also [11].

The role of the SGBEM in the present computational procedure is to provide a complete boundary-value solution from the given boundary data to calculate the elastic strain energy in these domains. Thus, at each time step and at each iteration of the functional minimisation in eqn (16), the SGBEM code calculates unknown tractions p along $\Gamma_C \cup \Gamma_D$ and unknown (fictitious) displacements v along $\Gamma_C \cup \Gamma_N$, assuming the displacement gap $[\![v]\!]$ on Γ_C to be known from the used minimisation procedure.

The implementation of SGBEM, deduced from the energetic principles as shown in [28, 29], guarantees the positive-definite character of the computed strain energy. The present system of BIEs written in the weighted formulation can be arranged in the following block form, cf. [28, 29]:

$$\begin{pmatrix} \phi_D^A \\ \psi_N^A \\ \phi_C^A \\ \psi_C^A \\ \phi_D^B \\ \psi_N^B \\ \phi_C^B \\ \psi_C^B \end{pmatrix}^T \begin{pmatrix} \mathbf{K}^A & \mathbf{M}^{AB} \\ \mathbf{M}^{AB*} & \mathbf{K}^B \end{pmatrix} \begin{pmatrix} p_D^A \\ v_N^A \\ p_C^A \\ v_C^A \\ p_D^B \\ v_N^B \\ p_C^B \\ v_C^B \end{pmatrix} = \begin{pmatrix} \phi_D^A \\ \psi_N^A \\ \phi_C^A \\ \psi_C^A \\ \phi_D^B \\ \psi_N^B \\ \phi_C^B \\ \psi_C^B \end{pmatrix}^T \begin{pmatrix} \mathbf{H}^A & 0 & -(\mathbf{M}^{AB})_{.4} \\ 0 & \mathbf{H}^B & 0 \end{pmatrix} \begin{pmatrix} g^A \\ f^A \\ g^B \\ f^B \\ w \end{pmatrix}, \tag{20}$$

With prescribed $w=[\![v]\!]$and with

$$\mathbf{K}^{\eta} = \begin{pmatrix} -\mathbf{U}^{\eta}_{\mathrm{DD}} & \mathbf{T}^{\eta}_{\mathrm{DN}} & -\mathbf{U}^{\eta}_{\mathrm{DC}} & \mathbf{T}^{\eta}_{\mathrm{DC}} \\ \mathbf{T}^{\eta*}_{\mathrm{ND}} & -\mathbf{S}^{\eta}_{\mathrm{NN}} & \mathbf{T}^{\eta*}_{\mathrm{NC}} & -\mathbf{S}^{\eta}_{\mathrm{NC}} \\ -\mathbf{U}^{\eta}_{\mathrm{CD}} & \mathbf{T}^{\eta}_{\mathrm{CN}} & -\mathbf{U}^{\eta}_{\mathrm{CC}} & \omega^{\eta}\frac{1}{2}\mathbf{I}^{\eta}_{\mathrm{CC}} + \mathbf{T}^{\eta}_{\mathrm{CC}} \\ \mathbf{T}^{\eta*}_{\mathrm{CD}} & -\mathbf{S}^{\eta}_{\mathrm{CN}} & \omega^{\eta}\frac{1}{2}\mathbf{I}^{\eta}_{\mathrm{CC}} + \mathbf{T}^{\eta*}_{\mathrm{CC}} & -\mathbf{S}^{\eta}_{\mathrm{CC}} \end{pmatrix}, \quad \omega^{\eta} = \begin{cases} -1 & \text{if } \eta = A, \\ 1 & \text{if } \eta = B, \end{cases}$$

$$\mathbf{H}^{\eta} = \begin{pmatrix} -\frac{1}{2}\mathbf{I}^{\eta}_{\mathrm{DD}} - \mathbf{T}^{\eta}_{\mathrm{DD}} & \mathbf{U}^{\eta}_{\mathrm{DN}} \\ \mathbf{S}^{\eta}_{\mathrm{DD}} & \frac{1}{2}\mathbf{I}^{\eta}_{\mathrm{NN}} - \mathbf{T}^{\eta*}_{\mathrm{NN}} \\ -\mathbf{T}^{\eta}_{\mathrm{CD}} & \mathbf{U}^{\eta}_{\mathrm{CN}} \\ \mathbf{S}^{\eta}_{\mathrm{CD}} & -\mathbf{T}^{\eta*}_{\mathrm{CN}} \end{pmatrix}, \qquad \mathbf{M}^{\mathrm{AB}} = \begin{pmatrix} 0 & 0 & 0 & 0 \\ 0 & 0 & 0 & 0 \\ 0 & 0 & 0 & \mathbf{I}^{\mathrm{AB}}_{\mathrm{CC}} \\ 0 & 0 & 0 & 0 \end{pmatrix}. \tag{21}$$

The above equations use the (weakly singular) Kelvin fundamental solution $U^{\eta}_{ij}(x,y)$, and the associated derivatives obtained by the differential traction operator – the strongly singular function $T^{\eta}_{ij}(x,y)$and the hypersingular function $S^{\eta}_{ij}(x,y)$. The compact form of BIEs in eqn (20) uses the notation

$$\omega^{\eta\mathrm{T}}_{q}\mathbf{Z}^{\eta}_{qr}w^{\eta}_{r} = \int_{\Gamma^{\eta}_{q}} \omega^{w}_{j}(y)\left(\int_{\Gamma^{\eta}_{r}} Z^{\eta}_{ji}(y,x)w^{\eta}_{i}(x)\mathrm{d}S(x)\right)\mathrm{d}S(y), \tag{22}$$

where ω stands for ϕ or ψ, while w stands for v or p, and further q and r stand for D, N, and C, and eventually Z^{η} stands for U^{η}, T^{η}, $T^{\eta}*$ or S^{η}, and where the inner integral can be regular, weakly singular, Cauchy principal value or Hadamard finite part integral. In the previous relations, $\mathbf{I}^{\eta}$ denotes the identity operator with the subscripts and superscripts specifying the part of the boundary where it is restricted to.

4 NUMERICAL RESULTS

The described model was used to analyse the compound structures made of steel and concrete. Two examples will be presented. In the first one, a standard four-point bending test is under consideration. Some numerical properties of the described model are presented, including convergence, influence of friction or of the bulk viscosity. The second example includes a T-shape connecting member to improve shearing connection between the two materials. In both examples, the elastic parameters are: E_s=210 GPa, v_s=0.3 for steel and E_c=38GPa, v_c=0.2 for concrete.

The geometrical model for the first example is shown in Fig. 2. The bottom layer is made from steel. The interface characteristics for the physical model based on eqns (3), (4), and (7) are $\kappa_{n0} = k_{s0} = 10\ \mathrm{TPam}^{-1}$, $\beta = 0.2195$, $\kappa_{c} = 100\ \mathrm{TPam}^{-1}$, $G^{\mathrm{I}}_{c} = 0.26\ \mathrm{kJm}^{-2}$, $G^{\mathrm{II}}_{c} = 1.0\ \mathrm{kJm}^{-2}$, if friction is considered then μ_0=0.8 and p=4, and if viscosity is considered then τ_r=0.01s or 0.1s. Pleasetic slip is not taken into account, hence σ_y is large.

As we intend to demonstrate numerically convergence properties, various discretisations have been used. The basic time step is τ=0:1 s which is refined by the factor 2 in subsequent discretisations. Accordingly, the space meshes are refined, where the smallest element is 10mm long and the longest one has 50 mm in the coarsest boundary element mesh.

The loading is prescribed in displacements as shown in Fig. 2 by the simple linear function $g(t)$=1t mms^{-1} up to maximal vertical shift 6 mm.

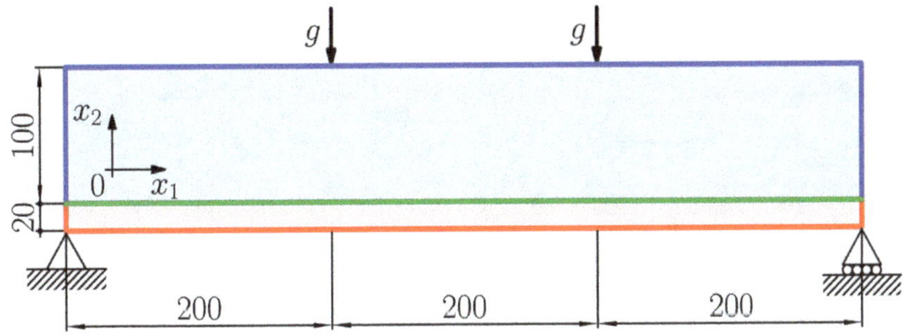

Figure 2: Geometry for the four-point bending example.

The results show how some global characteristics of the system are satisfied. First, satisfaction of the energy imbalance from eqn (18) is shown in Fig. 3. Naturally, refined discretisations, i.e. smaller τ, make the difference to be smaller independently of friction or used viscosity time-relaxation parameter τ_r. Second, a working diagram for the beam showing the relation of the applied force on the prescribed displacement at the point of the right applied load is shown in Fig. 4. Generally, a tendency of the force distribution for more refined discretisations can be guessed. Also the influence of friction in the part (a) is evident, while the difference between viscid and inviscid calculation is not significant, cf. [23].

Finally, the results of the interface variables for the frictional case and given τ_r are shown in Fig. 5 at the selected time instant. According to the graphs in Figs. 3(c) and 4(c), the extend of the crack represented by the domains where ζ =0 is appropriate. The displacements are almost the same which means that the boundary elements converge satisfactorily.

The second example includes a T-shape connecting steel element shown in Fig. 6. Here, we consider interface characteristics different from those above. Friction is considered with

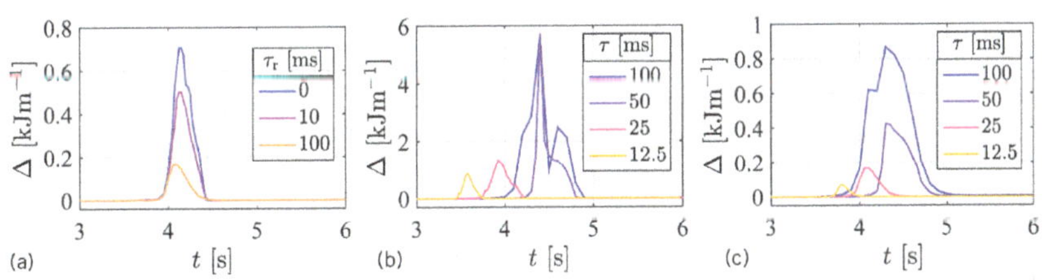

Figure 3: The energy imbalance of numerical solutions due to eqn (18): (a) various τ_r,τ=25 ms, μ_0=0.8; (b) τ_r=0 ms, various τ, no friction; (c) τ_r=100 ms, various τ,μ_0=0.8.

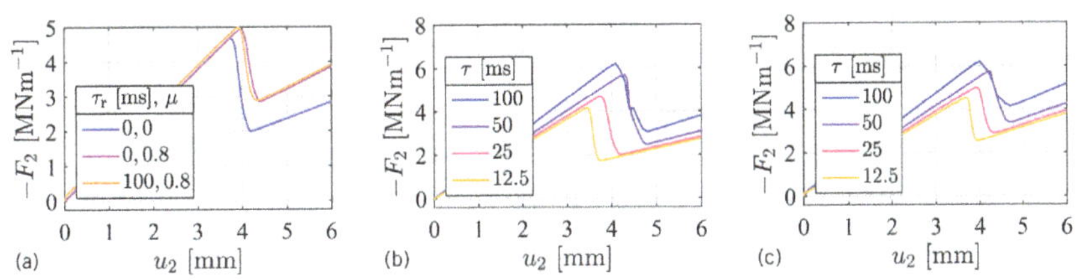

Figure 4: Working diagrams – the total vertical force F_2 applied at the point where acts one of the prescribed displacements u_2=g: (a) various τ_r; (b) τ_r=0 ms, various τ, no friction; (c) τ_r=100 ms, various τ, μ_0=0.8.

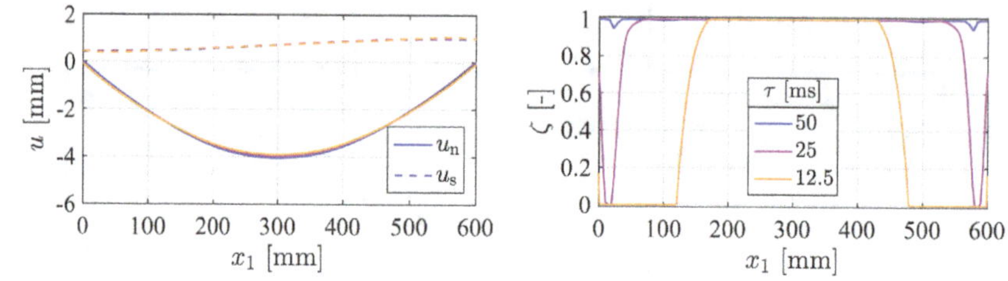

Figure 5: Distributions of interface variables for various discretisations (referred by τ) at the instant t=3.75s, τ_r=100 ms, μ_0=0.8.

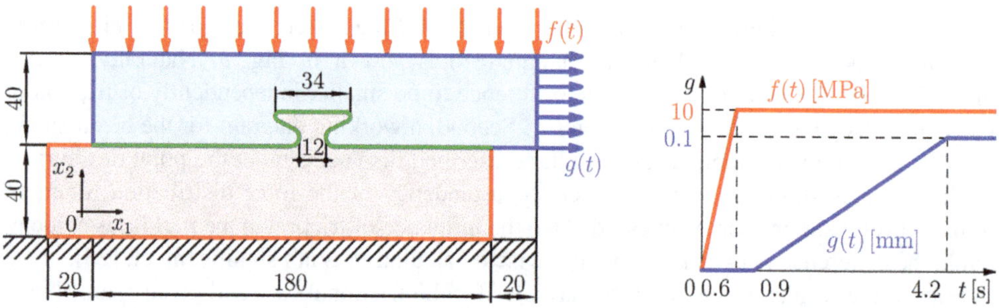

Figure 6: Geometry for the T-reinforcement example and the applied non-proportional loading.

μ_0=0.4 and the same p=4 and viscosity is neglected, τ_r=0 s. The interface damage characteristics include constant fracture energy G_c=5 Jm^{-2} and the interface plasticity is used with σ_y=1.25MPa and k_H=100GPam^{-1} to similate crack mode sensitivity.

The damage state of interface is described by the function, see [11]

$$\Phi(\zeta)=\frac{\lambda_1\exp(-\gamma^{-1}(\alpha\zeta+\varepsilon))-\lambda_2}{1-\lambda_2},\quad \alpha=0.999,\ \varepsilon=0.0005, \tag{23}$$

with $\lambda_1=\exp(\gamma^{-1}(\alpha+\varepsilon))$, $\lambda_2=\lambda_1\exp(-\gamma^{-1}(\varepsilon))$, and $\gamma(x)=\mathrm{e}^{-x}\left(1+x+\frac{1}{2}x^2\right)$ which, in terms of CZM, provokes stress-strain relation equivalent to the Ortiz-Pandolfi model [1]. The initial stiffness are k_{n0}=k_{s0}=2 TPam^{-1}, k_c=200TPam^{-1}.

The time step is set to τ=0·02 s. The boundary element mesh is non-uniform with refinements at the curved interface and at vicinity of subdomains' corner points, the smallest element is about 0.5 mm long.

The loading is considered non-proportional as it is shown in Fig. 6. First, the normal pressure is applied up to its maximal value of 10 MPa and subsequently the horizontal displacements are applied at the top layer up to the maximal value of 0.1 mm.

Some results are shown in Fig. 7. The displacements show the initial deformation after applying the normal pressure (a) and the final state of the structure (b). The evolution of the interface variables can be seen in the graphs below. The plastic slip starts to evolve first where the shearing stresses are important (around points G and D). At the vicinity of the points B, C, E and F the tangential stresses are small, so no plastic slip appears.

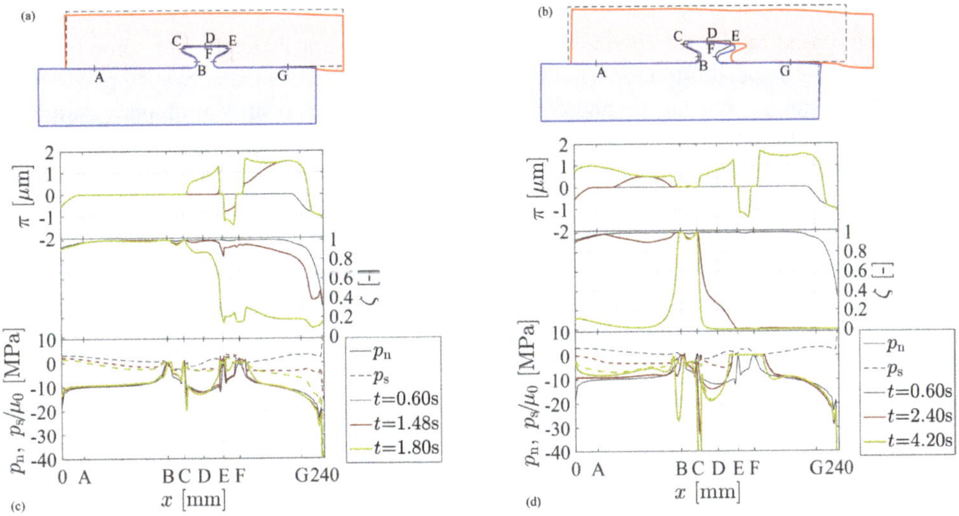

Figure 7: Results for the T-reinforcement example: (a) deformed shape at the instant t=0.6 s, (b) deformed shape at the instant t=4.2 s, (c) plastic slip, damage and interface stresses in the initial part of loading (ζ positive), (d) plastic slip, damage and interface stresses in the final part of loading (interface crack evolving). The tangential stresses are scaled in order to see the zone where friction takes place. The displacements are magnified by a factor 200.

The damage and subsequently the interface crack develops either due to shear around the point G or in the opening mode at the segment EF.

As the model incorporates also friction, the proposed friction function eqn (4) makes it significant only for substantially damaged interface (ζ close to 0). When a shearing crack appears at the zone close to point G, the influence of friction can be observed in the relation of the normal traction p_n and the tangential traction p_s where $|p_n| = |p_s|/\mu_0$.

Finally, the total force applied at the face where the displacement load is invoked is shown in Fig. 8. It documents the role of the T-shape reinforcement element as there is no weakening part in the force-displacement graph as we could see in Fig. 4.

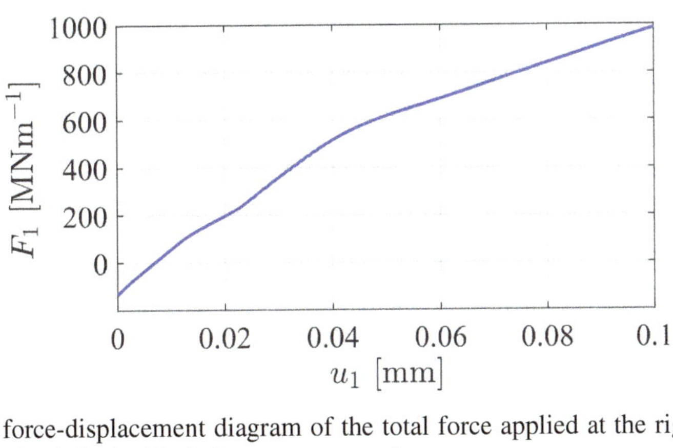

Figure 8: The force-displacement diagram of the total force applied at the right face of the upper body in the T-reinforcement example.

5 CONCLUSIONS

An energy-based model for solving problems of interface cracks applied to some types of steel-concrete composite structures has been presented. The model includes such phenomena as interface damage, plastic slip and contact with friction. The results of numerical simulations show capabilities of the model to provide satisfactory results either from the mathematical point of view or from the point of view of application in engineering. The former was documented by some convergence properties observed in the first solved example, the latter is advocated by the analysis of a steel T-connector subject to pressure and shear in the other example. Obtained data confirm the influence of all observed phenomena and their importance in practical calculations. The presented numerical implementation confirmed the expected behaviour of the model in accordance with the applied theory. Its applicability for steel-concrete composite materials was satisfactorily demonstrated, too.

ACKNOWLEDGEMENTS

Authors acknowledge support from Slovak Research and Development Agency by the grant APVV-15-0486 and from the Slovak Ministry of Education by the grant VEGA 1/0078/16.

REFERENCES

[1] Ortiz, M. & Pandolfi, A., Finite-deformation irreversible cohesive elements for three dimensional crack propagation analysis. *International Journal for Numerical Methods in Engineering*, **44**, pp. 1267–1283, 1999. https://doi.org/10.1002/(sici)1097-0207(19990330)44:9<1267::aid-nme486>3.3.co;2-z

[2] Park, K. & Paulino, G., Cohesive zone models: a critical review of traction-separation relationships across fracture surfaces. *Applied Mechanics Reviews*, **64(6)**, 2011. https://doi.org/10.1115/1.4023110

[3] Távara, L., Mantič, V., Salvadori, A., Gray, L. & París, F., Cohesive-zone-model formulation and implementation using the symmetric Galerkin boundary element method for homogeneous solids. *Computational Mechanics*, **51**, pp. 535–551, 2013. https://doi.org/10.1007/s00466-012-0808-5

[4] Leguillon, D., Strength or toughness? a criterion for crack onset at a notch. *European Journal of Mechanics and Solids*, **21**(1), pp. 61–72, 2002. https://doi.org/10.1016/s0997-7538(01)01184-6

[5] Cornetti, P., Mantič, V. & Carpinteri, A., Finite fracture mechanics at elastic interfaces. *International Journal of Solids and Structures*, **49**, pp. 1022–1032, 2012. https://doi.org/10.1016/j.ijsolstr.2012.01.002

[6] García, I., Mantič, V., Blázquez, A. & París, F., Transverse crack onset and growth in cross-ply $[0=90]_s$ laminates under tension: application of a coupled stress and energy criterion. *International Journal of Solids and Structures*, **51**, pp. 3844–3856, 2014.

[7] Kočvara, M., Mielke, A. & Roubíček, T., A rate-independent approach to the delamination problem. *Mathematics and Mechanics of Solids*, **11**, pp. 423–447, 2006. https://doi.org/10.1177/1081286505046482

[8] Roubíček, T., Kružík, M. & Zeman, J., Delamination and adhesive contact models and their mathematical analysis and numerical treatment. *Mathematical Methods and Models in Composites*, ed. V. Mantič, Imperial College Press, pp. 349–400, 2013. https://doi.org/10.1142/9781848167858_0009

[9] Raous, M., Cangemi, L. & Cocu, M., A consistent model coupling adhesion, friction and unilateral contact. *Computer Methods in Applied Mechanics and Engineering*, **177**(6), pp. 383–399, 1999.
https://doi.org/10.1016/s0045-7825(98)00389-2

[10] Frémond, M., Dissipation dans l'adhérence des solides. *CR Acad Sci, Paris, SérII*, **300**, pp. 709–714, 1985.

[11] Vodička, R., A quasi-static interface damage model with cohesive cracks: SQP–SGBEM implementation. *Engineering Analysis with Boundary Elements*, **62**, pp. 123–140, 2016.
https://doi.org/10.1016/j.enganabound.2015.09.010

[12] Vodička, R. & Mantič, V., A multilinear cohesive law within a quasi-static interface damage model. *Discrete and Continuous Dynamical Systems*, 2017. (accepted).

[13] Banks-Sills, L. & Askenazi, D., A note on fracture criteria for interface fracture. *International Journal of Fracture*, **103**, pp. 177–188, 2000.
https://doi.org/10.1023/a:1007612613338

[14] Mantič, V., Discussion on the reference length and mode mixity for a bimaterial interface. *Journal of Engineering Materials and Technology*, **130**, pp. 045501–1–2, 2008.
https://doi.org/10.1115/1.2975232

[15] Panagiotopoulos, C., Mantič, V. & Roubíček, T., BEM implementation of energetic solutions for quasistatic delamination problems. *Computational Mechanics*, **51**, pp. 505–521, 2013.
https://doi.org/10.1007/s00466-012-0826-3

[16] Vodička, R., Mantič, V. & Roubíček, T., Energetic versus maximally-dissipative local solutions of a quasi-static rate-independent mixed-mode delamination model. *Meccanica*, **49(12)**, pp. 2933–296, 2014.
https://doi.org/10.1007/s11012 014-0045-4

[17] Mantič, V. & París, F., Relation between SIF and ERR based measures of fracture mode mixity in interface cracks. *International Journal of Fracture*, **130**, pp. 557–569, 2004.
https://doi.org/10.1023/b:frac.0000049496.77533.d5

[18] Roubíček, T., Panagiotopoulos, C. & Mantič, V., Local-solution approach to quasistatic rate-independent mixed-mode delamination. *Mathematical Models and Methods in Applied Sciences*, **25(7)**, pp. 1337–1364, 2015.
https://doi.org/10.1142/s0218202515500347

[19] Wriggers, P., *Computational Contact Mechanics*. Springer: Berlin, 2006.

[20] Vodička, R., Mantič, V. & Roubíček, T., Quasistatic normal-compliance contact problem of visco-elastic bodies with Coulomb friction implemented by QP and SGBEM. *Journal of Computational and Applied Mathematics*, **315**, pp. 249–272, 2017.
https://doi.org/10.1016/j.cam.2016.10.010

[21] Roubíček, T., Panagiotopoulos, C. & Mantič, V., Quasistatic adhesive contact of visco-elastic bodies and its numerical treatment for very small viscosity. *Zeitschrift für Angewandte Mathematik und Mechanik*, **93**, pp. 823–840, 2013.
https://doi.org/10.1002/zamm.201200239

[22] Panagiotopoulos, C., Mantič, V. & Roubíček, T., A simple and efficient BEM implementation of quasistatic linear visco-elasticity. *International Journal of Solids and Structures*, **51**, pp. 2261–2271, 2014.
https://doi.org/10.1016/j.ijsolstr.2014.02.028

[23] Roubíček, T., Adhesive contact of visco-elastic bodies and defect measures arising by vanishing viscosity. *SIAM Journal on Mathematical Analysis*, **45**(1), pp. 101–126, 2013. https://doi.org/10.1137/12088286x
[24] Dostál, Z., *Optimal Quadratic Programming Algorithms*, volume 23 of *Springer Optimization and Its Applications*. Springer: Berlin, 2009.
[25] Vodička, R., Roubíček, T. & Mantič, V., General-purpose model for various adhesive frictional contacts at small strains. *Interf Free Bound*, 2017. (submitted).
[26] Benzeggagh, M. & Kenane, M., Measurement of mixed-mode delamination fracture toughness of unidirectional glass/epoxy composites with mixed-mode bending apparatus. *Composites Science and Technology*, **56**, pp. 439–449, 1996. https://doi.org/10.1016/0266-3538(96)00005-x
[27] Hutchinson, J. & Wu, T., Mixed mode cracking in layered materials. *Advances in Applied Mechanics*, **29**, pp. 63–191, 1992. https://doi.org/10.1016/s0065-2156(08)70164-9
[28] Vodička, R., Mantič, V. & París, F., Symmetric variational formulation of BIE for domain decomposition problems in elasticity – an SGBEM approach for nonconforming discretizations of curved interfaces. *CMES – Computer Modeling in Engineering & Sciences*, **17**(3), pp. 173–203, 2007.
[29] Vodička, R., Mantič, V. & París, F., Two variational formulations for elastic domain decomposition problems solved by SGBEM enforcing coupling conditions in a weak form. *Engineering Analysis with Boundary Elements*, **35**, pp. 148–155, 2011. https://doi.org/10.1016/j.enganabound.2010.05.002

ANALYSIS OF TWO COHESIVE ZONE MODELS FOR CRACK PROPAGATION IN NOTCHED BEAMS USING THE BEM

P.C. GONÇALVES[1], L. PALERMO JR.[2] & S.P.B. PROENÇA[3]
[1]Natural Resources Institute, Federal University of Itajuba, Brazil.
[2]School Civil Engineering, Architecture and Urban Design, University of Campinas, Brazil.
[3]São Carlos School of Engineering, University of São Paulo, Brazil.

ABSTRACT
Crack propagation in a single-edge notched beam is analyzed with the three-point bending test. Two constitutive laws that describe the material softening in the cohesive zone were tested, and their results were compared. The dual boundary element method (DBEM) is employed with the traction boundary integral equation using the tangential differential operator. A constitutive law was introduced in the system of equations, and the cohesive forces were directly computed at each loading step. The results are compared with the experimental and numerical results available in the literature.
Keywords: cohesive model, crack analysis, dual boundary element model, Plane problems, tangential differential operator

1 INTRODUCTION

A notched specimen of linear elastic material will concentrate stress in the region in front of the notch tip. The development of the damage zone in front of the notch tip is a consequence of these high stresses. The stress distributions in front of a notch, according to Ref. [1], are shown in Fig. 1a for a linear elastic material (curve a) and for a non-yielding material with a micro-cracked zone in front of the notch tip (curve b). Micro-cracks appear in the damage zone of concrete and other non-yielding materials. The material in the fracture or the micro-cracked zone is partly destroyed but still able to transfer stress. The crack surfaces just behind the fictitious crack tip are not completely separated, and tractions can be transferred across the relatively long extended crack trace, which is called the cohesive zone and is shown in Fig. 1b. The main assumption is that material softening occurs beyond the peak load in a narrow layer with a negligible volume behind the fictitious crack tip in which cohesive forces can stop crack opening. The crack in the cohesive zone can be represented with a two-parameter model named the fictitious crack model. The constitutive laws for the material in the cohesive zone can be defined in terms of stresses and strains accompanied by a layer thickening law or employing a traction-displacement relationship, as adopted in this study. Barenblatt [2] presented a cohesive model using the fictitious crack theory. Hillerborg *et al.* [3] proposed a function for the softening model related to mode I crack opening (Fig. 1c), which allowed for a finite element model (FEM) analysis of the problem, such as those presented by Petersson [1], Carpinteri [4] and Rots [5].

Several numerical methods besides the FEM have been used to perform crack analyses. The dual boundary element method (DBEM) is one of the most widely used due to its accuracy in computation of the stress intensity factors and the simplicity of adding more elements for crack propagation [6]. The crack analyses for cohesive materials were studied with the DBEM by Saleh and Aliabadi [7, 8], with the Galerkin multizone BEM by Chen *et al.* [9] and with the displacement discontinuity BEM by Gospodinov [10]. Karlis *et al.* employed the two dimensional gradient elasticity for crack analyses with the BEM in Ref. [11] and Leonel

DOI: 10.2495/CMEM-V6-N6-1057-1066

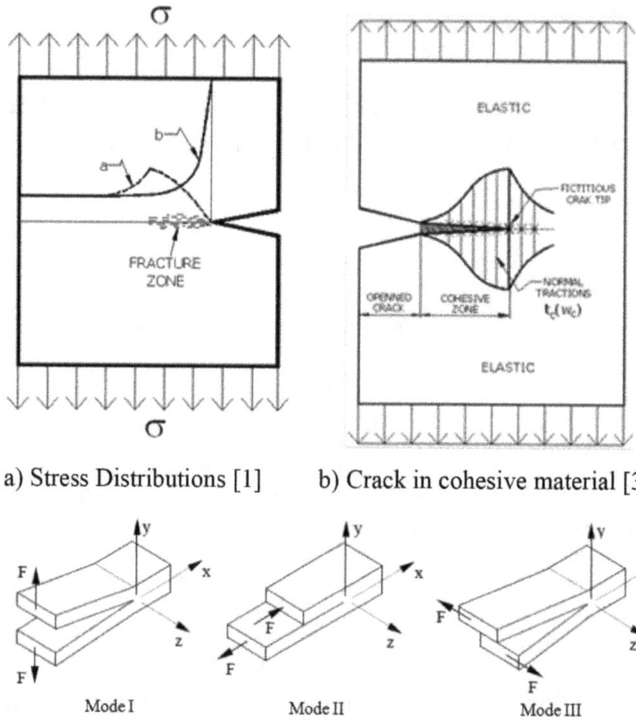

a) Stress Distributions [1] b) Crack in cohesive material [3]

c) Fracture modes

Figure 1: Crack in the cohesive zone.

and Venturini used the tangent operator in the non-linear analyses with a BEM for crack propagation in quasi-brittle materials in Ref. [12]. Tavara *et al.* performed a BEM analysis of crack propagation along the fiber–matrix interface of laminates using a linear elastic–brittle interface model in Ref. [13]. Trinh *et al.* employed the fast multipole symmetric Galerkin BEM to solve multizone and multicrack problems in Ref. [14], and Benedetti and Aliabadi employed the BEM in multiscale modeling for polycrystalline materials to analyze material degradation and fracture in Ref. [15].

The DBEM in this study used the tangential differential operator (TDO) in the traction boundary integral equation (BIE). The TDO, in conjunction with integration by parts, is one way to reduce the order of the singularity in kernels of Kelvin-type fundamental solutions and is applied to the hypersingular kernel of the traction BIE here. Kupradze [16] first presented an application with the TDO, and Sladek [17] employed the TDO in a curved crack solution. Bonnet [18] presented regularized formulations of gradients in potential problems by employing the TDO in the BIE, and in stress BIEs for elasticity problems including fracture mechanics formulations. The use of non-conformal interpolations with the TDO requires additional care as explained to the traction BIEs of elasticity for two-dimensional problems [19] and three-dimensional problems [20] and to the stress equations for plate bending including the effect of shear deformation, as in Ref. [21].

2 DUAL BOUNDARY INTEGRAL EQUATIONS FOR COHESIVE MATERIALS

The coincidence of two crack surfaces requires two different boundary integral equations (BIEs) for the solution and the dual boundary element method (DBEM) is the technique employed. The dual equations of the method are the displacement and the traction boundary integral equations:

$$\frac{1}{2}u_i(x')+\int T_{ij}(x',x)u_j(x)d(x)=\int U_{ij}(x',x)t_j(x)d(x) \tag{1}$$

$$\frac{1}{2}t_k(x')=n_a(x')C_{akim}\int\sigma_{ibj}(x',x)D_{mb}\left[u_j(x)\right]d(x)-n_a(x')\int\sigma_{jak}(x',x)t_j(x)d(x) \tag{2}$$

with

$$C_{akim}=\mu\left(\frac{2\nu}{1-2\nu}\delta_{ak}\delta_{im}+\delta_{ai}\delta_{km}+\delta_{am}\delta_{ki}\right)$$

$$\mathrm{D}_{bm}\left[f(x)\right]=n_b(x)f_{,m}(x)-n_m(x)f_{,b}(x)$$

where $U_{ij}(x`, x)$ and $T_{ij}(x`, x)$ are the displacement and traction, respectively, in the direction j at boundary point x due to a singular load in the direction i at the collocation point $x`$ based on the Kelvin solution for two-dimensional problems $u_i(x)$ and $t_i(x)$ are the displacement and traction at boundary point $x$, respectively, $n_a(x')$ is the direction cosine of the outward normal at collocation point $x`$, C_{akim} is the Hooke tensor for an isotropic material, σ_{ibj} is the stress σ_{bj} at boundary point x due to a singular load in the direction i at collocation point $x`$, $D_{bm}(\,)$ is the tangential operator, μ is the shear modulus, ν is the Poisson ratio, and δ_{ij} is the Kronecker delta.

The displacement BIE (eqn (1)) is applied to one of the crack surfaces, and the traction BIE (eqn (2)) is applied to the other surface to solve general mixed-mode crack problems with a single domain formulation. Although the integration path is still the same for coincident points on crack surfaces, the respective boundary integral equations are now distinct. The collocation point needed to perform the traction BIE and the strategy used to treat improper integrals are the key features of this formulation. The positions of the collocation points adopted in this study were inside the boundary elements to satisfy the continuity requirements for both BIEs. The left member in eqn (1) has a $1/r$ singularity, whereas the right member has a logarithmic singularity when the field point approaches the collocation point. Both integrals in the right member of eqn (2) have $1/r$ singularities because the TDO was used in the first integral.

The cohesive zone is the extension of the crack in which material softening beyond the peak load occurs in the narrow layer behind the fictitious crack tip (Fig. 1b). The cohesive zone is modeled as a crack region connecting coincident surfaces. The points that were originally coincidental on opposite sides of the crack line can separate into distinct points but remain connected by the cohesive zone material. Continued strain increases the separation between these two points and eventually leads to cracking. The simplest approximation for the σ-w curve in the cohesive zone is a single descending straight line, as shown in Fig ure 2a, where f_t is the tensile strength, and w_c is the maximum widening of the fracture zone when it is still able to transfer stress. A generalization for the σ–w curve can be made with straight-line approximations using an arbitrary number of break points with relations similar to those

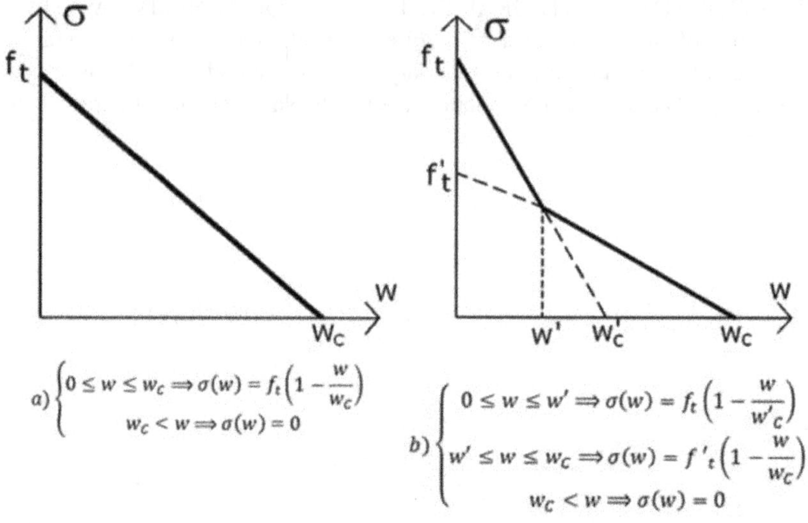

Figure 2: Approximation for a σ–w curve: a) Straight line, b) Two straight lines.

presented in Fig. 2b. The maximum principal stress criterion is used to determine the crack extension increment. When the maximum principal stress at the fictitious crack tip reaches the critical value of f_t, the tip will extend under further loading using one of the constitutive laws shown in Fig. 2. The tip advances in the direction perpendicular to the maximum principal stress at that point, and the extension is such that the maximum principal stress at the new tip position is equal to the critical f_t value during continued loading. The generalized plane stress problem was used in this study to model the beam in a three-point bending test. The maximum principal stress was obtained according to the following expression:

$$\sigma_{max} = \frac{1}{2}\left(\sigma_n + \sigma_t + \sqrt{(\sigma_n - \sigma_t)^2 + 4\tau_{nt}^{\ 2}}\right) \tag{3}$$

Figure 3 shows a crack containing a cohesive zone. The external boundary line of the problem is Γ_e, and the tangent direction used to compute the boundary integrals is shown. Γ_1 and Γ_4 are portions of the crack surface without cohesive forces (open crack), while Γ_2 and Γ_3 are

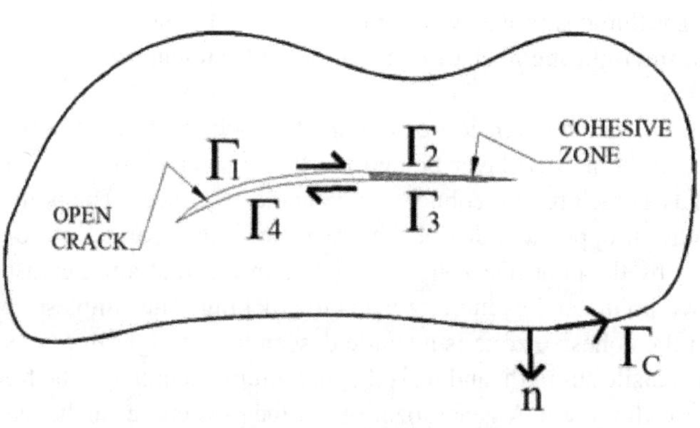

Figure 3: Crack inserted in a finite domain with boundary Γ_e.

portions containing cohesive forces. The cohesive zone has tractions on the crack surfaces of opposite sign and equal magnitude. The dual equations for the problem shown in Fig. 3 can be simplified when the directions and magnitudes of the cohesive tractions on crack surfaces Γ_2 and Γ_3 and the directions of integration on these surfaces are taken into account, yielding:

$$\frac{1}{2}u_i(x')+\int T_{ij}(x',x)u_j(x)d(x)=\int_e U_{ij}(x',x)t_j(x)d(x)+2\int_c U_{ij}(x',x)t_j^c(x)d(x) \quad (4)$$

$$\frac{1}{2}t_k(x')=n_a(x')C_{akim}\int\sigma_{ibj}(x',x)D_{mb}\left[u_j(x)\right]d(x)-n_a(x')\int_e\sigma_{jak}(x',x)t_j(x)d(x)$$

$$-2n_a(x')\int_c\sigma_{jak}(x',x)t_j^c(x)d(x) \quad (5)$$

In eqns (4) and (5), t^c_j was introduced as the traction in the direction j on one of the crack surfaces within the cohesive zone. The index c was introduced because surface Γ_2 or Γ_3 with tractions t^2 or t^3, respectively, can be used. $\Gamma=\Gamma_1+\Gamma_2+\Gamma_3+\Gamma_4+\Gamma_e$, and $\Gamma_2=\Gamma_3=\Gamma_c$.

3 NUMERICAL IMPLEMENTATION

The tractions on one of the crack surfaces in the cohesive zone (t_c) are the unknowns in the system of equations obtained from eqns (4) and (5), and the cohesive law is the additional equation used to find the solution. This strategy allows the direct computation of tractions in the cohesive zone at each incremental loading step.

$$\begin{bmatrix} H_{ee} & H_{eo} & H_{ez} & -G_{ez} \\ H_{oe} & H_{oo} & H_{oz} & -G_{oz} \\ H_{ze} & H_{zo} & H_{zz} & -G_{zz} \\ 0 & 0 & A & B \end{bmatrix}\begin{Bmatrix} u^e \\ u^o \\ u^z \\ t^c \end{Bmatrix}=\begin{bmatrix} G_{ee} & 0 \\ G_{oe} & 0 \\ G_{ze} & 0 \\ 0 & C \end{bmatrix}\begin{Bmatrix} t^e \\ w_f \end{Bmatrix} \quad (6)$$

Equation (6) summarizes the system of equations where the dual eqns (4) and (5) are converted to submatrices H_{ij} and G_{ij}, and the cohesive law is converted to submatrices A, B and C. The indices e, o and z are related to the boundary portions and correspond to the external boundary, the open crack boundary and the cohesive zone, respectively. Submatrices A and B relate the displacements and tractions in the normal and tangent directions, respectively, on the crack surfaces in the cohesive zone.

The softening criterion in the cohesive zone is for tractions and opening in the normal direction, and the tangential tractions are zero. The numerical algorithm is summarized next, considering the boundary portions shown in Fig. 3 and the straight line constitutive law shown in Fig. 2a:

a) The normal tractions (t_c) are less than or equal to f_t, i.e. no opening occurs in the cohesive zone, and opposite points along the crack surfaces have equal displacement. Submatrices B and C are zero, whereas submatrix A contains direction cosines that relate the displacements in the directions x_i, i.e. the openings in the normal and tangent directions are zero:

$$u_i^2 n_i^2 + u_i^3 n_i^3 = 0 \tag{7}$$

$$u_i^2 s_i^2 + u_i^3 s_i^3 = 0 \tag{8}$$

b) With subsequent loads after the normal traction has reached f_t, opening occurs in the cohesive zone according to the cohesive law as shown in Fig. 2a. The submatrices *B* and *C* are then modified to introduce the cohesive law in the line containing the opening in the normal direction. The tractions in the tangent direction become zero, and this condition is introduced in the line describing opening in the tangent direction. Equations (7) and (8) in corresponding lines of matrices *A* and *B* are replaced by:

$$\left(u_i^2 n_i^2 + u_i^3 n_i^3\right) - \frac{w_c}{f_t} n_i^c t_i^c = -w_c \tag{9}$$

$$s_i^c t_i^c = 0 \tag{10}$$

c) If the opening reaches w_c in subsequent loads, the tractions (t_c) must be zero at that point in the cohesive zone. The absence of tractions in the normal and tangent directions causes the tractions in the direction x_i to be zero. The tractions (t^c_i) at this point on the crack surfaces are eliminated from the system of equations. Equations (9) and (10) in the corresponding lines of matrices *A* and *B* are replaced by:

$$t_1^c = 0 \tag{11}$$

$$t_2^c = 0 \tag{12}$$

The break point value (*w'*) is introduced in the numerical algorithm when the constitutive law employs two straight lines, as shown in Fig. 2b. The constitutive law in eqn (9) uses the relation for either the first or second line in Fig. 2b, according to values of the normal opening (*w*) and the break point (*w'*). The first line is used in the subsequent loading after the normal traction has reached f_t and opening occurs in the cohesive zone. The second line for the constitutive law replaces the first line when the normal opening value (*w*) is greater than the break point value (*w'*). The tractions (t_c) must be zero when the normal opening (*w*) reaches w_c in the loading process and eqns (11) and (12) must be used instead of eqns (9) and (10).

The boundary element code was derived from that used in Refs. [22, 23]. Linear mapping functions were used to represent the displacements and tractions in the boundary elements. The same mapping function was used for conformal and non-conformal interpolations, with nodal parameters positioned at the ends of the elements. The collocation points were always positioned on the boundary line at the position (ξ') in the range (-1, 1): i) ξ'= -0.67 for continuous elements, and ii) ξ'= -0.67 and ξ'= +0.67 for discontinuous elements. Analytical expressions were used to evaluate the singular integrals in the Cauchy principal value, and the Gauss-Legendre scheme was used for regular integrals.

4 NUMERICAL EXAMPLE

Experimental and numerical analyses were performed in Ref. [1] for the notched beam under the three-point bending test shown in Fig. 4. The depth of the beam (d) is 0.2 m, the width is 0.05 m and the length (l) is 2.0 m. The adopted ratio between notched depth and beam depth (a/d) was 0.5. The Young's modulus (E) was 30 GPa, the Poisson's ratio (v) was 0.2 and the

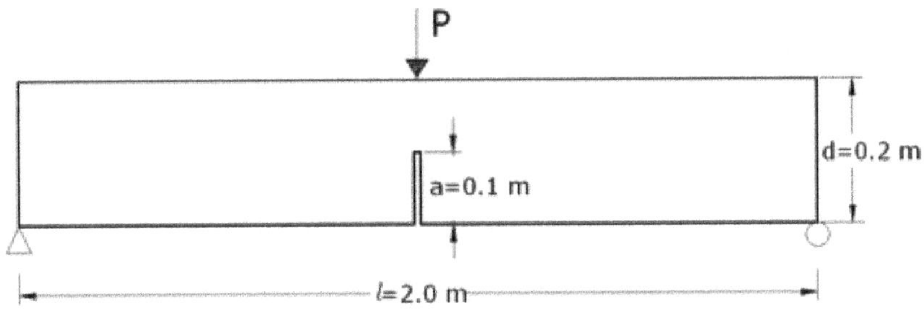

Figure 4: The notched beam subjected to three-point bending.

Table 1: Mechanical properties according to the constitutive laws in Ref. [1].

σ–w curve	G_f(Nm^1)	f_t'(MPa)	w_c (10^{-5} m)	w_c' (10^{-5} m)	w' (10^{-5} m)
Straight line (SL)	124	-	7.45	-	-
Two straight lines (C)	124	1.427	13.40	4.468	2.979

tensile strength f_t was 3.33 MPa. Table 1 presents the properties used in the analysis of the crack propagation in Ref. [1] for each of the constitutive laws shown in Fig ure 2.

Petersson presented results for crack propagation [1], which were obtained experimentally for two values of fracture energy (G_f): 115 Nm^1 and 137 Nm^1. Experimental results were compared with those obtained from a numerical analysis [1] with the FEM using 124 Nm^1 for G_f, which was the median value of the fracture energy values of the experiments. Two constitutive laws were considered in the numerical analyses of Ref. [1]: a) Straight line and b) Two straight lines. Petersson used the letter C to represent results from the constitutive law using two straight lines because he considered this law better for representing the concrete behavior. The breaking point value (w') used in Ref. [1] for the constitutive law with two straight lines was related to the following value of the normal traction:

$$w' = \frac{2}{9} w_C \xrightarrow{yields} \sigma = \frac{1}{3} f_t$$

Saleh and Aliabadi presented results in Ref. [7] for this problem using a DBEM with quadratic boundary elements; the constitutive law in the cohesive zone was the straight line and the mechanical properties were the same employed in Ref. [1]. The results obtained with the DBEM formulation presented here, using TDO and linear boundary elements, are shown in Fig. 5 and compared with those in Ref. [7]. Two types of constitutive laws were used with the present formulation: a single straight line (SL) and two straight lines (C). The same boundary element mesh was used in analyses with both constitutive laws. This analysis employed 196 boundary elements with two boundary nodes placed at each corner and at the fictitious crack tip.

According to Fig. 5, the peak load obtained with a straight line (blue curve) was greater than that obtained with two straight lines (green curve), which were both similar to the result obtained in Ref. [7] with a straight line constitutive law.

A comparison of the experimental and numerical results presented in Ref. [1] is shown in Fig. 6. The results obtained with the present DBEM formulation for both types of constitutive

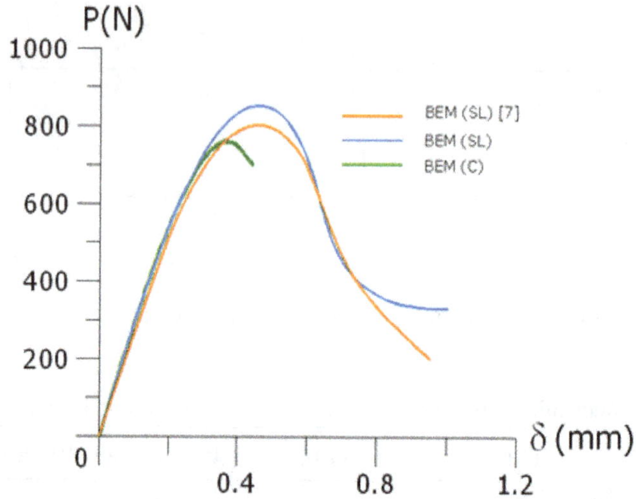

Figure 5: Comparison between the results obtained in this study and those in Ref. [7].

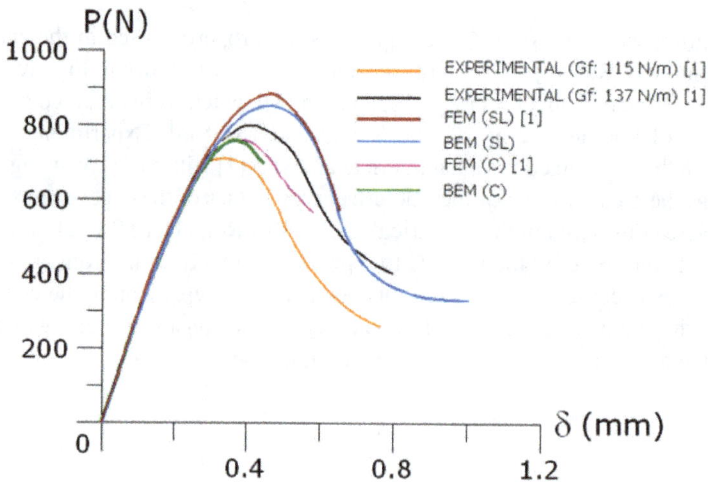

Figure 6: Comparison between the results obtained in this study and in Ref. [1].

laws are very similar to those in Ref. [1], obtained from an FEM analysis. Furthermore, the results obtained with the constitutive law using two straight lines demonstrate that using two straight lines is better to represent concrete behavior, as first noted by Petersson in Ref. [1]. In Fig. 6, the numerical results from the present BEM formulation and FEM [1], using two straight lines for the constitutive law and the median value of fracture energy (pink and green curves), were plotted between the curves obtained experimentally for the upper and the lower value of the fracture energy (G_f), which were 137 Nm^1 (black curve) and 115 Nm^1 (orange curve). On the other hand, numerical analyses using a straight line for the constitutive law with the median value for the fracture energy (124 Nm^1) presented a peak load (brown and blue curves) greater than that obtained with the experimental analysis (black curve) with the upper value for the fracture energy (137 Nm^1).

5 CONCLUSIONS

The results obtained in the crack propagation analyses with the present DBEM formulation match those presented in the literature, when either a straight line or two straight lines were used for the constitutive law of the cohesive zone models. Furthermore, the comparison with earlier results presented by Petersson [1] confirmed his conclusions, which were based on experimental tests and numerical analyses using an FEM. The model formulation used in this study, which was presented in Ref. [22], was more stable and computationally faster than the method presented in Ref. [24], as noted in Ref. [23].

REFERENCES

[1] Peterson, P.E., *Crack growth and development of Fracture Zone in Plain Concrete and similar materials*, Report Nº TVBM-1006, Division of Building Materials, Lund Institute of Technology, Lund Sweden, 1981.

[2] Barenblatt, G.I., The mathematical theory of equilibrium cracks in brittle fracture. *Advances in Applied Mechanics*, **7**, pp. 55–129, 1962. https://doi.org/10.1016/s0065-2156(08)70121-2

[3] Hillerborg, A., Modeer, M. & Petersson, P.E., Analysis of crack formation and crack growth in concrete by means of fracture mechanics and finite elements. *Cements Concrete Research*, **6**, pp. 773–782, 1976. https://doi.org/10.1016/0008-8846(76)90007-7

[4] Carpinteri, A., Post-peak and post-bifurcation analysis of cohesive crack propagation. *Engineering Fracture Mechanics*, **32**, pp. 265–278, 1989. https://doi.org/10.1016/0013-7944(89)90299-3

[5] Rots J.G., *Computational modeling of concrete fracture*. Ph.D. Thesis, Delft University of Technology, 1988.

[6] Brebbia, C.A., The birth of the boundary element method from conception to application. *Engineering Analysis with Boundary Elements*, **77**, pp. iii–x, 2017. https://doi.org/10.1016/j.enganabound.2016.12.001

[7] Saleh, A.L. & Aliabadi, M.H., Crack growth analysis in concrete using boundary element method. *Engineering Fracture Mechanics*, **51**(4), pp. 533–545, 1995. https://doi.org/10.1016/0013-7944(94)00301-w

[8] Saleh, A.L. & Aliabadi, M.H., Fracture Mechanics Analysis of crack on plain concrete using the boundary element methods. *Engineering Fracture Mechanics*, **69**, pp. 267–280, 2002.
https://doi.org/10.1016/s0013-7944(01)00089-3

[9] Chen T, Wang B, Cen Z, Wu Z., A symmetric Galerkin multi-zone boundary elemento method for cohesive crack growth. *Engineering Fracture Mechanics,* **63**, pp. 591–609, 1999.

[10] Gospodinov, G., Boundary element modelling of cohesive cracks using displacement discontinuity method. *Advances in Boundary Element Techniques*, pp. 19–26, 2003.

[11] Karlis, G.F., Tsinopoulos, S.V., Polyzos, D. & Beskos, D.E., Boundary Element analysis of mode I and mixed mode (I and II) crack problems of 2-D gradient elasticity. *Computer Methods in Applied Mechanics and Engineering*, **196**, pp. 5092–5103, 2007. https://doi.org/10.1016/j.cma.2007.07.006

[12] Leonel, E.D. & Venturini, W.S., Non-linear boundary element formulation with tangent operator to analyze crack propagation in quasi-brittle materials. *Engineering Analysis with Boundary Elements*, **34**, pp. 122–129, 2010. https://doi.org/10.1016/j.enganabound.2009.08.005

[13] Távara, L., Mantic, V., Graciani, E. & París, F., BEM analysis of crack onset and propagation along fiber–matrix interface under transverse tension using a linear elastic–brittle interface model. *Engineering Analysis with Boundary Elements*, **35**, pp. 207–222, 2011. https://doi.org/10.1016/j.enganabound.2010.08.006

[14] Trinh, Q.T., Mouhoubi, S., Chazallon, C. & Bonnet, M., Solving multizone and multicrack elastostatic problems: a fast multipole symmetric Galerkin boundary element method approach. *Engineering Analysis with Boundary Elements*, **50,** pp. 486–495, 2015.
https://doi.org/10.1016/j.enganabound.2014.10.004

[15] Benedetti, I. & Aliabad, M.H., Multiscale modeling of polycrystalline materials: a boundary element approach to material degradation and fracture. *Computer Methods in Applied Mechanics and Engineering*, **289**, pp. 429–453, 2015.
https://doi.org/10.1016/j.cma.2015.02.018

[16] Kupradze, V.D., Three-dimensional problems of the mathematical theory of elasticity and thermoelasticity. North Holand, 1979.

[17] Sladek, J. & Sladek, V., Three-dimensional curved crack in an elastic body. *International Journal of Solids and Structures*, **19**, pp. 425–436, 1983.
https://doi.org/10.1016/0020-7683(83)90053-7

[18] Bonnet, M., *Boundary Integral Equation Methods for Solids and Fluids*, John Wiley & Sons Ltd, 1999.

[19] Palermo, Jr., L., Almeida, L.P.C.P.F. & Gonçalves, P.C., The use of the tangential differential operator in the dual boundary element equation. *Structural Durability & Health Monitoring*, **2**(2), pp. 123–130, 2006.

[20] Palermo, Jr., L. & Almeida, L.P.C.P.F., On the use of the tangential differential operator in the traction boundary integral equation of the dual boundary element method for three dimensional problems. *ICCES*, **7**(2), pp. 83–87, 2008.

[21] Palermo, Jr., L., The tangential differential operator applied to a stress boundary integral equation for plate bending including the shear deformation effect. *Engineering Analysis with Boundary Elements*, **36**, pp. 1213–1225, 2012.
https://doi.org/10.1016/j.enganabound.2012.02.010

[22] Gonçalves, P.C., *Aplicação do método dos elementos de contorno no estudo da propagação de fissura discreta para modelos coesivos* (PhD thesis, written in Portuguese), 136p., Faculdade de Engenharia Civil, Arquitetura e Urbanismo, Universidade Estadual de Campinas, 2015.

[23] Gonçalves, P.C., Palermo, L. & Proenca, S.P.B., A boundary element formulation for crack analyses incorporating a cohesive zone model. *International Journal of Computational Methods and Experimental Measurements*, **5**(3), pp. 231–240, 2017.
https://doi.org/10.2495/cmem-v5-n3-231-240

[24] Gonçalves, P.C., Figueiredo, L.G., Palermo Jr., L. & Proença, S.P.B., *The Dual-Boundary-Element Formulation Using the Tangential Differential Operator and Incorporating a Cohesive Zone Model for Elastostatic Cracks.* 12th International Conference Boundary Element and Meshless Techniques, Brasília, 2011.

SIMULATION OF SOUND STRUCTURE INTERACTIONS BY THE COUPLED FEM/BEM

ZAI-YOU YAN
College of Aerospace Engineering, Nanjing University of Aeronautics and Astronautics, P.R. China.

ABSTRACT
Sound transmission through thin elastic shell with different fluids on the inside and outside is simulated using the in-house program based on the coupled finite element and boundary element method. The structure dynamics is simulated using the finite element method and the acoustic fields are simulated using the boundary element method. To avoid the non-uniqueness problem existing in the exterior acoustic boundary element method, Burton and Miller formulation is employed. The hyper-singular boundary integral is dealt with a regularization relationship. To validate this approach, a case with analytical solutions is simulated.
Keywords: boundary element method, Burton and Miller's formulation, finite element method, sound structure interaction

1 INTRODUCTION

Acoustic radiation and scattering from fluid-loaded elastic shells may be encountered in the aeronautical and naval industries as well as in underwater acoustics. It is well known that the presence of fluid modifies considerably the resonance characteristics of the structure. In the mean time, the propagation of sound in fluids is altered by the presence of the elastic structure, which causes serious noise problems. The radiated noise from a vibrating structure is important for underwater-related applications. The scattering of acoustic waves from such structures contains information relating to the geometry and composition of the structure, which makes it possible to identify the structure by the remote sensing. Therefore, it is of considerable interest to predict the acoustic fields both radiated and scattered by a submerged vibrating structure.

Acoustic radiation and scattering from submerged elastic structures have been studied by many researchers over the past seventy years. Numerous popular textbooks and monographs [1–3] have been devoted to this important subject. The overwhelming majority of investigations dealing with problems in structural acoustics have focused on the steady state response of the fluid-structure-coupled system. Therefore, it is natural that solutions have been carried out primarily in the frequency domain. Most of the problems addressed are about structures of revolution immersed in an unbounded acoustic fluid medium. The problem of sound radiation and scattering from shells of arbitrary shape with different fluids on both the inside and outside has been investigated very few. Analytical approaches to fluid-structure interaction problems are almost invariably concerned with spherical or infinite circular cylindrical shells subjected to axisymmetric excitations for which the classical method of separation of variables is available. For example, Junger [4] investigated the scattering of plane acoustic waves by air-filled, immersed in fluid elastic spherical and cylindrical shells. Goodman and Stern [5] investigated a submerged elastic spherical shell excited by plane acoustic waves using the exact theory of three-dimensional elasticity.

When analyzing the sound radiated or scattered by submerged elastic shells of more complicated shapes, it is almost indispensable to use numerical codes that can handle the complexity of the structure in question. For a complex structure subjected to known applied

DOI: 10.2495/CMEM-V6-N6-1067-1078

forces, the finite element method (FEM) has become an accepted, well-proven, and highly successful analysis tool. Therefore, for the sound/structure interaction model, it is almost consistent to formulate the structural dynamic equations *via* FEM techniques. When applying to the case of interior problems where the fluid is inside the structure, FEM also gives satisfactory solutions. However, in the case of exterior problem in which fluid occupies an unbounded domain, the FEM is inefficient. When FEM has been applied to unbounded exterior problems, the domain has to be truncated and radiating boundary conditions have to be enforced. In addition, they are limited by computer memory and runtime considerations as well. To deal with unbounded exterior acoustic field problems, the boundary element method (BEM) based on the utilization of the Helmholtz integral equation (HIE) has been the most popular numerical tool. Boundary integral formulations have long been recognized as an elegant and computationally economical method of modeling infinite fluid loading upon an elastic shell. The BEM method has several advantages over a FEM treatment of the acoustic problem, including a reduction of dimensionality of the problem by one, and an automatic satisfaction of the radiation condition. The elegance of this method is the mathematical simplicity of the resulting integral expressions. In the past few decades, numerical methods based on BEM had been employed to address the general acoustic radiation and scattering from elastic bodies.

One potential shortcoming of the classical BEM in acoustics is that the exterior boundary integral formulation shares the well-known difficulty of nonuniqueness of solution at certain characteristic frequencies. A well-known formulation to overcome the nonuniqueness problem is the method proposed by Burton and Miller [6].

To deal with the fluid-structure interaction problems, many researchers have developed FEM/BEM coupled schemes, using various boundary integral equations (BIEs) for the unbounded fluid domain together with a FEM to model the elastic target. The use of the same elements in both the FEM and BEM allows one to match the continuity at the fluid structure interface. Now, coupled FEM/BEM method is the most popular numerical tool for sound/structure interaction problems. Luke and Martin [7] gave a mathematical investigation about the coupled FEM and BEM method. They proved the existence, uniqueness theorems and so on. The first reference about coupled FEM/BEM method dated back to the late seventies. In 1977, Zienkiewicz *et al.* [8] proposed the coupled of the standard FEM to those derived from the BIEs. Subsequently, Wilton [9] formulated a coupled FEM/BEM system in terms of the acoustic variables. A comparison between the two different approaches for the coupled FEM/BEM, i.e. removal of the structural variables versus removal of the acoustic variables was made by Mathews [10]. Also using the submerged spherical shell, Mathews concluded that combining the equations in terms of the acoustic variables was both computational efficient and more accurate. Wu and Dandapani [11] used the coupled FEM and multi-domain boundary element method to model sound transmission through thin structures with different fluids on the inside and outside. As an example, they calculated the sound transmission through thin elastic spherical shell, which was submerged in water, filled with air, and subjected to a sound point source located at the center of the shell. Jeans and Mathews [12] presented a unique coupled FEM/BEM method for the elastoacoustic analysis of fluid-filled thin shells. They concluded that except for problems having a significant density difference between the internal and external acoustic fields, for example, air and water, their formulation was suitable.

This paper numerically investigates the sound transmission through thin elastic shell with different fluids on both the inside and outside using numerical method. To deal with such kind of structural-acoustics coupled problems, a numerical model based on the coupled finite/boundary element method (FEM/BEM) is developed, in which the elastic target is modeled using the finite element method and the acoustic equations are solved using the boundary element method. Finally, the finite element model and the boundary models are completely coupled through the continuity conditions on the fluid structure interfaces. Example with analytical solution is simulated to verify the developed method.

2 FINITE ELEMENT MODEL FOR SHELLS

The general shell element employed here had been described in Ref. [13]. The elements are eight-nodded isoparametric quadrilateral elements. Strains normal to the midsurface are neglected, and lines originally normal to the midsurface are constrained to remain straight during deformation. The effect of transverse shear deformation has been included. Therefore, the rotation of the midsurface normal vector is an independent variable.

According to Hamilton's principle, the finite element governing equation [10] for the dynamic fluid-structure interaction system is given by

$$[-\omega^2 M - i\omega C + K]\{U\} = \{F_I\} + \{F_A\} \tag{1}$$

where M, K and C are the mass, stiffness and damping matrices, respectively. U denotes the displacement and ω is the circular frequency. F_A represents the known applied excitation forces and F_I represents the interaction forces generated by the acoustic fluid acting on the fluid-structure interaction surfaces. The vector of interaction force can be defined through the structure coupled matrix $L_s(L_s^i$ and $L_s^e)$ and the nodal acoustic pressures $\{\varphi^i\}$ and $\{\varphi^e\}$, that is

$$\{F_I\} = L_s^i\{\varphi^i\} - L_s^e\{\varphi^e\} \tag{2}$$

where $\{\varphi^i\}$ and $\{\varphi^e\}$ are the interior and exterior surface acoustic pressures, respectively. L_s^i and L_s^e are the structure coupled matrices on the interior surface and exterior surface. The coupled matrix of element is defined as

$$L_s = \iint_S [N]_f^T\{\vec{n}\}[N]dS \tag{3}$$

where $[N]_f^T$ is the shape function matrix about displacement in the finite element method. $[N]$ is the shape function matrix about acoustic pressure in the boundary element method and $\{\vec{n}\}$ is the vector of the direction cosines of normal vector. Detail discussion of the coupled matrix will be presented in Ref. [14].

Previous to describe the general shell element, eight-node isoparametric element displayed in Fig. 1 will be introduced first. Isoparametric elements are formulated using an natural coordinate system $\xi\eta$, which is defined by element geometry and not by the element orientation in the global coordinate system xy. Coordinates $\xi\eta$ (and ζ, if the element is three-dimensional) are attached to the element and are scaled so that sides of a quadrilateral element shown in Fig. 1 are defined by $\xi=-1$, $\xi=1$, $\eta=-1$ and $\eta=1$, as shown in Fig. 2. Shape functions [14] for quadratic eight-node isoparametric curvilinear quadrilateral elements are given as follows:

$$N_1 = -\frac{1}{4}(1-\xi)(1-\eta)(1+\xi+\eta),\ N_2 = \frac{1}{2}(1-\xi^2)(1-\eta)$$

$$N_3 = \frac{1}{4}(1+\xi)(1-\eta)(\xi-\eta-1),\ N_4 = \frac{1}{2}(1-\xi)(1-\eta^2)$$

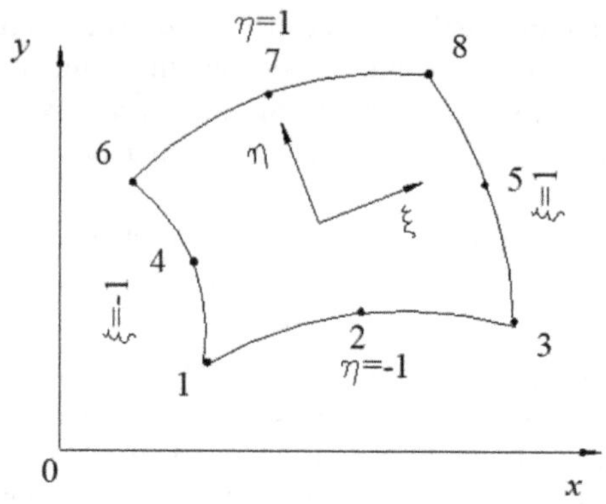

Figure 1: Curvilinear quadrilateral element.

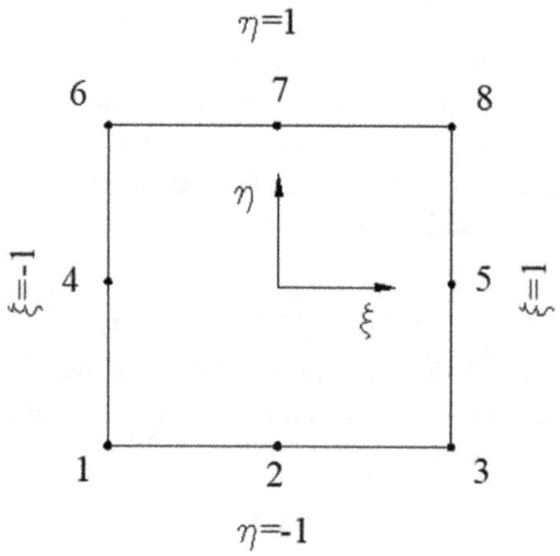

Figure 2: Parent element.

$$N_5 = \frac{1}{2}(1+\xi)(1-\eta^2),\ N_6 = \frac{1}{4}(1-\xi)(1+\eta)(-1-\xi+\eta)$$

$$N_7 = \frac{1}{2}(1-\xi^2)(1+\eta),\ N_8 = \frac{1}{4}(1+\xi)(1+\eta)(\xi+\eta-1)$$

Therefore, the shape function matrix $[N]$ can be expressed as follows

$$[N]=[N_1 \quad N_2 \quad N_3 \quad N_4 \quad N_5 \quad N_6 \quad N_7 \quad N_8] \tag{4}$$

For thin shell, the solid element can be reduced to an eight-node isoparametric surface element, as shown in Fig. 3. Note that the mid-surface normal needs not to be kept normal during deformation; it may take angles of other 90 degrees with the deformed mid-surface. In this way, the ability to model transverse deformation is retained. Consider a local curvilinear coordinate system $o\xi\eta\zeta$ on each element. Let ξ and η be curvilinear coordinates on the middle surface. $\zeta=1$ represents upper surface S_u and $\zeta=-1$ represents lower surface S_l.

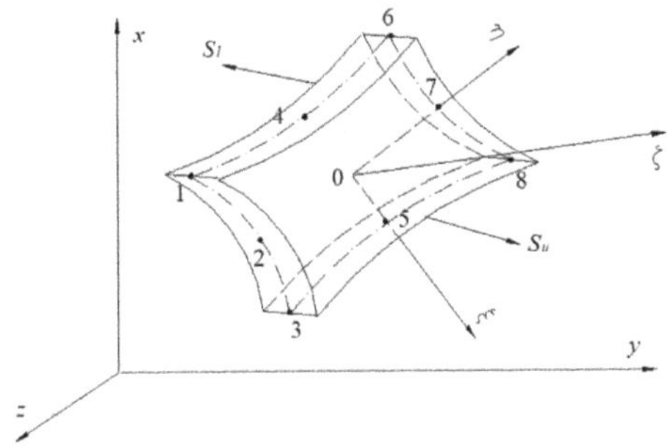

Figure 3: Eight-node isoparametric mid-surface element.

A unit normal vector on middle surface through node i is given by

$$\vec{V}_{3i} = \begin{Bmatrix} l_{3i} \\ m_{3i} \\ n_{3i} \end{Bmatrix} = \frac{1}{h_i}\left(\begin{Bmatrix} x_i \\ y_i \\ z_i \end{Bmatrix}_u - \begin{Bmatrix} x_i \\ y_i \\ z_i \end{Bmatrix}_l \right) \tag{5}$$

where, l_{3i}, m_{3i} and n_{3i} are the direction cosines of $\vec{V}_{3i}$ in the global coordinate system and h_i is the thickness of the shell structure.

Then, the global coordinates of arbitrary point in the solid element (see Fig. 3) can be obtained using the shape functions eqn (4) as

$$\begin{Bmatrix} x \\ y \\ z \end{Bmatrix} = \sum_{i=1}^{8} N_i(\xi,\eta)\left(\begin{Bmatrix} x_i \\ y_i \\ z_i \end{Bmatrix} + \frac{h_i}{2}\zeta \vec{V}_{3i} \right) \tag{6}$$

Suppose the mid-surface normal line rotates about two different axes, which are perpendicular to each other and parallel to the mid-surface. Let $\vec{V}_{1i}$ and $\vec{V}_{2i}$ are the unit vectors of these two axes respectively.

The corresponding rotation angles of normal vector $\vec{V}_{3i}$ around the axes $\vec{V}_{1i}$ and $\vec{V}_{2i}$ are β_i and α_i. Then the rotation vector can be expressed by

$$\vec{\omega}_i = \beta_i \vec{V}_{1i} + \alpha_i \vec{V}_{2i} \tag{7}$$

Consequently, the displacements of arbitrary point on the normal vector $\vec{V}_{3i}$ can be expressed using the mid-surface nodal displacements u_i, v_i, w_i and rotation vector $\vec{\omega}_i$ of mid-surface normal line. By the shape function interpolation, the displacements of arbitrary point in the solid element are

$$\begin{Bmatrix} u \\ v \\ w \end{Bmatrix} = \sum_{i=1}^{8} N_i(\xi,\eta)\left(\begin{Bmatrix} u_i \\ v_i \\ w_i \end{Bmatrix} + \zeta[\phi_i]\begin{Bmatrix} \alpha_i \\ \beta_i \end{Bmatrix} \right) \tag{8}$$

where ϕ_i is defined as follows

$$[\phi_i] = \begin{bmatrix} \phi_{11i} & \phi_{12i} \\ \phi_{21i} & \phi_{22i} \\ \phi_{31i} & \phi_{32i} \end{bmatrix} = \frac{h_i}{2}\begin{bmatrix} l_{1i} & -l_{2i} \\ m_{1i} & -m_{2i} \\ n_{1i} & -n_{2i} \end{bmatrix} = \frac{h_i}{2}[\vec{V}_{1i} \quad -\vec{V}_{2i}] \tag{9}$$

Equation (8) can be expressed in standard form as

$$\begin{Bmatrix} u \\ v \\ w \end{Bmatrix} = [N]_f \begin{Bmatrix} \{\delta_1\} \\ \{\delta_2\} \\ \vdots \\ \{\delta_8\} \end{Bmatrix} = \sum_{i=1}^{8} [N_i]_f \{\delta_i\} \tag{10}$$

where $\{\delta_i\} = [u_i \quad v_i \quad w_i \quad \alpha_i \quad \beta_i]^T$ is the vector for degrees of freedom at node i. The matrix $[N]_f$ is defined as the shape function matrix for displacements.

$$[N]_f = \begin{bmatrix} N_1 & 0 & 0 & N_1\zeta\phi_{111} & N_1\zeta\phi_{121} & \cdots & N_8 & 0 & 0 & N_8\zeta\phi_{118} & N_8\zeta\phi_{128} \\ 0 & N_1 & 0 & N_1\zeta\phi_{211} & N_1\zeta\phi_{221} & \cdots & 0 & N_8 & 0 & N_8\zeta\phi_{218} & N_8\zeta\phi_{228} \\ 0 & 0 & N_1 & N_1\zeta\phi_{311} & N_1\zeta\phi_{321} & \cdots & 0 & 0 & N_8 & N_8\zeta\phi_{318} & N_8\zeta\phi_{328} \end{bmatrix}_{3\times 40} \tag{11}$$

The element stiffness matrix can be written in block form. That is each element stiffness matrix, $[k]_{40\times 40}$, can be partitioned into sixty-four $[k_{ij}]_{5\times 5}$ element sub-matrices. These sub-matrices will be calculated in terms of the following integral equation [14]

$$[k_{ij}] = \int_{-1}^{1}\int_{-1}^{1}\int_{-1}^{1} [B_i]^T [D][B_j] |J| d\xi d\eta d\zeta \tag{12}$$

The expressions for the matrices B and D can be found in the reference [14]. J represents the Jacobian matrix.

The element consistent mass matrix can be expressed by using the shape functions for displacements as

$$[m] = \int_{-1}^{1}\int_{-1}^{1}\int_{-1}^{1} [N]_f^T \rho [N]_f \, |J| d\xi d\eta d\zeta \tag{13}$$

The structure damping matrix C is implemented numerically through the use of the following hysteric relationship [12].

$$C = \gamma K \tag{14}$$

Here γ is assumed as the structural loss factor and K is the structure stiffness matrix. In general, γ is chosen to be 0.01.

3 BOUNDARY ELEMENT MODEL FOR ACOUSTICS

An elastic thin shell is defined on the closed surface S. The shell submerged in an infinite fluid with density ρ^e in the exterior domain E and contains a fluid with density ρ^i in the interior domain D. The fluids on the inside and outside are assumed to be inviscid and compressible. The surface of the shell is assumed smooth. The normal vector is defined to point into the exterior domain, as shown in Fig. 4.

A harmonic time dependency term of $e^{i\omega t}$ is assumed and the wave number k can be defined as

$$k = \omega / c \tag{15}$$

where ω is the circular frequency and c is the sound velocity

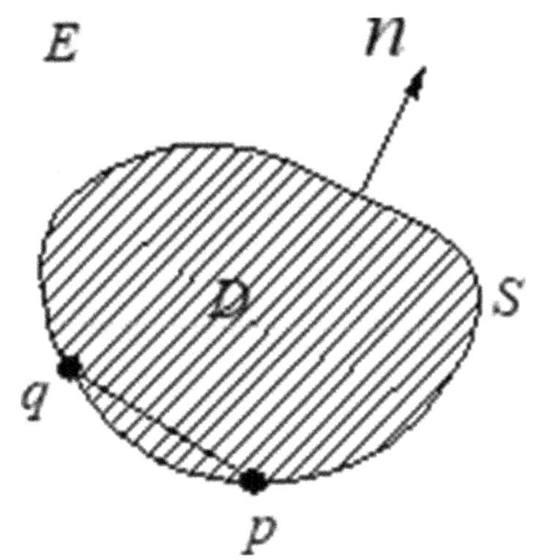

Figure 4: Notation for submerged fluid-filled shell.

Assuming there are a point sound source in the interior domain and a plane acoustic wave in the exterior domain. Then, the acoustic pressure φ^e in the exterior domain satisfies the homogeneous reduced wave equation

$$(\nabla^2 + k^2)\varphi^e(p) = 0 \tag{16}$$

The point sound source with strength Ω is located at the point p_0 in the interior domain. Then, the incident acoustic pressure φ_I^i produced by the point sound source can be expressed as

$$\varphi_I^i(r) = \frac{\Omega}{4\pi}\frac{e^{-ikr}}{r} \tag{17}$$

The acoustic problems are defined using the reduced wave equations and boundary conditions on the surfaces of the shell. The partial derivatives of the exterior surface acoustic pressure φ^e and interior surface acoustic pressure φ^i with the surface normal displacements u_n^e and u_n^i satisfy the following relations.

$$\frac{\partial \varphi^e}{\partial n} = \omega^2 \rho^e u_n^e \tag{18}$$

$$\frac{\partial \varphi^i}{\partial n} = \omega^2 \rho^i u_n^i \tag{19}$$

Besides the above boundary continuity conditions, the pressure difference between the exterior acoustic pressure φ^e and the incident plane acoustic wave pressure φ_I^e is subjected to the Sommerfeld radiation condition.

The Helmholtz integral equation for exterior acoustic pressure φ^e is

$$c^e(p)\varphi^e(p) = \iint_S \left[\varphi^e(q)\frac{\partial G(p,q)}{\partial n_q} - G(p,q)\frac{\partial \varphi^e(q)}{\partial n_q} \right] dS_q + \varphi_I^e(p) \tag{20}$$

And the interior acoustic pressure φ^i can be expressed in the form of boundary integral equation as

$$c^i(p)\varphi^i(p) = \iint_S \left[\varphi^i(q)\frac{\partial G(p,q)}{\partial n_q} - G(p,q)\frac{\partial \varphi^i(q)}{\partial n_q} \right] dS_q - \varphi_I^i(p) \tag{21}$$

where $c^e(p)$ and $c^i(p)$ are the dimensionless solid angles on the exterior and interior surfaces respectively.

In operator notations, the surface Helmholtz integral equations can be given as,

$$\left[-\frac{1}{2}I + M_k \right]\varphi^e = L_k \frac{\partial \varphi^e}{\partial n} - I\varphi_I^e \tag{22}$$

$$\left[\frac{1}{2}I + M_k \right]\varphi^i = L_k \frac{\partial \varphi^i}{\partial n} - I\varphi_I^i \tag{23}$$

where the integral operators M_k and L_k are defined as

$$M_k \mu = \iint_S \mu \frac{\partial G}{\partial n} dS,\ L_k \mu = \iint_S \mu G dS$$

Burton and Miller [6] had shown that a combination of the surface Helmholtz integral equation (22) and its differentiation will provide a unique solution for the entire frequency range. It can be represented in operator notation as

$$\left[-\frac{1}{2}I + M_k + \alpha N_k \right]\varphi^e = \left[L_k + \alpha \left[\frac{1}{2}I + M_k^T \right] \right] \frac{\partial \varphi^e}{\partial n} - I\varphi_I^e - \alpha I \frac{\partial \varphi_I^e}{\partial n} \tag{24}$$

where α, the coupling constant, is chosen to be strictly complex. Generally, α is taken to be $-i/k$. i is an imaginary unit. The integral operators N_k and M_k^T are defined as

$$N_k \mu = \iint_S \mu(q) \frac{\partial^2 G(p,q)}{\partial n_q \partial n_p} dS_q,\ M_k^T \mu = \iint_S \mu(q) \frac{\partial G(p,q)}{\partial n_p} dS_q$$

The hypersingular integral in the operator N_k is dealt with the regularization relationship developed in Refs. [14, 15]. Since this regularization relationship reduces the hyersingular integral to weakly singulars, C^0 element is enough to compute the hyersingular integral.

4 THE COUPLED FEM/BEM MODEL

There are two kinds of continuity boundary conditions on the fluid structure interfaces. One is the kinetic boundary condition and the other is the dynamic boundary condition.

4.1 Kinetic continuity boundary condition

The partial derivatives of the surface acoustic pressures φ^e and φ^i with the surface normal displacements u_n^e and u_n^i satisfy eqns (18) and (19). The nodal values of normal displacements $\{u_n\}$ and displacements $\{U\}$ have the following relationship

$$\{u_n\} = Q\{U\} \tag{25}$$

where Q is defined as the kinetic coupled matrix. It can be explicitly expressed as

$$Q_{t\times 5t} = \begin{bmatrix} [\vec{n}_1]^T[I_{3\times 3} \quad \zeta[\phi_1]] & 0 & \cdots & 0 \\ 0 & [\vec{n}_2]^T[I_{3\times 3} \quad \zeta[\phi_2]] & \cdots & 0 \\ \vdots & \vdots & \ddots & \vdots \\ 0 & 0 & \cdots & [\vec{n}_t]^T[I_{3\times 3} \quad \zeta[\phi_t]] \end{bmatrix} \tag{26}$$

where t is the total number of node. Matrix Q is not a square one. It can be stored using one-dimensional compacted storage.

$$Q = \begin{cases} Q^e & \zeta = 1 \\ Q^i & \zeta = -1 \end{cases}$$

Then, the kinetic continuity boundary condition can be expressed as

$$\frac{\partial \varphi^e}{\partial n} = \omega^2 \rho^e Q^e \{U\} \tag{27}$$

$$\frac{\partial \varphi^i}{\partial n} = \omega^2 \rho^i Q^i \{U\} \tag{28}$$

4.2 Dynamic boundary condition

For problems of sound transmission through a fluid-filled, submerged shell structure, the coupled nodal forces $\{F_l\}$ satisfy the eqn (2). The element-coupled matrix L can be partitioned into 64 sub-matrices L_{ij} of 5 by 1.

$$L_{ij} = \iint_S [N_{,i}]_f^T [\vec{n}] N_{,j} dS \tag{29}$$

where

$$[N_{,i}]_f^T [\vec{n}][N_{,j}] = \begin{bmatrix} l_3 N_{,i} N_{,j} \\ m_3 N_{,i} N_{,j} \\ n_3 N_{,i} N_{,j} \\ a_i \zeta N_{,i} N_{,j} \\ b_i \zeta N_{,i} N_{,j} \end{bmatrix}_{5\times 1}, \; i, j = 1, 2, \cdots, 8 \tag{30}$$

and

$$a_j = l_3 \phi_{11j} + m_3 \phi_{21j} + n_3 \phi_{31j}, \; j = 1, 2, \cdots, 8$$

$$b_j = l_3 \phi_{12j} + m_3 \phi_{22j} + n_3 \phi_{32j}, \; j = 1, 2, \cdots, 8$$

The structure-coupled matrix L_s is not a square one. Assuming the total node number is t, because there are five degrees of freedom at each node, the dimension of structure matrix L_s is $5t \times t$. The matrices L^e and L^i with the matrix L satisfy the following relation

$$L = \begin{cases} L^e & \zeta = 1 \\ L^i & \zeta = -1 \end{cases}$$

Then, combing the eqns (1), (2), (21) and (24), sound transmission through thin elastic shell with different fluids on both the inside and outside can be solved.

5 NUMERICAL EXAMPLE

In order to validate the correctness of the present coupled FEM/BEM method, a fluid-filled submerged elastic spherical shell excited by a point sound source located at the center is analyzed. In this example, the fluid mediums inside and outside the spherical shell are air and sea water, respectively. The spherical shell is discretized 96 surface element. The external radius of the spherical shell is 1.01 m and the internal radius is 1.0 m. Therefore, the relative thickness of the spherical shell is 1%. It is a thin shell. The density of the spherical shell is $\rho = 7.81\times10^3 kg/m^3$. The elasticity modulus is $E = 2.07\times10^{11}$Pa and Poisson's ratio is $\mu = 0.3$. The air density is $\rho^a = 1.21kg/m^3$ and sound velocity in air is $c^a = 346m/s$. While, the sea water density is $\rho^w = 1030kg/m^3$ and sound velocity in sea water is $c^w = 1500m/s$.

The frequency range is from 50 to 2000Hz. Figure 5 shows the comparison of the exterior surface acoustic pressure amplitude obtained using the coupled FEM/BEM and the corresponding analytical solutions [11]. The numerical results agree with the corresponding analytical solutions throughout the frequency range. Figure 6 shows the

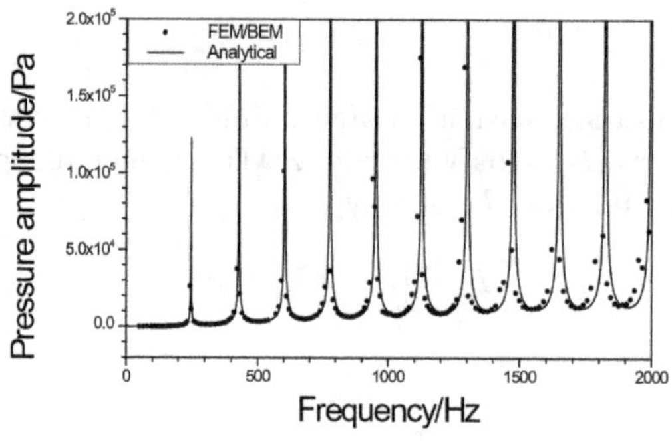

Figure 5: Frequency response of exterior surface acoustic pressure amplitude.

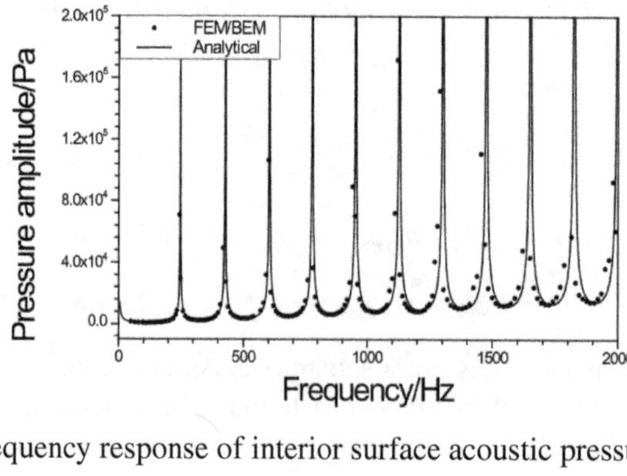

Figure 6: Frequency response of interior surface acoustic pressure amplitude.

comparison of the interior surface acoustic pressure amplitude obtained using coupled FEM/BEM and the corresponding analytical solutions. Clearly, the numerical results agree with the corresponding analytical solutions.

6 CONCLUSIONS

A coupled finite element method and boundary element method are developed to model sound transmission through thin elastic shell with different fluids on the inside and outside of the shell. Burton & Miller's composite formulation is employed to avoid the nonuniqueness problem existing in the exterior acoustic boundary element method. To validate the present method, sound transmission through an air-filled and submerged in sea water spherical shell is simulated.

ACKNOWLEDGEMENT

This project was supported by NSFC (11172132) and NSFC (51405225); the Priority Academic Program Development of Jiangsu Higher Education Institutions.

REFERENCES

[1] Junger, M.C. & Feit, D., *Sound Structure and Their Interaction*, 2nd ed., MIT, Cambridge, MA, 1986.

[2] Mason, W.P. & Thurston, R.N., *Physical Acoustics*, Vol. XV, London: Academic Press, INC, pp.191-285, 1981.

[3] Ciskowski, R.D. & Brebbia, C.A., *Boundary Element Methods in Acoustics*, Southampton Boston: Computational Mechanics Publications, pp. 61–74, pp. 109–129, 1991.

[4] Junger, M.C., Sound scattering by thin elastic shells. *The Journal of the Acoustical Society of America*, **24**, pp. 366–373, 1952.
https://doi.org/10.1121/1.1906905

[5] Goodman, R.R. & Stern, R., Reflection and transmission of sound by elastic spherical shells. *The Journal of the Acoustical Society of America*, **34**, pp. 338–344, 1962.
https://doi.org/10.1121/1.1928120

[6] Burton, A.J. & Miller, G.F., The application of integral equation methods to the numerical solution of some exterior boundary value problems. *Proceedings of the Royal Society A: Mathematical, Physical and Engineering Sciences*, **A323**, pp. 201–210, 1971.
https://doi.org/10.1098/rspa.1971.0097

[7] Luke, C.J. & Martin, P.A., Fluid-solid interaction: acoustic scattering by a smooth elastic obstacle. *SIAM Journal on Applied Mathematics*, **55**(4), pp. 904–922, 1995.
https://doi.org/10.1137/s0036139993259027

[8] Zienkiewicz, O.C., Kelly, D.W. & Bettess, P., The coupling of the finite element method and boundary solution procedures. *International Journal for Numerical Methods in Engineering*, **11**, pp. 355–375, 1977.
https://doi.org/10.1002/nme.1620110210

[9] Wilton, D.T., Acoustic radiation and scattering from elastic structures. *International Journal for Numerical Methods in Engineering*, **13**, pp. 123–138, 1978.
https://doi.org/10.1002/nme.1620130109

[10] Mathews, I.C., Numerical techniques for three dimensional steady-state fluid-structure interaction. *The Journal of the Acoustical Society of America*, **79**, pp. 1317–1325, 1986.
https://doi.org/10.1121/1.393711

[11] Wu, T.W. & Dandapani, A., A boundary element solution for sound transmission through thin panels. *Journal of Sound and Vibration*, **171**, pp. 145–157, 1994. https://doi.org/10.1006/jsvi.1994.1110

[12] Jeans, R.A. & Mathews, I.C., A unique coupled boundary element/finite element method for the elastoacoustic analysis of fluid-filled thin shells. *The Journal of the Acoustical Society of America,* **94**, pp. 3473–3479, 1993. https://doi.org/10.1121/1.407201

[13] Cook, R.D., *Concepts and Applications of Finite Element Analysis*, John Wiley & Sons, Inc: New Yorks, pp. 99–118, 191–207, 1974.

[14] Yan, Z.Y., *Sound Transmission Through Thin Elastic Shell with Different Fluids on Both the Inside and Outside*. Ph. D Thesis, Shanghai Jiao Tong University, P.R. China, 2000.

[15] Yan, Z.Y., Hung, K.C. & Zheng, H., Solving the hypersingular boundary integral equation in three dimensional acoustics using a regularization relationship. *The Journal of the Acoustical Society of America,* **113**, pp. 2674–2683, 2003. https://doi.org/10.1121/1.1560164

EXPLORING RICHTMYER–MESHKOV INSTABILITY PHENOMENA AND THE LINKS BETWEEN SURFACE PERTURBATIONS AND SHOCKED-INDUCED MASS EJECTION

SHI YINA, WANG PEI, ZHAO QIANG & FENG QIJING
Institute of Applied Physics and Computational Mathematics, Beijing, P. R. China.

ABSTRACT

This work investigates the mass ejected from surface perturbations as the shockwave reaches the AL-vacuum interface, which originates from unstable Richtmyer–Meshkov (RMI) impulse phenomena. The main purpose is to explore the relationships between the shockwave impulse and the geometric properties of surface perturbations, and how those relationships drive the total ejected mass, directionality and velocity distribution. We discuss in detail different types of surface geometry (sinusoidal, square-wave, chevron and semicircle), as well as the wavelengths and amplitudes of surface perturbation. The time evolutions of micro-jet ejection are simulated using a hydrodynamic Lagrangian-Remapping Eulerian method. The calculated results show that primary jetting ejection can be formed from the different shapes, and with increasing wavelength, the ejection mass keeps an increase while the jet head-velocity decreases. However, not all periodic perturbations behave similarly, and masses ejected from irregular surface cannot be normalized to its cross-sectional areas. The square-wave surface may yield pronounced, velocity-enhanced secondary jetting, which is a result of collision of primary jets.

Keywords: L-R two-step Eulerian method, mass ejection, metal, Richtmyer–Meshkov instability, surface perturbation

1 INTRODUCTION

Metals under shock-loaded conditions can lead to complex phenomena depending on the properties of material and shock conditions. We know that in one-dimensional (1D) geometry, the reflection of a planar shock front from a parallel, perfectly planar free surface accelerates this surface to approximately twice the particle velocity induced by the incident shock. In practice, material surface characteristics such as pits, bumps, voids, grooves, or scratches disrupt this 1D ideal case. When shock pressures are typically on the order of a few hundred thousand atmospheres, but gas pressures are about an atmosphere, the metal will release to a very low pressure at the surface as the shockwave reflects back into the metal whether there is a vacuum or a gas across the interface. Of interest is what happens to metals when a shock wave encounters different perturbed metal/vacuum interfaces. The phenomenon, which rapid ejection of micron-scale fragments usually happens from material surface, has attracted much attention since being discovered by Asay [1, 2]. These ejected high-speed particles often have negative effects on optical and electrical measurements at metal surfaces, and are relevant to inertial confinement fusion.

This process of mass ejection has been extensively studied under explosive and impact loading. Previous research to understand ejecta sources has in most cases focused on experiment characterization and originated from a Richtmyer–Meshkov instability (RMI). The microjetting observed behind those perturbed surfaces may become one kind of main ejection modes. Experimental techniques include piezoelectric pins, optical shadowgraph, X-ray radiography and velocity measurements, *et al.* Some significant efforts have been dedicated to studying on the microjetting mechanism. The effects of groove shape, shock wave rise-time

DOI: 10.2495/CMEM-V6-N6-1079-1086

and shock pressure on micro-jet are investigated, including continuum level and molecular dynamics simulations [3, 4]. Recently, some measured results on ejecta particle size distributions are given [5], which are also predicted by percolation theory and MD simulations. Models aiming at a prediction of ejection velocity and mass are from early, empirical descriptions to modern theories based on the nonlinear RMI [6–9]. It is shown that the mass, size, and velocity of the ejecta vary depending closely on the surface status, initial shock conditions and the material properties.

In this work, we investigate the mass ejected from surface perturbations as the shockwave reaches the AL-vacuum interface. We discuss in detail the underlying mechanisms of surface microjetting related to different types of surface geometry (sinusoidal, square-wave, chevron groove and semicircle) as well as the wavelengths and amplitudes of surface perturbation. The time evolutions of micro-jet ejection are simulated using a hydrodynamic Lagrangian-Remapping Eulerian method. The effects of shockwave impulse as well as surface geometric properties on total ejected mass, directionality and velocity distribution are analyzed.

2 MECHANISM AND MODELING OF MASS EJECTION

2.1 Methodology and modeling

The simulation sample is metal aluminium with surface perturbation of initial wavelength λ and amplitude h. Shock waves are generated by assigning the flyer plate an initial velocity to impact the target. The length of the flyer plate is much longer than that of the target to delay the effects of release fan originated from the free surface of the flyer plate. Here, the impact velocity is 3.4 km/s, and the impact pressure of aluminum samples is about 30 GPa. The initial configurations of our simulations are illustrated in Fig. 1. A proper propagation distance is set between the flyer plate and the bottom of periodic perturbation to ensure the formation of a supported shock wave before it encounters the tip.

The elastic-plastic hydrodynamic Lagrangian-Remapping two-step Euler method is used for the simulation of aluminium micro-jet ejection. The hydrocode uses dimensional split algorithm with Lagrangian re-map and is second order accurate in both time and space. A Young's or VOF approach is employed to capture the interface. The strengths of the code lie in its ability to capture large-scale material deformation, wave propagation and interaction generated from impact phenomena. For the aluminium material, the plastic stress–strain

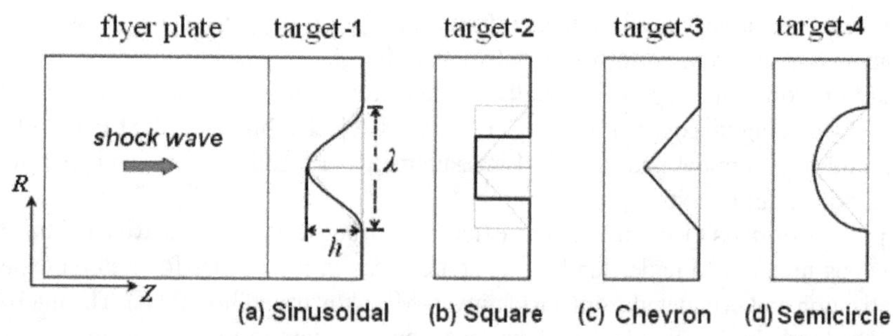

Figure 1: The shapes of initial surface perturbation. (a) Sinusoidal; (b) Square-wave; (c) Chevron and (d) Semicircle.

relation is modelled by the von Misses yield criterion, and the yield strength of the material σ_Y is determined by the Johnson–Cook (JC) model, which can be expressed as

$$S_{rr}^2 + S_{zz}^2 + S_{\theta\theta}^2 + 2S_{rz}^2 \leq (2/3)\sigma_Y^2. \tag{1}$$

$$\sigma_Y = \left[A + B(\varepsilon^p)^n\right]\cdot\left[1 + C\ln(\dot{\varepsilon}^p/\dot{\varepsilon}_0)\right]\cdot\left[1-(T^*)^m\right]. \tag{2}$$

where parameters A, B, C, m and n are determined by experiments. ε^p and $\dot{\varepsilon}^p$ are the equivalent plastic strain and strain-rate, $\dot{\varepsilon}_0 = 1.0\ s^{-1}$, $T^* = (T-T_r)/(T_m-T_r)$, T_m is the melting temperature and T_r is the room temperature,. The equation of state (EOS) is adopted by the Mie-Gruneisen (MG) equation as

$$p = \frac{\rho_0 c_0^2 \mu\left[1 + (1-\frac{\gamma_0}{2})\mu - \frac{a}{2}\mu^2\right]}{\left[1-(S_1-1)\mu - S_2\frac{\mu^2}{\mu+1} - S_3\frac{\mu^3}{(\mu+1)^2}\right]^2} + (\gamma_0 + a\mu)E,\quad \rho \geq \rho_0. \tag{3}$$

$$p = \rho_0 c^2 \mu + (\gamma_0 + a\mu)E,\quad \rho < \rho_0. \tag{4}$$

where $\mu = \rho/\rho_0 - 1$, ρ_0 is the initial density, c_0 is the initial sound velocity, and a, γ_0 are the Gruneisen coefficients. S_1, S_2 and S_3 are the parameters that relate the shock speed to the particle velocity, and E is the initial internal energy.

2.2 Mechanisms of microjetting

We first carry out the simulations of microjetting ejection from chevron perturbation surface of $h = 2.0\ \mu$m, $\lambda = 2.31\ \mu$m, to illustrate jetting mechanisms. The chevron groove is one kind of common surface profile obtained from a machining process. When a shock wave breaks out at a groove tip, a series of rarefaction waves propagate into the target interior. The matters near the groove free surface acquire R-component velocities and move toward the axis of symmetry, as shown in Fig. 2(a). Consequently, the matters above and below the axis collide

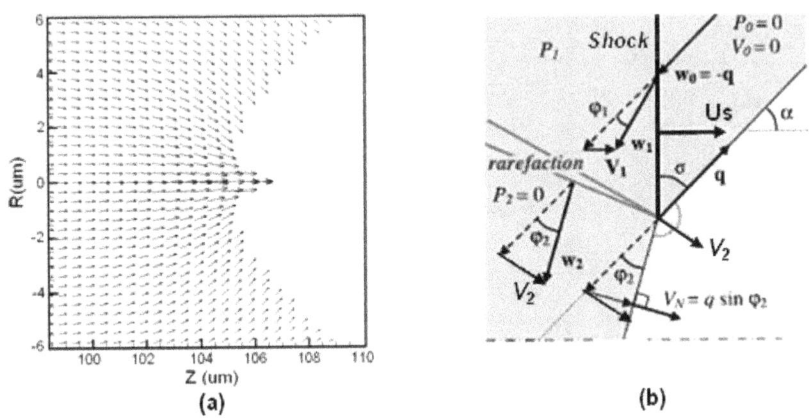

Figure 2: (a) Velocity vector plot near the tip of the chevron groove; (b) Two-dimensional analytical description in the steady-state regular case.

around the axis, along with the central, fast, forward-moving matters, forming a high-speed jet along the shock direction.

The main features governing wave reflection from the groove surface can be approached by shock wave theory. Figure 2(b) is the two-dimensional description of the interaction of a planar shock wave (shock pressure P_1, shock velocity U_s, particle velocity V_1) with a planar free surface at an incidence angle σ. In the regular case, the shocked state is below the sonic point, and the flow is supersonic, which allows steady-state conditions where velocities and angles remain constant during the interaction. By neglecting pressure decay during the interaction of the planar shock with the groove edge, the resulting jet velocity can be estimated. In Fig. 2(b), $q = U_s/\sin\sigma$ is the shock apparent velocity along the interface, $\alpha = 90^0\text{-}\sigma$ is the groove half-angle, φ_2 is the deviation angle after full release and V_2 is the collapse velocity along the bisecting direction, which can be derived from its normal component.

A spatial distribution of the jet velocity (V_R or u_z) is plotted in Fig. 3(b), showing acceleration near the groove, reduced acceleration further away, and deceleration within the jet head. The acceleration is close to linear, and u_z is symmetric about the axis of symmetry, while V_R shows negative gradients along the z-axis near the groove tip [Fig. 3(a)].

3 RESULTS AND DISCUSSION

3.1 Surface geometry effects on mass ejection

With the above model and methods, we conducted a variety of numerical simulations to explore the underlying mechanisms of surface microjetting related to different types of surface geometry, including sinusoidal, square-wave, chevron and semicircle. Figure 4 is simulation results of the micro-jet morphology evolution from different perturbation surface of $h = 2.0$ mm, $\lambda = 4.0$ mm. It shows that primary jetting ejection can be formed from these different shapes, and the growth in lateral and longitudinal dimensions followed by necking. The former is accompanied by pronounced mass accumulation, and then, the necking is the dominant phenomenon. Both mass accumulation and necking are due to the velocity gradi-

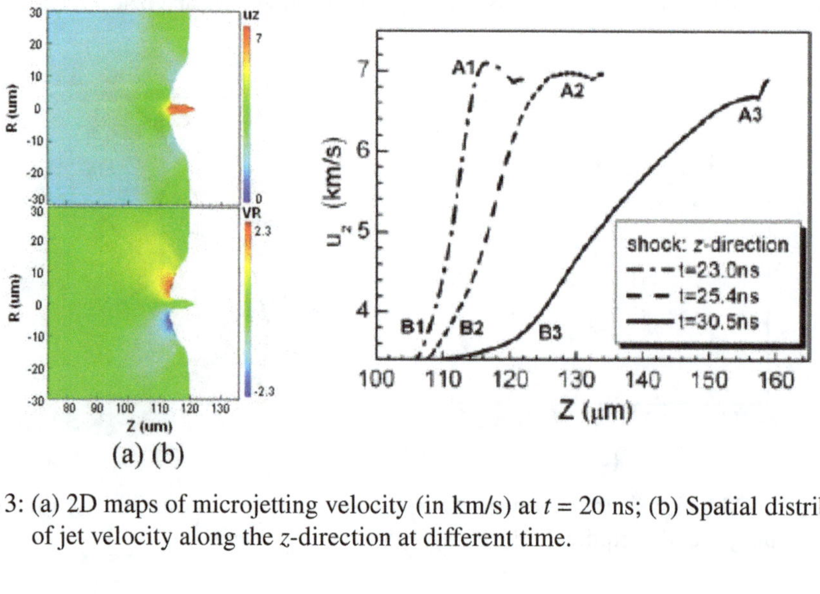

(a) (b)

Figure 3: (a) 2D maps of microjetting velocity (in km/s) at $t = 20$ ns; (b) Spatial distribution of jet velocity along the z-direction at different time.

ents along jet stretching direction, as seen in Fig. 3(b): the gradient values in jet different region give rise to two competing phenomena: mass accumulation in the head (A1, A2 and A3) and necking (B1, B2 and B3) behind it. The decrease in their amplitudes as increasing time leads to reduced mass accumulation and increased necking at later times. Especially, for the square-wave perturbation, two jetting stages can be clearly identified: in the first stage seen as Fig. 4(b), two jets form at the left and right corners within a short period, and the jetting becomes relatively stable; in the second stage, the first two jets collide, leading to a stable, velocity-enhanced, secondary jet. The jet head velocity evolutions are compared in Fig. 5. Secondary jet from square-wave perturbation surfaces can greatly enhance jetting velocity ratio.

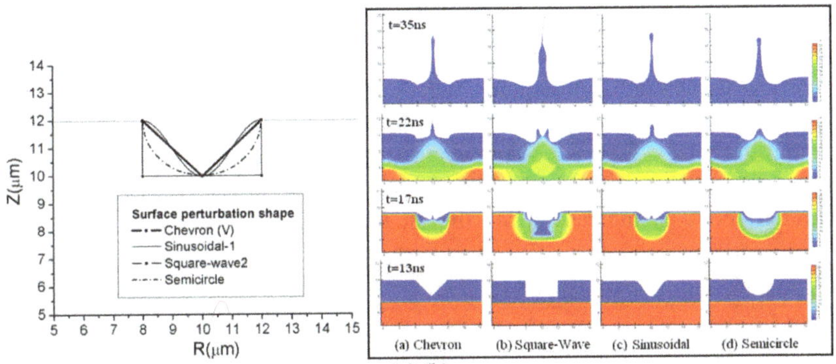

(a) (b)

Figure 4: Jet evolution for different surface shapes with $h = 2.0$ um, $\lambda = 4.0\ \mu$m. (a) Surface perturbation shape; (b) Simulation results at $t = 35$ ns.

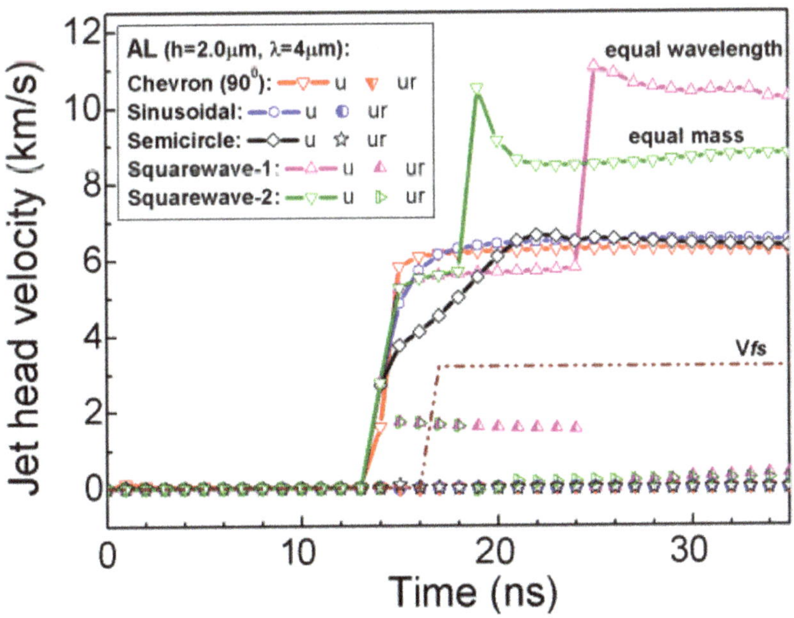

Figure 5: Effect of initial perturbation shapes on jet head velocity evolution.

3.2 Perturbation wavelength and amplitude dependences of mass ejection

Here, we also carried out some simulations to investigate the effects of perturbation wavelengths and amplitudes on mass ejection. We consider the four cases of different perturbation wavelength and amplitude with the same impact velocity. The ratios of amplitude and wavelength are 0.25, 0.5 and 1.0, respectively. Figure 6 is simulation results of the micro-jet morphology evolution under the above perturbation parameter conditions. The calculated results are compared between the standard sinusoidal and square-wave surface.

The jet head velocity evolutions and mass accumulation are shown in Fig. 6. It can be seen that for the sinusoidal surface, with increasing wavelength, the maximum and asymptotic values of jet velocity (*u*, *u*z) decrease, while the time durations to reach the asymptotic or stable velocity increase. The reason is that, at smaller wavelengths, V_R is larger so the jet head gains more acceleration and reaches the peak velocity within a shorter amount of time. For the chevron and semicircle shapes, there are the similar jetting features. However, the jet head velocity evolution is very different from the square-wave surface in the early stage and later secondary jetting stages. Figure 7 compares the jet mass accumulation for the four different wavelength and amplitude ratios. The results indicate that with increasing wavelength, the ejection mass keeps an increase while the jet head-velocity decreases. The square-wave

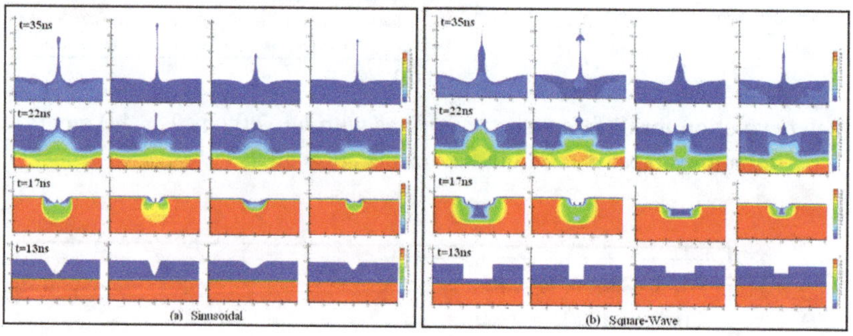

Figure 6: Microjetting evolutions for different ratio of perturbation amplitude and wavelength. (a) Sinusoidal shape; (b) Square-wave shape.

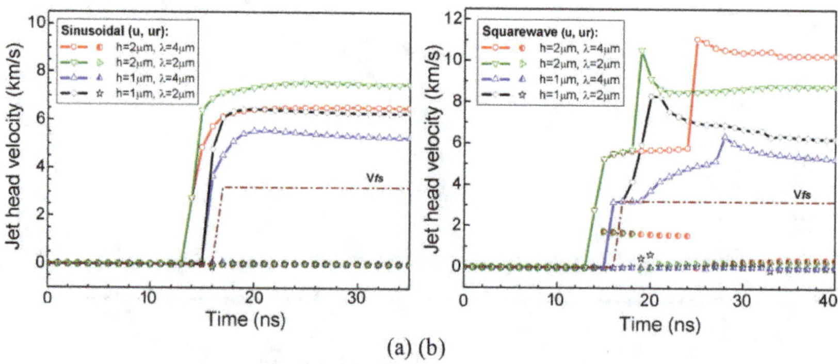

(a) (b)

Figure 7: Effect of initial surface perturbation parameters on jet head velocity evolutions. (a) Sinusoidal shape; (b) Square-wave shape.

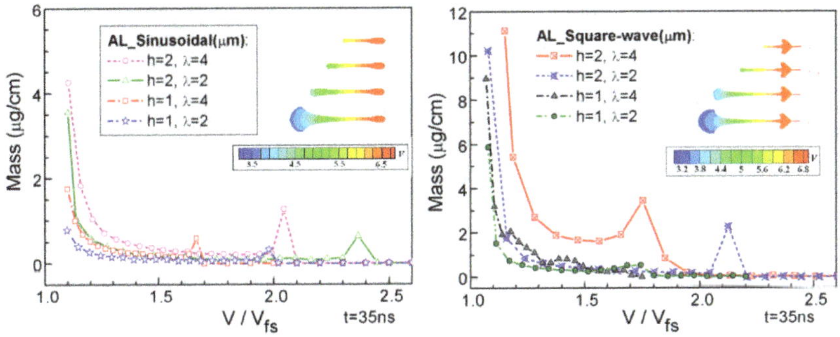

(a) (b)

Figure 8: Ejection mass vs. velocity for different ratio of perturbation amplitude and wavelength. (a) Sinusoidal shape; (b) Square-wave shape.

surface may yield pronounced, velocity-enhanced secondary jetting, and the total ejecta mass increases more significantly, which is a result of collision of primary jets.

4 CONCLUSION

This work investigates the mass ejected from surface perturbations as the shockwave reaches the AL-vacuum interface, which originates from unstable Richtmyer–Meshkov impulse phenomena. Using a hydrodynamic L-R two-step Eulerian method, we explore in detail the underlying mechanisms of surface microjetting related to different types of surface geometry (sinusoidal, square-wave, chevron groove and semicircle) as well as the wavelengths and amplitudes of surface perturbation. The calculated results show that primary jetting ejection can be formed from different perturbation surfaces, and with increasing wavelength, the ejection mass keeps an increase while the jet head-velocity decreases. The fundamental mechanism for jet formation is that transverse collision induces high stress concentrations, and the resulting strong stress gradient near the metal/vacuum interface. However, not all periodic perturbations behave similarly, and masses ejected from irregular surface cannot be normalized to its cross-sectional areas. For the square-wave surface, secondary jetting ejection can be clearly identified: in the first stage, relatively stable primary jets; in the second stage, the first two jets collide, leading to a stable, velocity-enhanced, secondary jet. The total ejecta mass increases more significantly, which is a result of collision of primary jets.

ACKNOWLEDGEMENTS

This work was supported by the National Natural Science Foundation of China (No.11371067) and the Science and Technology Development Foundation of China Academy of Engineering Physics (No.2015B0201044).

REFERENCES

[1] Walsh, J.M., Shreffler, R.G. & Willig, F.J., Limiting conditions for jet formation in high velocity collisions. *Journal of Applied Physics*, **124**(3), pp. 349–359, 1953. https://doi.org/10.1063/1.1721278

[2] Asay, J.R., Material ejection from shock-loaded free surfaces of aluminum and lead. Report No. SAND-76-0542, 1976. https://doi.org/10.2172/7136578

[3] Dyachkov, S.A., Parshikov. A.N. & Zhakhovsky, V.V., Shock-produced ejecta from tin: comparative study by molecular dynamics and smoothed particle hydrodynamics methods. *Journal of Physics: Conference Series*, **653**, pp. 012043, 2015.
https://doi.org/10.1088/1742-6596/653/1/012043
[4] Durand, O. & Soulard, L., Large-scale molecular dynamics study of jet breakup and ejecta production from shock-loaded copper with a hybrid method. *Journal of Applied Physics*, **111**, pp. 044901, 2012.
https://doi.org/10.1063/1.3684978
[5] Sorenson, D.S., Minich, R.W., Romero, J.L., Tunnell, T.W. & Malone, R.M., Ejecta particle size distributions for shock loaded Sn and AL Metals. *Journal of Applied Physics*, **92**(10), pp. 5830–5836, 2002.
https://doi.org/10.1063/1.1515125
[6] Han, C.S., A semi-empirical equation for estimating the micro-jet ejection from shocked free surface. *Chinese Journal of High Pressure Physics*, **3**(3), pp. 234–240, 1989.
[7] Buttler, W.T., Oro, D.M., Preston, D.L., Mikaelian, K.O., Cherne, F.J., Hixson, R.S., Mariam, F.G., Morris, C., Stone, J.B., Terrones, G. & Tupa, D., Unstable Richtmyer-Meshkov growth of solid and liquid metals in vacuum. *Journal of Fluid Mechanics*, **703**, pp. 60–84, 2012.
[8] Shi, Y.N., Qin, C. S., Liu, C. & Feng, Q.J., Study on fragment size of micro-jet from shocked metal surface with groove defect. *Acta ArmamentarII*, **34**, pp. 280–286, 2013.
[9] Dimonte, G., Terrones, G., Cherne, F.J. & Ramaprabhu, P., Ejecta source model based on the nonlinear Richtmyer-Meshkov instability. *Journal of Applied Physics*, **113**, pp. 024905, 2013.
https://doi.org/10.1063/1.4773575

BOUNDARY-DOMAIN INTEGRAL METHOD FOR VORTICITY TRANSPORT EQUATION WITH VARIABLE VISCOSITY

J. RAVNIK & J. TIBAUT
University of Maribor, Faculty of Mechanical Engineering, Maribor, Slovenia.

ABSTRACT

In this paper, we derive a boundary-domain integral formulation for the vorticity transport equation under the assumption that the viscosity of the fluid, through which the vorticity is transported by diffusion and convection, is spatially changing. The vorticity transport equation is a second order partial differential equation of a diffusion-convection type.

The final boundary-domain integral representation of the vorticity transport equation is discretized using a domain decomposition approach, where a system of linear equations is set-up for each sub-domain, while subdomains are joint by compatibility conditions. The validity of the method is checked using several analytical examples. Convergence properties are studied yielding that the proposed discretization technique is second order accurate for constant and variable viscosity cases.

Keywords: boundary-domain element method, variable material properties, vorticity transport equation,

1 INTRODUCTION

Convection-diffusion type partial differential equations govern the physical mechanisms of transport phenomena - transport of momentum, energy and mass. In many engineering or natural circumstances, the flow velocity or the diffusion coefficient vary with time and location. Variation may be due to the unsteady nature of the underlying phenomena or due to temperature or pressure dependence of diffusion coefficient. In terms of the boundary element method, this type of problems can in the case of the constant velocity field and constant diffusivity be described by pure boundary integral equations by using the diffusion-convection fundamental solution.

Several researchers have considered variable velocity fields. Driessen and Dohner [7] proposed finite element - boundary element method for advection-diffusion problems with variable advective fields and in finite domains. DeSilva *et al.* [6] considered 2D energy transport. Wrobel and DeFigueiredo [18] and Rap *et al.* [11] developed a dual reciprocity boundary element formulation for convection-diffusion problems with variable velocity fields.

Variable diffusivity was considered by Grzhibovskis *et al.* [8], Chkadua *et al.* [5], Al Jawary [1–3] and Ang *et al.* [4]. Ravnik and Skerget [12] proposed a boundary-domain integral formulation for diffusion-convection equations with variable coefficient and velocity and considered the energy equation with variable material properties, [13]. Boundary-domain integral method for compressible fluid flow was proposed by Škerget and co-workers [15, 16].

In this paper we, consider a spatially variable fluid viscosity and develop the boundary-domain integral method for the solution of vorticity transport equation in 3D. The vorticity transport equation governs the advective and diffusive movement of vorticity in fluid flows. It is an inhomogeneous equation due to the presence of the vortex twisting and stretching term, which accounts for inherent the 3D nature of fluid flows. The dual reciprocity method [10], which is traditionally used to handle the domain contribution steaming from non-homogeneous parts of the governing equation, is in this work replaced by a

DOI: 10.2495/CMEM-V6-N6-1087-1096

domain decomposition approach, which results in sparse integral matrices and thus makes DBIM computationally affordable.

2 GOVERNING EQUATIONS

We consider a domain Ω with a boundary Γ filled with an incompressible fluid. The flow of the fluid is assumed to be steady. Let $\vec{r} \in \Omega$ denote the position in the domain. If the fluid velocity field is denoted by $\vec{\upsilon} = \vec{\upsilon}(\vec{r},t)$ then the mass conservation law for incompressible fluids states that the divergence of the velocityfield vanishes, i.e.

$$\vec{\nabla} \cdot \vec{\upsilon} = 0 \tag{1}$$

The steady fluid momentum transport equation for this case is

$$(\vec{\upsilon} \cdot \vec{\nabla})\vec{\upsilon} = -\frac{1}{\rho_0}\vec{\nabla} p + \vec{\nabla} \cdot \underline{\tau} + \vec{S}_m, \tag{2}$$

where p is the pressure, $\underline{\tau}$ is the stress tensor and $\vec{S}_m$ accounts for any additional body forces, which may act as a sink or source of momentum.

Taking into account the Newton's law of viscosity, which is a linear relationship between stress and rate of strain, and considering incompressibility, the stress tensor may be written as $\underline{\tau} = 2\mu \underline{\dot{\epsilon}}$; where $\underline{\dot{\epsilon}}$ is the rate of strain tensor and μ is the kinematic viscosity of the fluid. The rate of strain tensor is defines as $\dot{\epsilon}_{ij} = \frac{1}{2}\left(\frac{\partial v_i}{\partial x_j} + \frac{\partial v_j}{\partial x_i}\right)$. Calculating the divergence of the stress tensor gives the following expression. $\vec{\nabla} \cdot \underline{\tau} = \mu\vec{\nabla} \cdot (2\underline{\dot{\epsilon}}) + 2\underline{\dot{\epsilon}} \cdot \vec{\nabla}\mu$ Due to incompressibility $\vec{\nabla} \cdot \vec{\upsilon} = 0$, we can simplify it by showing that $\mu\vec{\nabla} \cdot (2\underline{\dot{\epsilon}}) = \mu\nabla^2\vec{v}$. Thus, the final expression for the momentum transport equation is

$$(\vec{\upsilon} \cdot \vec{\nabla})\vec{\upsilon} = -\frac{1}{\rho_0}\vec{\nabla} p + \mu\nabla^2\vec{\upsilon} + 2\underline{\dot{\epsilon}} \cdot \vec{\nabla}\mu + \vec{S}_m, \tag{3}$$

where the fluid viscosity may vary with time and location.

To derive the velocity-vorticity formulation, we introduce vorticity, which is defined at the curl of the velocity field, $\vec{\omega} = \vec{\nabla} \times \vec{v}$. Vorticity is by definition solenoidal $\vec{\nabla} \cdot \vec{\omega} = 0$. The kinematics equation is a vector elliptic partial differential equation od Poisson type, which links the velocity and vorticity fields for every point in space and time. For an incompressible fluid, it can be stated as (Ravnik [14]):

$$\nabla^2\vec{\upsilon} + \vec{\nabla} \times \vec{\omega} = 0. \tag{4}$$

The vorticity transport equation is derived by taking a curl of the momentum equations. First, the following relationships are used $(\vec{\upsilon} \cdot \vec{\nabla})\vec{\upsilon} = \frac{1}{2}\vec{\nabla}\upsilon^2 - \vec{\upsilon} \times \vec{\omega}, \nabla^2\vec{\upsilon} = -\vec{\nabla} \times \vec{\omega}$, to rewrite the momentum equation into

$$\frac{1}{2}\vec{\nabla}\upsilon^2 - \vec{\upsilon} \times \vec{\omega} = -\frac{1}{\rho_0}\vec{\nabla} p - \mu\vec{\nabla} \times \vec{\omega} + 2\underline{\dot{\epsilon}} \cdot \vec{\nabla}\mu + \vec{S}_m, \tag{5}$$

Introducing the curl considering that the curl of a gradient is zero, we obtain

$$\vec{\nabla} \times (\vec{\upsilon} \times \vec{\omega}) = \vec{\nabla} \times (\mu\vec{\nabla} \times \vec{\omega}) - \vec{\nabla} \times (2\underline{\dot{\epsilon}}\vec{\nabla}\mu + \vec{S}_m), \tag{6}$$

Next, we can use

$$\vec{\nabla}\times(\vec{\upsilon}\times\vec{\omega})=(\vec{\omega}\cdot\vec{\nabla})\vec{\upsilon}-(\vec{\upsilon}\cdot\vec{\nabla})\vec{\omega} \tag{7}$$

$$\vec{\nabla}\times(\mu\vec{\nabla}\times\vec{\omega})=\mu\vec{\nabla}\times(\vec{\nabla}\times\vec{w})-(\vec{\nabla}\cdot\vec{\omega})\times\vec{\nabla}\mu, \tag{8}$$

$$\vec{\nabla}\times(\vec{\nabla}\times\vec{\omega})=-\nabla^2\vec{\omega}, \tag{9}$$

to write the final form of the vorticity transport equation for an incompressible fluid with variable viscosity:

$$(\vec{\upsilon}\cdot\vec{\nabla})\vec{\omega}=(\vec{w}\cdot\vec{\nabla})\vec{\upsilon}+\mu\nabla^2\vec{\omega}+\vec{\nabla}\times\vec{S}_m+(\vec{\nabla}\times\vec{\omega})\times\vec{\nabla}\mu+\vec{\nabla}\times(2\,\underline{\dot{\in}}\cdot\vec{\nabla}\mu). \tag{10}$$

The last two terms in equation (10) are due to variable viscosity. They vanish in the case of constant viscosity and the equation simplifies to a standard expression known for an incompressible fluid with constant material properties.

3 BOUNDARY-DOMAIN INTEGRAL FORMULATION

Vorticity transport equation (10) can be viewed as a Poisson type equation of the following format: $\nabla^2\vec{\omega}=\vec{b}$, where $\vec{b}$ represents all other terms. The boundary-domain integral formulation of a Poisson type equation is (Wrobel [17]):

$$\begin{aligned}\mu c\left(\vec{\xi}\right)\vec{\omega}\left(\vec{\xi}\right)+\int_\Gamma \mu\vec{\omega}\vec{\nabla}u^*\cdot\vec{n}d\Gamma=\int_\Gamma \mu\vec{q}u^*\,d\Gamma\\ -\int_\Omega\left((\vec{\upsilon}\cdot\vec{\nabla})\vec{\omega}-(\vec{\omega}\cdot\vec{\nabla})\vec{\upsilon}\right)u^*\,d\Omega\\ +\int_\Omega\left((\vec{\nabla}\times\vec{w})\times\vec{\nabla}\mu\right)u^*\,d\Omega\\ +\int_\Omega\left(\vec{\nabla}\times(2\,\underline{\dot{\in}}\cdot\vec{\nabla}\mu+\vec{S}_m)\right)u^*\,d\Omega\end{aligned} \tag{11}$$

where $\vec{\xi}$ is a point in the domain and $u^*=u^*\left(\vec{\xi},\vec{r}\right)=1/\left(4\pi\,|\,\vec{\xi}-\vec{r}\,|\right)$ is the fundamental solution for the diffusion operator. The last domain integral on the right hand side features a curl of body forces. This term may be reformulated using the following expression

$$\vec{\nabla}\times\left(\left(2\,\underline{\dot{\in}}\cdot\vec{\nabla}\mu+\vec{S}_m\right)u^*\right)=\left(\vec{\nabla}\times\left(2\,\underline{\dot{\in}}\cdot\vec{\nabla}\mu+\vec{S}_m\right)\right)u^*-\left(2\,\underline{\dot{\in}}\cdot\vec{\nabla}\mu+\vec{S}_m\right)\times\vec{\nabla}u^* \tag{12}$$

And the Gauss theorem

$$\int_\Omega\vec{\nabla}\times\left(\left(2\,\underline{\dot{\in}}\cdot\vec{\nabla}\mu+\vec{S}_m\right)u^*\right)d\Omega=-\int_\Gamma\left(2\,\underline{\dot{\in}}\cdot\vec{\nabla}\mu+\vec{S}_m\right)u^*\times\vec{n}d\Gamma \tag{13}$$

so that the calculation of the curl in the last term of equation (11) is avoided:

$$\begin{aligned}\mu c\left(\vec{\xi}\right)\vec{\omega}\left(\vec{\xi}\right)+\int_\Gamma \mu\vec{\omega}\vec{\nabla}u^*.\vec{n}d\Gamma=\int_\Gamma \mu\vec{q}u^*\,d\Gamma\\ -\int_\Omega\left(\left(\vec{\upsilon}\cdot\vec{\nabla}\right)\vec{\omega}-\left(\vec{\omega}\cdot\vec{\nabla}\right)\vec{\upsilon}\right)u^*\,d\Omega\\ +\int_\Omega\left(\left(\vec{\nabla}\times\vec{\omega}\right)\times\vec{\nabla}\mu\right)u^*\,d\Omega\\ -\int_\Gamma\left(2\,\underline{\dot{\in}}\cdot\vec{\nabla}\mu+\vec{S}_m\right)u^*\times\vec{n}d\Gamma\\ +\int_\Omega\left(2\,\underline{\dot{\in}}\cdot\vec{\nabla}\mu+\vec{S}_m\right)\times\vec{\nabla}u^*\,d\Omega\end{aligned} \tag{14}$$

The first domain integral on the right-hand side of equation (14) includes derivatives of the velocity and vorticity field functions. In order to avoid calculation of derivatives, we use algebraic relations to move the derivative from the unknown field function to the fundamental solution. Let us first write the first domain integral alone for j^{th} component of vorticity only:

$$\int_\Omega \left(\left(\vec{\upsilon} \cdot \vec{\nabla} \right) \omega_j - \left(\vec{\omega} \cdot \vec{\nabla} \right) \upsilon_j \right) u^* \, d\Omega \tag{15}$$

Due to the solenoidality of the velocity and vorticity fields, we may use $\left(\vec{w} \cdot \vec{\nabla} \right) \upsilon_j = \vec{\nabla} \cdot \left(\vec{\omega} \upsilon_j \right)$ and $\left(\vec{\upsilon} \cdot \vec{\nabla} \right) \omega_j = \vec{\nabla} \cdot \left(\vec{\upsilon} \omega_j \right)$ to transform equation (15) into

$$\int_\Omega \left(\vec{\nabla} \cdot \left(\vec{\upsilon} \omega_j - \vec{\omega} \upsilon_j \right) \right) u^* \, d\Omega \tag{16}$$

In order to move the derivative towards the fundamental solution, the following algebraic relation $\vec{\nabla} \cdot \left(u^* \left(\vec{\upsilon} \omega_j - \vec{\omega} \upsilon_j \right) \right) = u^* \vec{\nabla} \cdot \left(\vec{\upsilon} \omega_j - \vec{\omega} \upsilon_j \right) + \left(\vec{\upsilon} \omega_j - \vec{\omega} \upsilon_j \right) \cdot \vec{\nabla} u^*$ is used to obtain two integrals

$$\int_\Omega \vec{\nabla} \cdot \left(u^* \left(\vec{\upsilon} \omega_j - \vec{\omega} \upsilon_j \right) \right) d\Omega - \int_\Omega \left(\vec{\upsilon} \omega_j - \vec{\omega} \upsilon_j \right) \cdot \vec{\nabla} u^* \, d\Omega \tag{17}$$

The first integral may be converted to a boundary integral using a Gauss divergence theorem. Thus, the final form of the first domain integral on the right-hand side of equation (14) for j^{th} vorticity component without derivatives of field functions may be stated as:

$$\int_\Gamma \vec{n} \cdot \left(u^* \left(\vec{\upsilon} \omega_j - \vec{\omega} \upsilon_j \right) \right) d\Gamma - \int_\Omega \left(\vec{\upsilon} \omega_j - \vec{\omega} \upsilon_j \right) \cdot \vec{\nabla} u^* \, d\Omega \tag{18}$$

The final integral expression for *j*-th component of vorticity is

$$\begin{aligned} \mu c(\vec{\xi}) \omega_j(\vec{\xi}) + \int_\Gamma \mu \omega_j \vec{\nabla} u^* \cdot \vec{n} \, d\Gamma = & \int_\Gamma \mu q_j u^* \, d\Gamma \\ & - \int_\Gamma \vec{n} \cdot \left(u^* \left(\vec{\upsilon} \omega_j - \vec{\omega} \upsilon_j \right) \right) d\Gamma \\ & + \int_\Omega \left(\vec{\upsilon} \omega_j - \vec{\omega} \upsilon_j \right) \cdot \vec{\nabla} u^* \, d\Omega \\ & + \int_\Omega \left(\left(\vec{\nabla} \times \vec{\omega} \right) \times \vec{\nabla} u \right)_j u^* \, d\Omega \\ & - \int_\Gamma \left(\left(2 \underline{\underline{\dot{\epsilon}}} \cdot \vec{\nabla} \mu + \vec{S}_m \right) u^* \times \vec{n} \right)_j d\Gamma \\ & + \int_\Omega \left(\left(2 \underline{\underline{\dot{\epsilon}}} \cdot \vec{\nabla} \mu + \vec{S}_m \right) \times \vec{\nabla} u^* \right)_j d\Omega \end{aligned} \tag{19}$$

4 DISCRETIZATION

The governing equation (19) features, due to the inhomogeneous and non-linear nature of the underlying flow problem, boundary and domain integrals. Thus, we are required to discretize the boundary and the domain into boundary elements and domain cells. We choose quadrilateral boundary elements (Fig. 1) and hexahedral domain cells (Fig. 2) with quadratic interpolation for function (u) and linear interpolation for flux (q). Interpolation of function and flux on each boundary element is done using a local coordinate system (ξ, η). Node

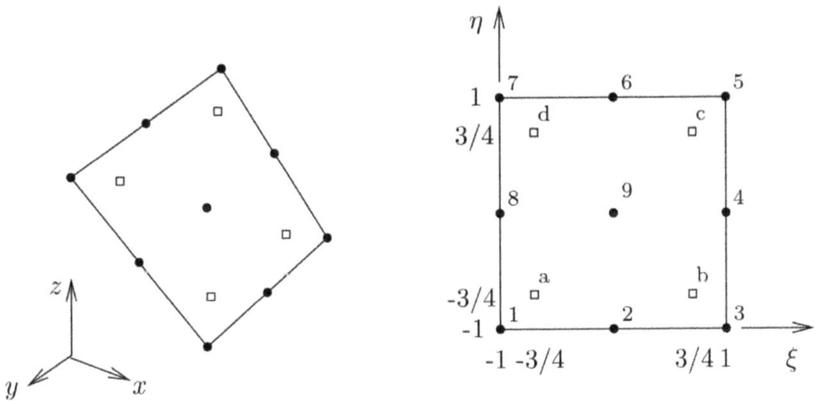

Figure 1: Linear interpolation for flux (squares) and quadratic interpolation for function (circles) over a surface. Left $\mathbb{R}^3$ space, right local coordinate system.

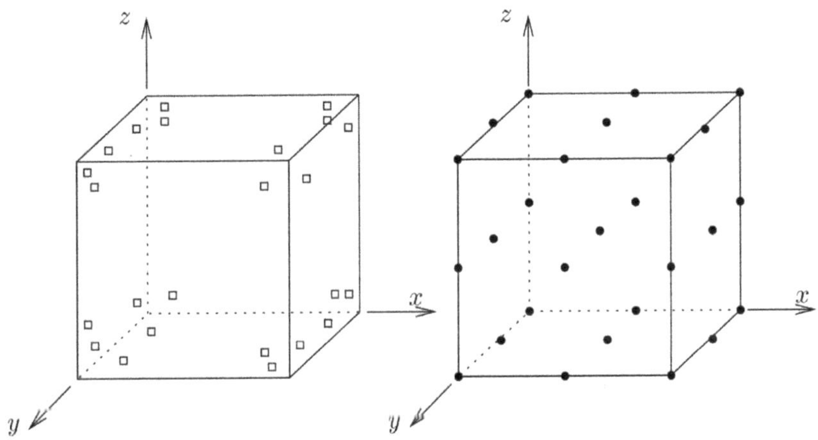

Figure 2: A computational domain consisting of one domain cell and six boundary elements. Distribution of nodes is shown: left flux, right function.

distribution is shown on Fig. 1 and the Lagrange φ_i and ϕ_i shape functions are used to achieve interpolation:

$$u(\xi,\eta)=\sum_{i=1}^{9}\varphi_i(\xi,\eta)u_i \quad q(\xi,\eta)=\sum_{i=1}^{4}\phi_i(\xi,\eta)q_i \tag{20}$$

Interpolation of function within domain cells is also quadratic using 27 nodes and the Lagrange Φ_i shape functions are used to achieve interpolation:

$$u(\xi,\eta,\zeta)=\sum_{i=1}^{27}\Phi_i(\xi,\eta,\zeta)u_i \tag{21}$$

The geometry of the boundary element is defined by 8 corner nodes, thus each surface is defined by 4 nodes (numbers 1,3,5,7 in Fig. 1). One may find the location of flux nodes (a,b,c,d) by the following transformation

$$\begin{vmatrix} x_a & y_a & z_a \\ x_b & y_b & z_b \\ x_c & y_c & z_c \\ x_d & y_d & z_d \end{vmatrix} = \frac{1}{64}\begin{vmatrix} 49 & 7 & 1 & 7 \\ 7 & 49 & 7 & 1 \\ 1 & 7 & 49 & 7 \\ 7 & 1 & 7 & 49 \end{vmatrix} \cdot \begin{vmatrix} x_1 & y_1 & z_1 \\ x_3 & y_3 & z_3 \\ x_5 & y_5 & z_5 \\ x_7 & y_7 & z_7 \end{vmatrix} \tag{22}$$

Based on this transformation shape functions for flux interpolation are

$$\begin{aligned} \phi_1 &= \frac{4}{9}\left(\xi - \frac{3}{4}\right)\left(\eta - \frac{3}{4}\right), \quad \phi_2 = -\frac{4}{9}\left(\xi + \frac{3}{4}\right)\left(\eta - \frac{3}{4}\right) \\ \phi_3 &= \frac{4}{9}\left(\xi + \frac{3}{4}\right)\left(\eta + \frac{3}{4}\right), \quad \phi_4 = -\frac{4}{9}\left(\xi - \frac{3}{4}\right)\left(\eta + \frac{3}{4}\right) \end{aligned} \tag{23}$$

Based on the proposed discretization scheme, in order to write the discrete version of equation (19) the following type of integrals have to be calculated

$$[H] = \int_\Gamma \varphi_i \vec{\nabla} u^* \cdot \vec{n} d\Gamma, \qquad [G] = \int_\Gamma \phi_i u^* d\Gamma, \tag{24}$$

$$[\vec{A}] = \int_\Gamma \varphi_i \vec{n} u^* \cdot d\Gamma, \qquad [\vec{D}] = \int_\Omega \Phi_i \vec{\nabla} u^* d\Omega, \qquad [B] = \int_\Omega \Phi_i u^* d\Omega \tag{25}$$

These integrals depend on the domain geometry, computational mesh and the choice of the source point location. Due to the non-local nature of the fundamental solution, all integrals are non-zero.

Calculation of the free coefficient $c(\vec{\xi})$ When a rigid body movement is applied, $u = 1$, $q = 0$, we see that the sum of all H matrix elements for one source point must be equal to 0, thus we may use this fact to calculate $c(\vec{\xi})$. If the source point is located on the surface, we know that c = 1/2, also if the source point is inside of the element then c = 1. Those two are used to check the accuracy of calculated integrals.

4.1 Setting up the system of equations

We use a collocation scheme to set up linear equations at unknown nodes in the computational mesh. Since domain integrals are present in the governing integral equation this leads to full matrices of integrals and unreasonably high computational demands. To avoid this, we employ a domain-decomposition strategy to set up a sparse system of linear equations.

The strategy is as follows. We consider each domain element an individual subdomain. Such a subdomain includes 26 function nodes on the surface (8 in the corners, 6 at the middle of the surfaces and 12 at the middle of edges) and 24 flux nodes (4 on each side). In order to set up a system of equations, the source point is set in all of those nodes (24+26=50). Additionally, the source point is set into a node in the centre of the domain element, where the function value may be obtained explicitly from known boundary values. Thus all in all, we have 51 equations for each element.

Since neighbouring elements share nodes and since boundary conditions on the outer boundaries of the domain are prescribed, we obtain an over-determined system of equations. An example of setting up the system of linear equations in this manner is shown in Fig. 3.

The system of linear equations obtain in this way is solved in a least squares manner using the Paige and Saunders [9] solver.

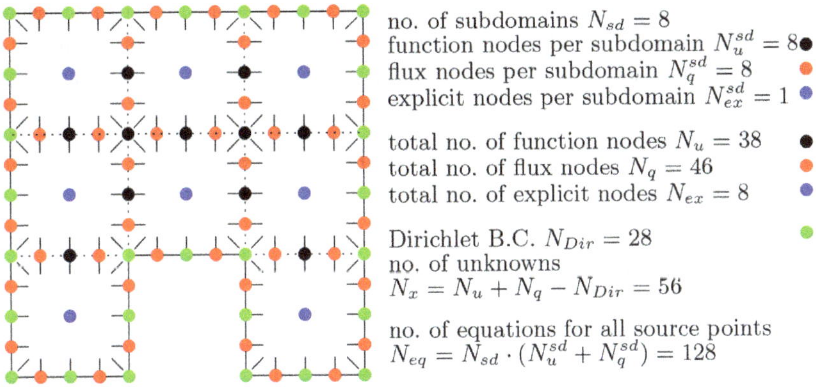

Figure 3: An example of building the system of linear equations in the proposed domain-decomposition scheme. A 2D cross-section of a domain is shown. Dirichlet boundary conditions are assumed. Each equation is shown with a short line originating from the source point location and pointing into a subdomain for each it is written.

The computational demands of the proposed scheme scale linearly with the number of nodes. In the worst case (corner node) each node is connected to eight subdomains and thus renders approximately 51·8 non-zero elements in the system matrix. If n is the number of all nodes, then the number of non-zero matrix elements is $51{\cdot}8n$. For large n this number of non-zero elements is significantly smaller that the n^2 storage requirements of the standard approach, which does not utilise domain decomposition.

5 VALIDATION

In order to validate the integral formulation of the vorticity transport equation, we solved a flow problem featuring a fluid with variable viscosity and momentum sources. We consider pressure drop driven developed flow between to large parallel horizontal plates. The vertical walls are assumed to be very far, so the flow is two-dimensional. The computational domain is shaped as a box with the flow in the x direction. At the inlet, a uniform flow field is prescribed ($\vec{\upsilon} = (1,0,0)$), while at the outlet a developed flow profile is prescribed ($d\upsilon_x / d_x = 0,\ \upsilon_y = \upsilon_z = 0$). The fluid slips freely at the vertical side walls and does not slip at the top and bottom walls. We consider three cases: a constant viscosity case, a case where viscosity is a linear function of position and a case where viscosity changes quadratically. In order for the flow to have a parabolic profile in all cases, we assume that the fluid is under the influence of momentum sources. Table 1 lists all cases.

Table 1: List of cases for testing the vorticity transport equation. The analytical solution of Navier-Stokes equation for all cases is $\vec{\upsilon} - = \left(6\left(z-z^2\right),0,0\right), \vec{\omega} = \left(0,6\left(1-2z\right),0\right)$

case	Viscosity μ	momentum sources $\vec{S}_m$
A	1	(0,0,0)
B	$1+z$	$(-6(4z-1),0,0)$
C	$1+z^2$	$(12(1-3z)z,0,0)$

In order to asses the influence of the variable viscosity on the flow field results, we measure the difference between the solution obtained in the constant viscosity case A and the cases with linear and quadratic dependence of viscosity. Relative RMS norm is used, defined as

$$\| v_x \| = \frac{\sum_i (v_{x,A} - v_{x,B})^2}{\sum_i v_{x,A}^2}, \quad \|\omega_y\| = \frac{\sum_i (\omega_{y,A} - \omega_{y,B})^2}{\sum_i \omega_{y,A}^2} \tag{26}$$

where i includes all nodes along the vertical outlet pro le. We consider several meshes with equidistantly placed nodes. The meshes have 5, 9, 17, 33 and 65 nodes in each direction. For a domain of size 1, the element sizes are between h= 0.2 and 0.0154.

Figure 4 displays the norms and thus compares the simulated velocity and vorticity fields for cases with and without variable viscosity. For the coarsest mesh, the difference is large, while for all others the differences are lower and at approximately the same level. Thus, we can conclude, that the proposed integral formulation and the numerical method are able to

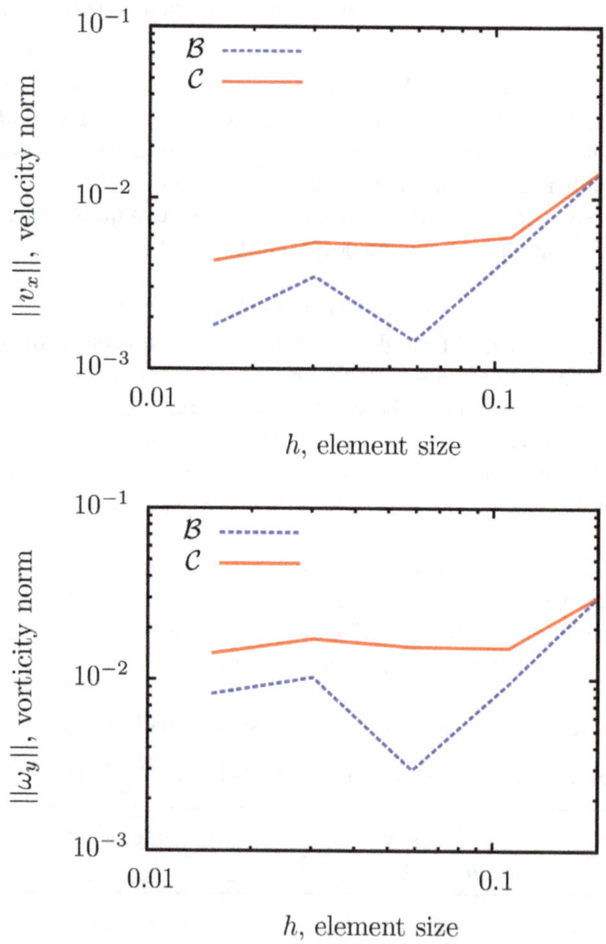

Figure 4: Velocity (top panel) and vorticity (bottom panel) norms expressing the difference between simulation of constant viscosity problem and variable viscosity problem.

capture the physical effect of variable viscosity field in fluid flows. The norms for vorticity are higher than the norms for velocity. This is due to the fact, that vorticity is, by definition, a derivative of velocity, thus lower accuracy is expected.

6 CONCLUSIONS

We developed a boundary-domain integral formulation of the vorticity transport equation with variable viscosity. The contribution of the variable viscosity is present in additional integral terms which feature the viscosity gradient. These terms vanish when the viscosity is constant and the formulation simplifies to the constant viscosity case. Decomposition of the viscosity into a constant and variable part, as was proposed by other authors, was not used.

In order to efficiently employ the boundary-domain integral method on large computational grids, we propose the use of domain decomposition, which leads to over-determined systems of linear equations.

The formulation proposed in this work will be used in the future as a part of a Navier-Stokes fluid flow and heat transfer solver.

REFERENCES

[1] AL-Jawary, M.A., Ravnik, J., Wrobel, L.C. & Skerget, L., Boundary element formulations for numerical solution of two-dimensional diffusion problems with variable coefficients. *Computers and Mathematics with Applications*, **64**(8), pp. 2695–2711, 2012. https://doi.org/10.1016/j.camwa.2012.08.002

[2] AL-Jawary, M.A. & Wrobel, L.C., Numerical solution of two-dimensional mixed problems with variable coefficients by the boundary-domain integral and integro-differential equation methods. *Engineering Analysis with Boundary Elements*, **35**, pp. 1279–1287, 2011.
https://doi.org/10.1016/j.enganabound.2011.06.004

[3] AL-Jawary, M.A. & Wrobel, L.C., Radial integration boundary integral and integro-differential equation methods for two-dimensional heat conduction problems with variable coefficients. *Engineering Analysis with Boundary Elements*, **36**(5), pp. 685–695, 2012. https://doi.org/10.1016/j.enganabound.2011.11.019

[4] Ang, W.T., Kusuma, J. & Clements, D.L., A boundary element method for a second order elliptic partial differential equation with variable coefficients. *Engineering Analysis with Boundary Elements*, **18**(4), pp. 311–316, 1996.
https://doi.org/10.1016/s0955-7997(97)83178-5

[5] Chkadua, O., Mikhailov, S.E. & Natroshvili, D., Analysis of direct boundary-domain integral equation for a mixed BVP with variable coefficient, I: equivalence and invertibility. *Journal of Integral Equations and Applications*, **21**, pp. 499–543, 2009.
https://doi.org/10.1216/jie-2009-21-4-499

[6] DeSilva, S.J., Chan, C.L., Chandra, A. & Lim, J., Boundary element method analysis for the transient conduction convection in 2-D with spatially variable convective velocity. *Applied Mathematical Modelling*, **22**(12), pp. 81–112, 1998.
https://doi.org/10.1016/s0307-904x(98)00010-9

[7] Driessen, B.J. & Dohner, J.L., A finite element - boundary element method for advection diffusion problems with variable advective fields and infinite domains. *International Journal of Heat and Mass Transfer*, **44**, pp. 2183–2191, 2001.
https://doi.org/10.1016/s0017-9310(00)00226-x

[8] Grzhibovskis, R., Mikhailov, S. & Rjasanow, S., Numerics of boundary-domain integral and integro-differential equations for BVP with variable coefficient in 3D. *Computational Mechanics*, **51**, pp. 495–503, 2013. https://doi.org/10.1007/s00466-012-0777-8

[9] Paige, C.C. & Saunders, M.A., LSQR: An algorithm for sparse linear equations and sparse least squares. *ACM Transactions on Mathematical Software*, **8**, pp. 43–71, 1982. https://doi.org/10.1145/355984.355989

[10] Partridge, P.W., Brebbia, C.A. & Wrobel, L.C., *The dual reciprocity boundary element method.* Computational Mechanics Publications Southampton, U.K.; Boston : Computational Mechanics Publications; London; New York, 1992.

[11] Rap, A., Elliott, L., Ingham, D.B., Lesnic, D. & Wen, X., DRBEM for Cauchy convection-diffusion problems with variable coefficients. *Engineering Analysis with Boundary Elements*, **28**(11), pp. 1321–1333, 2004. https://doi.org/10.1016/j.enganabound.2004.06.003

[12] Ravnik, J. & Skerget, L., Integral equation formulation of an unsteady diffusion–convection equation with variable coefficient and velocity. *Computers and Mathematics with Applications*, **66**, pp. 2477–2488, 2014. https://doi.org/10.1016/j.camwa.2013.09.021

[13] Ravnik, J., Skerget, L., Tibaut, J. & Yeigh, B.W., Solution of energy trans-port equation with variable material properties by BEM. *International Journal of Computational Methods and Experimental Measurements*, **5**(3), pp. 337–347, 2017. https://doi.org/10.2495/cmem-v5-n3-337-347

[14] Ravnik, J., Skerget, L. & Zunic, Z., Velocity-vorticity formulation for 3D natural convection in an inclined enclosure by BEM. *International Journal of Heat and Mass Transfer*, **51**(17–18), pp. 4517–4527, 2008. https://doi.org/10.1016/j.ijheatmasstransfer.2008.01.018

[15] Skerget, L. & Ravnik, J., BEM simulation of compressible fluid flow in an enclosure induced by thermoacoustic waves. *Engineering Analysis with Boundary Elements*, **33**(4), pp. 561–571, 2009. https://doi.org/10.1016/j.enganabound.2008.08.003

[16] Skerget, L. & Samec, N., BEM for the two-dimensional plane compress-ible fluid dynamics. *Engineering Analysis with Boundary Elements*, **29**, pp. 41–57, 2005. https://doi.org/10.1016/j.enganabound.2004.08.007

[17] Wrobel, L.C., *The Boundary Element Method.* John Willey & Sons, LTD, 2002.

[18] Wrobel, L.C. & DeFigueiredo, D.B., A dual reciprocity boundary element formulation for convection-diffusion problems with variable velocity fields. *Engineering Analysis with Boundary Elements*, **8**, pp. 312–319, 1991. https://doi.org/10.1016/0955-7997(91)90044-t

ACCELERATION OF BEM WITH THE CROSS APPROXIMATION FOR DETERMINATION OF BOUNDARY VORTICITY

JAN TIBAUT[1], LEOPOLD ŠKERGET[2] & JURE RAVNIK[1]
[1]Faculty of Mechanical engineering, University of Maribor, Slovenia.
[2]Wessex Institute, Ashurst Lodge, Southampton.

ABSTRACT

In this paper, we present a fast boundary element method (BEM) algorithm for the solution of the velocity-vorticity formulation of the Navier-Stokes equations. The Navier-Stokes equations govern incompressible fluid flow, which is inherently nonlinear and when discretizised by BEM requires the discretization of the domain and calculation of domain integrals. The computational demands of such method scale with $O(N^2)$, where N is the number of boundary nodes. To accelerate the solution process and reduce the computational demand, we present two different approaches, the subdomain method and an approximation procedure with hierarchical structure. Several approximation techniques exist, such as multipole approximation methods FMM (fast multiple method), SVD (singular value decomposition method), wavelet transform method and a cross approximation method. In this paper, we present the cross approximation method in combination with the hierarchical H-structure. The cross approximation method can reduce the computational demands from $O(N^2)$ to $O(N \log N)$. There are many forms of the cross approximation, like the algebraic cross approximation and the hybrid cross approximation. Here, we applied the algebraic cross approximation form. The main advantage is that we did not need to evaluate the integral and then to change it with a degenerate kernel function. The cross approximation algorithm was used to solve the kinematics equation for unknown boundary vorticity values. Results show that an increasing of the compression rate has a negative influence on the solution accuracy. On the other hand, the solution accuracy increases with computational grid density. Tests were performed using the 3D lid-driven cavity test case with Reynolds numbers up to 1000. Solution accuracy was similar for all Reynolds numbers considered. In conclusion, the tests showed that our implementation of the algebraic cross approximation for the acceleration of the solution of the kinematics equation can be applied to decrease the computational demands and to accelerate the BEM.

Keywords: adaptive cross approximation, boundary element method, boundary-domain integral method, fast algorithms, hierarchical structure, H-matrix, kinematics equation, lid-driven cavity, velocity-vorticity formulation

1 INTRODUCTION

The boundary element method (BEM) is a numerical method for solving partial differential equations (PDE). The numerical method is based on the Greens second theorem. Thus, only the discretization of the boundary is necessary [1]. However, due to the non-local nature of the fundamental solution, the resulting system of linear equations is fully populated. Due to this, the computational demands of the method scale with the square of the number of boundary unknowns $O(N^2)$.

To reduce computational demand, a number of accelerating procedures were proposed by diverse authors, which aim to reduce the cost from $O(N^2)$ to $O(N \log N)$ or even to $O(N)$. We present the cross approximation algorithm in combination with a hierarchical structure for the acceleration of the solution process. The hierarchical structure or H-structure is a procedure, which divides the full matrix formulation into smaller matrices, based on the shape of the numerical domain. A geometrical condition filters the matrix parts on which an approximation algorithm is applied and the matrix parts which are not admissible. Börm *et al.* [2] presented the application of hierarchical matrices and the bounding box and bisection method

DOI: 10.2495/CMEM-V6-N6-1097-1107

for building cluster trees. Hackbusch [3] wrote the arithmetic based on H-matrices. A different hierarchical structure is an H^2-matrix formulation. Börm [4] used the H^2-matrix formulation to construct efficient approximations of discretized integral formulation.

There are different authors who present different forms and variants of the cross approximation method. Bebendorf [5] introduced the fully pivoting adaptive cross approximation algorithm (ACA). Rjasanow and Steibach [6] employed the partly pivoting adaptive cross approximation algorithm (ACA+). Tamayo [7] presented a special algebraic form named the multilevel adaptive cross approximation (MLACA). The MLACA is a procedure that divides the matrix into a number of levels and applies the ACA algorithm on each level. Bebendorf [8] used the cross approximation algorithm on a collocation BEM and Rjasanov [9] applied it for Galerkin BEM. Börm and Grasedyck [10] introduced the hybrid cross approximation of integral operators, where they used the H-matrix representation of finite element stiffens matrix.

Many authors used the cross approximation method in different applications. In all publications, the approximation procedure was introduced to accelerate the BEM. The most often used form of the cross approximation is the adaptive cross approximation. Schröder *et al.* [11] used the adaptive cross approximation algorithm to solve the electromagnetic field for a known current distribution. Kurz and Rjasanow [12] used the ACA algorithm to accelerate a coupled BEM and finite element numerical method to solve the distribution of a symmetric electromagnetic field. Grytsenko and Galybin [13] used the ACA to solve a large number of cracks on a plate with the, singular integral method. Maerten [14] used the adaptive cross approximation to solve a 3D elasticity problem. Wei *et al.* [15] solved the Laplace equation to predict the temperature distribution in a 2D domain by combining ACA with the singular boundary method.

The purpose of this work is to implement the cross approximation algorithm with the H-structure to speed up a BEM-based fluid flow solver. The fluid flow solver [16, 17] has been developed for the velocity-vorticity formulation of the Navier-Stokes equations. Our main motivation was to decrease the memory storage demands. We tested the application and assessed its usefulness using the lid-driven cavity benchmark test case [18].

2 GOVERNING EQUATIONS

The fluid flow is defined by two equations, the kinematic and the transport vorticity equation. The two equations are formulated in the velocity-vorticity formulation of the Navier-Stokes equations. Let $\vec{v}$ denote the velocity field and $\vec{\omega} = \vec{\nabla} \times \vec{v}$ the vorticity field. With this consideration the kinematics equation may be specified from the continuity equation as [16]:

$$\nabla^2 \vec{v} + \vec{\nabla} \times \vec{\omega} = 0. \tag{1}$$

For a non-compressible fluid, the velocity field and the vorticity field are divergence free. Eqn (1) represents a connection between the velocity and vorticity vector field. The fluid movement governs the vorticity transport equation, written in a non-dimensional form, from the momentum equation as:

$$\frac{\partial \vec{\omega}}{\partial t} + \left(\vec{v} \cdot \vec{\nabla}\right)\vec{\omega} = \left(\vec{\omega} \cdot \vec{\nabla}\right)\vec{v} + \frac{1}{Re}\nabla^2 \vec{\omega}, \tag{2}$$

where $Re = \frac{vl}{\nu}$ is the Reynolds number and l the characteristic length scale. The fluid viscosity ν and fluid density ρ are considered constant.

The system of equations that is formed with the discretization process is solved with an iterative algorithm proposed by Ravnik *et al.* [17]. In this paper, we present an accelerated version of the estimation of boundary vorticity. In the second step, the vorticity momentum equation is solved with the sub-domain BEM to solve the domain velocity. Domain decomposition cannot be used for the solution of the kinematics equation due to the Biot-Savart law. Because of this, we used the cross approximation technique in combination with the H-structure to solve the boundary vorticity in eqn (1).

2.1 The kinematics equation for the boundary vorticity

The kinematic equation solves the vorticity field of the fluid flow, which is on the edge of the domain. Let us consider a domain Ω with a position vector, $\vec{r} \in \mathbb{R}^3$ and with a boundary $\Gamma = \delta\Omega$. The integral form of the kinematics eqn (1) without derivatives of the velocity and vorticity fields takes the following form [17]:

$$c(\vec{\xi})\vec{v}(\vec{\xi}) + \int \vec{v}\vec{\nabla}u^* d = \int \vec{v} \times (\vec{n} \times \vec{\nabla}) u^* d + \int (\vec{\omega} \times \vec{\nabla} u^*) d', \tag{3}$$

where $u^*(\vec{r}, \vec{\xi}) = \dfrac{1}{4\pi|\vec{r} - \vec{\xi}|}$ is the fundamental solution of the Laplace equation and $\vec{\xi}$ is the source point. In order to use the kinematics equation to solve the boundary vorticity values, we must rewrite eqn (3) in a tangential form by multiplying the system with a normal in the source point $\vec{n}(\vec{\xi})$. This yields the following integral equation,

$$c(\vec{\xi})\vec{n}(\vec{\xi}) \times \vec{v}(\vec{\xi}) + \vec{n}(\vec{\xi}) \times \int_\Gamma \vec{v}\vec{\nabla}u^* \cdot \vec{n} d\Gamma$$

$$= \vec{n}(\vec{\xi}) \times \int_\Gamma \vec{v} \times (\vec{n} \times \vec{\nabla}) u^* d\Gamma + \vec{n}(\xi) \times \int_\Omega (\vec{\omega} \times \vec{\nabla} u^*) d\Omega. \tag{4}$$

The discretization process, divides the domain into domain elements $\Omega = \sum_{i=1}^{d} \Omega_i$ and the boundary into boundary elements $\Gamma = \sum_{i=1}^{b} \Gamma_i$, where d is the number of domain elements and b the number of boundary elements. With this, we can write the discrete form of the kinematic equation. Each domain part Ω_i is defined with 27 nodes ϕ_i and each boundary part Γ_i is defined with 9 nodes Φ_i. A quadratic interpolation function was used for the boundary and domain nodes.

For each collocation node at the boundary $\vec{\xi}_f$, we calculate the following integrals over boundary element c:

$$h_{fcl} = \delta_{fcl} c(\xi_f) + \sum_{l=1}^{9} \int f n_i \frac{\partial u^*}{\partial x_i} d_c, \tag{5}$$

$$h_{fcl}^{ij} = \sum_{l=1}^{9} \int_c f \left[n_i \frac{\partial u^*}{\partial x_j} - n_j \frac{\partial u^*}{\partial x_i} \right] d_c. \tag{6}$$

and the following integrals over domain elements e

$$d_{fel}^{i} = \sum_{l=1}^{27} \int_{\Omega_e} \phi_l \frac{\partial u^*}{\partial x_i} d\Omega_e . \tag{7}$$

The indexes i, j in eqns (5)–(7) stand for the coordinates x, y, z. For the integral in eqn (6), $i \neq j$. The character δ_{fcl} is the Kronecker delta, which is equal to 1 when source point and the grid point are equal.

In order to separate the unknown boundary vorticity values, the vorticity vector can be written as a sum of boundary and domain vorticity as $\{\omega_i\} = \{\omega_i\}_\Gamma + \{\omega_i\}_{\Omega\setminus\Gamma}$. Vector $\{\omega_i\}_\Gamma$ includes the boundary vorticity and vector $\{\omega_i\}_{\Omega\setminus\Gamma}$ includes the vorticity at nodes in the domain. Hence, we can write the system of equations like this:

$$\begin{aligned}
&\left([n_i][D_i]+[n_j][D_j]+[n_k][D_k]\right)\{\omega_i\}_\Gamma \\
&= \left([n_j][H_{ki}]-[n_k][H_{ij}]\right)\{v_i\}+\left([n_j][H]-[n_k][H_{jk}]\right)\{v_k\} \\
&-\left([n_i][H]+[n_j][H_{jk}]\right)\{v_j\}+[n_k][D_x]\{\omega_k\}_\Gamma+[n_j][D_i]\{\omega_j\}_\Gamma \\
&+[n_i][D_i]\{\omega_i\}_\Gamma-\left([n_j][D_j]_{\Omega\setminus\Gamma}+[n_k][D_k]_{\Omega\setminus\Gamma}\right)\{\omega_i\}_{\Omega\setminus\Gamma} \\
&+[n_j][D_i]_{\Omega\setminus\Gamma}\{\omega_j\}_{\Omega\setminus\Gamma}+[n_k][D_i]_{\Omega\setminus\Gamma}\{\omega_k\}_{\Omega\setminus\Gamma}
\end{aligned} \tag{8}$$

Indexes i, j, k in eqn (8) are the coordinates i=x, j=y, k=z. From eqn (8), we write a system of three linear equations. In the first equation, the combination of indexes is i, j, k in the second the combination indexes is j, k, i and in the last one the combination is k, i, j.

Matrices $[n_x]$, $[n_y]$, $[n_z]$ are diagonal matrices that contain the components of the normal vector $\vec{n} = \{n_x, n_y, n_z\}$. The size of matrices $[H]$ and $[H_{ij}]$ is $n \times n$, where n is the number grid nodes on the boundary. The first domain matrices $[D_i]$ have the size $n \times n$, while the second $[D_i]_{\Omega\setminus\Gamma}$ are of the size $n \times m$, where i is the coordinate x, y, z and m is the number of domain nodes.

As the size of $[D_i]_{\Omega/\Gamma}$ domain matrices is one order of magnitude greater than all other matrices, we implemented the cross approximation algorithm for these matrices.

3 APPROXIMATION OF THE DOMAIN MATRIX

Let us consider one of the domain integral matrices $[D_i]_{\Omega\setminus\Gamma}$ and denote it as D. Its size is $n \times m$ and its entries are real numbers $D \in \mathbb{R}^{n \times m}$. To approximate the domain matrix D first, an H-matrix structure has to be built. The H-matrix structure is built from the block cluster tree. The block cluster tree is a combination of two cluster trees. The first cluster tree is built from the domain elements and the second one is built from the boundary elements. From the two cluster trees, a block cluster tree is formed. Each cluster of the block cluster tree is tested on a geometrical condition, which is based on the shape of the domain. Considering the geometrical condition, we can split the matrix $D_{n \times m}$ into smaller matrices $\hat{D}\,|_{\hat{n} \times \hat{m}}$. Matrix parts that fulfil the geometrical condition can be approximated.

3.1 Construction of the block cluster tree

A block cluster tree is a combination of the domain and boundary clusters. The computational domain is split into smaller boxes, which we will denote as clusters. In each level of the domain cluster tree, the number of clusters increases and the size of the boxes decreases. Clusters that are at the end of the cluster tree are called leaves. The cluster tree that is built from the computational domain is denoted as T_I. The total number of levels p defines the depth of a cluster tree. The following algorithm below was used to build the domain cluster tree T_I:

- for all levels i=1 to p
- for all clusters $I_k^{(i)}$ at the level, i
- find $I_k^{(i)}$'s eight neighbouring clusters and join them into a new cluster at level $i+1$

For each domain cluster in T_I cluster tree, we define a bounding box $Q_{dm} \subseteq \mathrm{IR}^3$, which includes all domain nodes in the cluster as proposed by Börm [2]. Similarly, for each boundary cluster in the T_J cluster tree, we define a bounding box $Q_{bn} \subseteq \mathbb{R}^3$ from the boundary nodes in the cluster. In Fig. 1, we present a block cluster of the block cluster tree.

3.2 H structure formulation

From the block cluster tree, we build an H-matrix, by applying a geometrical condition. The block cluster tree includes several parts. We decide which cluster is admissible by defining a geometrical condition. The geometrical condition is:

$$min\left\{dim\left(Q_{dm}\right), dim\left(Q_{bn}\right)\right\} \leq \eta \, dist\left(Q_{dm}, Q_{bn}\right). \tag{9}$$

Expressions $dim\left(Q_{dm}\right) = \left|\vec{r}_A - \vec{r}_B\right|$ and $dim\left(Q_{bn}\right) = \left|\vec{r}_C - \vec{r}_D\right|$ are the diameters of boundary and domain bounding boxes. The distance $dist\left(Q_{dm}, Q_{bn}\right)$ is the minimal distance between nodes in the boundary and domain clusters. All expressions are illustrated in Fig. 1. Parameter η is the admissibility parameter, which is defined by the user. This parameter enables the user to vary the geometrical condition and to control the number of admissible block clusters.

Block clusters which fulfil the geometrical condition, are admissible. Thus, an approximation procedure can be applied. However, block clusters, which do not meet the geometrical condition, are split into smaller blocks.

Leaves, which meet the geometrical condition in eqn (10), are admissible leaves. Leaves that do not fulfil the geometrical condition are not admissible and they have to be maintained

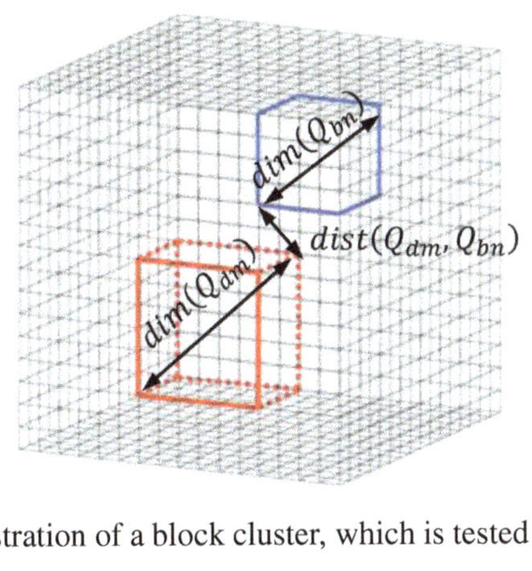

Figure 1: An illustration of a block cluster, which is tested for admissibility.

in their original form. In our simulations, we used $\eta = 1$ based on the recommendation given by Wei *et al.* [15].

3.3 Cross approximation

Cross approximation is a method that approximates matrices $\hat{D}_{\hat{n}\times\hat{m}}$, from the full matrix formulation to an RK-form. Every matrix that has a rank k>1, can be written into the RK-matrix formulation using the approximation rank r in the following way [2]:

$$\tilde{D}_{r(\hat{n}+\hat{m})} \approx \hat{A}\hat{B}^T, \hat{A} \in \mathbb{R}^{\hat{n}\times r}, \hat{B} \in \mathbb{R}^{r\times\hat{m}}. \tag{10}$$

To store the *RK*-matrix into memory and perform matrix-vector multiplications, we need $O(\hat{n}\hat{m})$ computational resources [10]. The approximation rank r is the number of rows in matrix A and the number of columns in matrix B^T. In this paper, we present the full pivoting cross approximation method. Full pivoting means that the algorithm evaluates each element in the matrix and picks the highest valued element, to determine a cross, which is the building block of the RK-matrix formulation. The cross approximation algorithm with full pivoting is written bellow:

1. $\hat{R}_0 = \hat{D}_{\hat{n}\times\hat{m}}$
2. $The\ loop\ u = 0,1,2,3,\ldots,r$

   $(2.1)\left(i^*,j^*\right)^u = ArgMax\left|\hat{R}^u\right|$

   $(2.2)\gamma^{u+1} = \left(\hat{R}^u_{i^*,j^*}\right)^{-1}$

   $(2.3)a^{u+1} = \gamma^{u+1}\hat{R}^u_{i,j^*}, b^{u+1} = \left(\hat{R}^u_{i^*,j}\right)^T$

   $(2.4)\hat{R}^{u+1} = \hat{R}^u - a^{u+1}b^{u+1}$

In point (1) of the algorithm, we define the residual matrix. The second point (2) is a start of a loop that performs the approximation. In eqn (2.1), the largest element in the matrix $\hat{R}^u$ is found, with the full pivoting procedure. The full pivoting algorithm needs to perform $O(\hat{n}\hat{m})$ pivoting steps. To reduce the number of pivoting steps Rjasanow and Steibach [6] proposed a partly pivoting cross approximation method (ACA+). However, the partly pivoting algorithm demands more computational resources. Because of this, the algorithm is less desirable. Next in eqn (2.2), the reverse of the largest element $\hat{R}^u_{i^*,j^*}$ is calculated. In eqn (2.3) vectors a^{u+1} and b^{u+1} are defined. These two vectors build a cross around the element $\hat{R}^u_{i^*,j^*}$. Eqn (2.4) calculates a new residual matrix $\hat{R}^{u+1}$.

We propose to use a user defined compression factor α, which defines the approximation rank r in the following way:

$$r = \alpha \cdot k\ , \tag{11}$$

where k is the rank of the matrix $\hat{D}_{r(\hat{n}+\hat{m})}$. Based on r, we can calculate the compression ratio φ

$$\varphi = \frac{\sum_i \hat{n}_i \cdot \hat{m}_i + \sum_j r_j\left(\hat{n}_j + \hat{m}_j\right)}{n \cdot m} \tag{12}$$

which, is the memory ratio of the approximated matrix versus the original matrix. The counter is the sum of the matrix elements in admissible block clusters and leaves $\tilde{D}_{r(\hat{n}+\hat{m})}$ plus the elements in inadmissible matrix parts $\hat{D}_{\hat{n}\times\hat{m}}$. And the denominator is the number of elements in the original matrix $D_{n\times m}$.

4 NUMERICAL MODEL AND EXPERIMENTS

Flow in a 3D lid-driven cavity is one of the standard test cases used in the development of flow solvers. The domain as well as the boundary conditions are unambiguously defined and do not change with the Reynolds number.

We tested the proposed algorithm using the lid-driven cavity problem [18]. The domain was a unit cube. We sought steady state flow solutions using three different grid densities, having 25^3, 41^3 and 49^3 nodes. The grid was structured and non-uniform. The size of hexahedral elements changed from the boundary to the middle elements by a length ratio 1:6.5.

The side and bottom walls of the cube are no-slip walls. The top wall is moving with a velocity, which is determined from the Reynolds number. The boundary vorticity is obtained by the solution of the accelerated kinematics equation, except for the $\omega_x = 0$ on the left and right wall, $\omega_y = 0$ on the front and back wall and $\omega_z = 0$ on the top and bottom walls. The flow was solved for three Reynolds numbers Re=100, 400 and 1000. Boundary conditions are shown in Fig. 2. The linear system of equations was solved with the LU-decomposition method. Iterations were stopped when the difference between subsequent iteration of all field functions dropped below 10^{-6}. The maximum number of iterations for each run was 8000.

The cross approximation was used on the matrices $[D_x]_{\Omega/\Gamma}$, $[D_y]_{\Omega/\Gamma}$, $[D_z]_{\Omega/\Gamma}$ in eqn (8). In order to measure the influence of the approximation on the result, find its dependence on the flow structure, and grid density, we used two norms. We measured the difference between boundary vorticity values as:

$$RMS_\omega = \left(\sum_{j=x,y,z} \frac{\sum_i \left(\omega_{ji} - \omega_{Aji}\right)^2}{\sum_i \left(\omega_{Aji}\right)^2} \right)^{\frac{1}{2}} \tag{13}$$

where ω_{ji} is the jth component of vorticity in i^{th} boundary node calculated without an approximation and the ω_{Aji} its approximated counterpart.

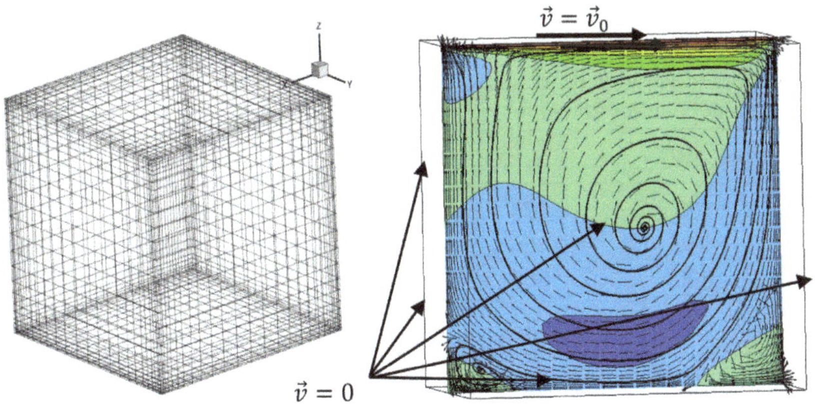

Figure 2: Flow structure and boundary conditions for the lid driven cavity at Re = 400 on the right and a 25^3 grid on the left.

5 RESULTS AND DISCUSSION

In this section, we illustrate the impact of the cross approximation method on the solution of the boundary vorticity. The Reynolds number was changed in an interval from 100 to 1000 to see how a different compression rate a effect the norm RMS_ω. All simulations were performed on an Intel Xenon 64-bit processor with 64 GB of memory space. Table 1 is showing the amount of memory that is needed to store one of the matrices $[D_i]_{\Omega/\Gamma}$. We observed that the space to store the original matrix formulation increases quadratically with the number of boundary unknowns, while the approximated formulation exhibits a quasilinear dependence.

In Table 2, we observed the number of iterations ($Niter_a$) that was necessary to obtain a solution of the kinematics equation. We noticed that with an increasing grid density and Reynolds number the number of iterations has increased. This is because the non-linearity of the problem increases with a higher Reynolds number and a better-resolved grid. However, the compression rate did not have an impact on the number of iterations, if a was between 1.0 and 0.5. When the compression rate was lowered below 0.5, the number of iterations started to increase. Thus, the compression rate has an influence on the number of iterations to solve the boundary vorticity.

In Fig. 3, we illustrate how the norm RMS_ω is dependent of the compression ratio φ and different Reynolds numbers. The RMS increases with a higher compression ratio. At a given Reynolds number, we observe that the norms are lower for a fine grid compared to course grid. This means that the simulation with a fine grid is able to use less memory storage to reach the same accuracy of results compared to a simulation on a coarse grid. Thus, the difference in the boundary vorticity is dependent of the grid density.

Using the norm, we focused on the impact of the cross approximation algorithm on the solution of the boundary vorticity values. The approximation of the boundary values effects

Table 1: Memory demands for storing one of the matrices [Di]Ω/Γ

	$ORG_{a=1.0}$	$RAM_{a=0.5}$[MB]	$RAM_{a=0.2}$[MB]	$RAM_{a=0.1}$[MB]
25^3	320	239	122	82
41^3	4345	3083	1833	1424
49^3	8828	5859	3171	2294

Table 2: Number of iteration to solve the kinematics equation with the proposed algorithm at different compression ratios a.

Re=100	$ORG_{a=1.0}$	$Niter_{a=0.8}$	$Niter_{a=0.5}$	$Niter_{a=0.2}$	$Niter_{a=0.1}$
25^3	728	728	728	728	821
41^3	766	766	765	769	841
49^3	800	800	799	830	939
Re=1000	$ORG_{a=1.0}$	$Niter_{a=0.8}$	$Niter_{a=0.5}$	$Niter_{a=0.2}$	$Niter_{a=0.1}$
25^3	2000	2184	2184	1896	8000
41^3	2401	2401	2401	2204	2331
49^3	2561	2561	2561	2679	3748

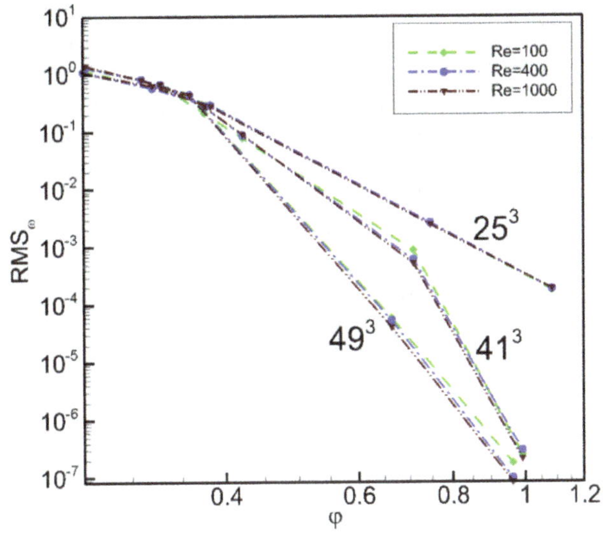

Figure 3: Influence of the Reynolds number and the grid density on the accuracy of the solution of the boundary vorticity values RMS$_{\omega}$.

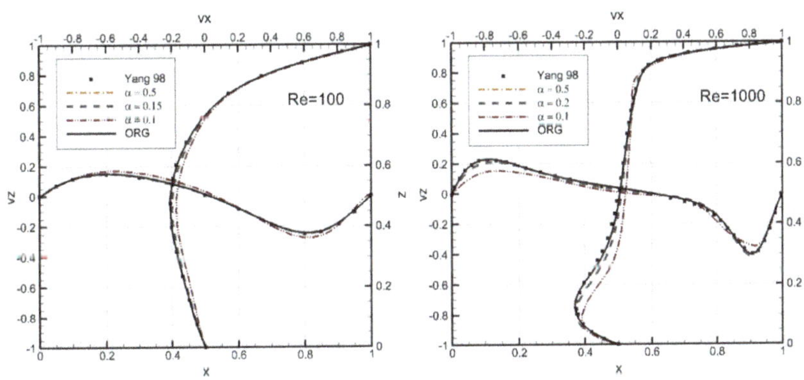

Figure 4: Velocity profiles vx(z) and vz(x) at the y=0.5 plane, for different compression ratio and Reynolds number. The grid density was 49^3.

the hole velocity field. In Fig. 4, we show the velocity profile through the centre of the cavity. We compared the benchmark results of Yang [18] with our method. We observed very good agreement between the benchmark result and our results obtained without the approximation method. When the approximation method is used, the difference in the profile increases. We also observed that the difference rises with a higher Reynolds number, and declines with a finer grid density.

6 CONCLUSION

In this study, we developed an accelerated algorithm based on BEM, for solving the kinematics equation. The developed algorithm used the H-matrix structure and the cross approximation method to decrease the computational demand of the BEM-based solver for incompressible fluid flow. We have shown that the cross approximation algorithm can be successfully applied to solve the kinematics equation. The implementation was tested using the 3D lid-driven cavity test case with the Reynolds number up to 1000 using several computational grids.

Results showed that the chosen solution accuracy of a selected grid density was dependent on the compression rate. However, also the number of iterations that were necessary to obtain a solution increased with a lower compression rate. A finer grid allows more compression. We plan to control the inaccuracy introduced into the boundary vorticity by the approximation by specifying the rank of the rank of the approximated matrix parts.

However, the elliptic integral kernel, which is in the integral form of the kinematics equation, is not very well suited for approximation. We can conclude that the usage of the cross approximation enables the usage of the flow solver on a finer grid at a reduced computational cost and thus allows the simulation of more complex flow structures.

ACKNOWLEDGEMENTS

The authors acknowledge the financial support from the Slovenian Research Agency (research core funding No. P2-0196).

REFERENCES

[1] Wrobel, L.C., *The boundary element method*, John Willey and Sons, 2002.

[2] Börm, S., Grasedyck, L. & Hackbusch, W., Introduction to hierarchical matrices with applications. *Engineering Analysis with Boundary Elements*, **27**(5), pp. 405–422, 2003. https://doi.org/10.1016/s0955-7997(02)00152-2

[3] Hackbusch, W., A sparse matrix arithmetic based on H -matrices. *Computing*, **108**, pp. 89–108, 1999. https://doi.org/10.1007/s006070050015

[4] Börm, S., Approximation of integral operators by H2-matrices with adaptive bases. *Computing*, **74**(3), pp. 249–271, 2005. https://doi.org/10.1007/s00607-004-0106-y

[5] Bebendorf, M., Approximation of boundary element matrices. *Numerische Mathematik*, **86**(4), pp. 565–589, 2000. https://doi.org/10.1007/pl00005410

[6] Rjasanow, S. & Steinbach, O., *The fast solution of boundary integral equations*, Springer Science+Business Media, 2007.

[7] Tamayo, J.M., Heldring, A. & Rius, J.M., Multilevel adaptive cross approximation (MLACA), *IEEE Antennas and Propagation Society International Symposium Antennas and Propagation,* pp. 2008–2011, 2009.

[8] Bebendorf, M. & Rjasanow, S., Adaptive low-rank approximation of collocation matrices. *Computing*, **70**(1), pp. 1–24, 2003. https://doi.org/10.1007/s00607-002-1469-6

[9] Rjasanow, S., Adaptive cross approximation of dense matrices, *IABEM 2002 Symposium*, pp. 1–12, 2002.

[10] Börm, S. & Grasedyck, L., Hybrid cross approximation of integral operators, *Numerische Mathematik*, **101**(2), pp. 221–249, 2005. https://doi.org/10.1007/s00211-005-0618-1

[11] Schröder, A., Brüns, H.-D. & Schuster, C., Fast evaluation of electromagnetic fields using a parallelized adaptive cross approximation. *Antennas and Propagation, IEEE Transactions on*, **62**(5), pp. 2818–2822, 2014. https://doi.org/10.1109/tap.2014.2303819

[12] Kurz, S., Rain, O. & Rjasanow, S., Application of the adaptive cross approximation technique for the coupled BE-FE solution of symmetric electromagnetic problems. *Computational Mechanics*, **32**(4), pp. 423–429, 2003.
https://doi.org/10.1007/s00466-003-0511-7

[13] Grytsenko, T. & Galybin, A.N., Numerical analysis of multi-crack large-scale plane problems with adaptive cross approximation and hierarchical matrices. *Engineering Analysis with Boundary Elements*, **34**(5), pp. 501–510, 2010.
https://doi.org/10.1016/j.enganabound.2009.12.001

[14] Maerten, F., Adaptive cross-approximation applied to the solution of system of equations and post-processing for 3D elastostatic problems using the boundary element method. *Engineering Analysis with Boundary Elements*, **34**(5), pp. 483–491, 2010.
https://doi.org/10.1016/j.enganabound.2009.10.016

[15] Wei, X., Chen, B., Chen, S. & Yin, S., An ACA-SBM for some 2D steady-state heat conduction problems. *Engineering Analysis with Boundary Elements*, **71**, pp. 101–111, 2016.
https://doi.org/10.1016/j.enganabound.2016.07.012

[16] Škerget, L., Hriberšek, M. & Žunic, Z., Natural convection flows in complex cavities by BEM. *International Journal of Numerical Methods for Heat & Fluid Flow*, 13(6), pp. 720–735, 2003.
https://doi.org/10.1108/09615530310498394

[17] Ravnik, J., Škerget, L. & Zunič, Z., Velocity-vorticity formulation for 3D natural convection in an inclined enclosure by BEM. *International Journal of Heat and Mass Transfer*, **51**(17–18), pp. 4517–4527, 2008.
https://doi.org/10.1016/j.ijheatmasstransfer.2008.01.018

[18] Yang, J., Yang, S., Chen, Y. & Hsu, C., Implicit weighted ENO schemes for the three-dimensional incompressible Navier – Stokes equations. *Journal of Computational Physics*, **487**, pp. 464–487, 1998.
https://doi.org/10.1006/jcph.1998.6062

MFS ANALYSIS OF THE VIBRATION FILTERING EFFECT OF PERIODIC STRUCTURES IN ELASTIC MEDIA

LUÍS GODINHO[1], PAULO AMADO-MENDES[1], PEDRO ALVES-COSTA[2] & CARLOS ALBINO[1]
[1]ISISE, Department Civil Engineer, University of Coimbra, Rua Luis Reis Santos, Coimbra, Portugal.
[2]Construct-FEUP, Department Civil Engineer, Faculty of Engineering of the University of Porto, R. Dr. Roberto Frias, Porto, Portugal.

ABSTRACT

Phononic crystals have been extensively studied, and their capacity to attenuate the propagation of sound waves at specific frequency bands is well known and documented in the literature. However, few studies exist concerning the behaviour of such structures in the context of elastic media, with the purpose of attenuating the transmission of vibrations. Applying this concept can be quite interesting, and may allow new vibration control devices to be developed, tailored at specific applications. As an example, buried periodic structures may be used to control elastic wave propagation in the ground, and thus to help reducing the vibrations that can reach sensible structures.

In this work, the authors make use of a 2.5D numerical model based on the Method of Fundamental Solutions (MFS) to analyse this complex problem, considering the case of arrays of elastic inclusions buried in a homogeneous medium, fully considering the complete elastodynamic interaction between the inclusions and the host medium. Due to the geometric periodicity of the analysed problem, the numerical formulation can be simplified, particularly in what concerns the calculation of the system matrix, and significant computational gains can be obtained. The results of a numerical study concerning the behaviour of a sequence of embedded inclusions within an elastic material, when subject to the incidence of waves with different frequencies, is here presented, and the interpretation of the involved phenomena is described in order to clarify the main wave propagation features in the presence of multiple elastic inclusions. The computed results are promising, clearly revealing the existence of band gaps where large attenuation occurs, although limitations related to the existence of guided waves traveling along the inclusions are also identified.
Keywords: elastodynamics, MFS, periodic structures, vibration filtering.

1 INTRODUCTION

Many important problems, such as those related to human comfort, make vibration mitigation a relevant worldwide societal challenge. New generation of mitigation measures include the so-called metamaterials, which, in essence, are artificial materials engineered to have properties not directly found in nature. Among these metamaterials, periodically arranged arrays of scatterers constitute a particular case, usually designated as "phononic crystals" or "sonic crystals". Application of these metamaterials has been tried in acoustic shielding, as described, for example, by Umnova *et al.* [1] or by Martinez-Sala *et al.* [2]. Many of the published works in that field indicate that these techniques can be effective, and seem to be quite promising for acoustical applications. These results also make it possible to expect that similar concepts may be used with success in vibration mitigation using buried metamaterial-like structures, as can be the case, for example, of buried sonic-crystal-like structures.

Recent works concerning "seismic" protection by periodic structures are already found in the literature. Kim and Das [3] analysed the effect of a periodic structure in the creation of a shadow zone to seismic waves. Alagoz and Alagoz [4] presented numerical studies concerning the effect of what they called "seismic crystals", indicating an effective attenuation of seismic waves. Possibly the first experimental evidence that buried periodic structures can efficiently attenuate vibrations in specific frequency bands is due to Brûle *et al.* [5]; these authors have shown experimental results of a seismic test carried out using seismic waves

DOI: 10.2495/CMEM-V6-N6-1108-1119

generated by a monochromatic vibrocompaction probe, revealing a modification of the seismic energy distribution in the presence of a buried metamaterial, in agreement with numerical simulations. Preliminary work by the authors of the present paper has also demonstrated the attenuating properties of arrays of scatterers (see [6]), although in a very limited number of scenarios. It looks like this concept for vibration protection based on buried metamaterials may constitute a powerful promising future solution.

Beyond the physical phenomena themselves, one of the challenges of this type of problem is related to their accurate computational simulation. Previous works in the field of wave propagation (as for example [7, 8]) made use of a number of techniques, which can be extended and adapted to treat the problem of metamaterials in solid media. Methods such as the BEM (Boundary Element Method) or the MFS (Method of Fundamental Solutions) seem to be appropriate choices for that purpose. Conceptually, one possibility is related to 2.5D analysis, in which the geometry remains constant along one direction, although with a purely 3D source. These models [9] are very efficient and can be of interest to study the propagation of elastic waves in the presence of arrays of scatterers; they may serve as an excellent base for understanding the underlying physics of such complex phenomena.

The present paper intends to provide a numerical strategy based on the use of a 2.5D MFS implementation, and tries to physically interpret the wave propagation phenomena that occur around solid buried inclusions embedded in a host homogeneous soil. Due to the geometric periodicity and geometrically repetitive nature of the analysed problem, the numerical implementation can be significantly simplified, particularly in what concerns the calculation of the system matrix; indeed, a large number of repeated blocks occur in this matrix, whose calculation just needs to be performed once, and so significant computational gains can be obtained. A number of numerical simulations is here used to illustrate the main features of wave propagation in this type of scenarios, using different arrays of scatterers and considering different elastic properties.

2 MATHEMATICAL FORMULATION

2.1 2.5D Problem Formulation

The problem to be solved consists of an elastic medium within which a set of elastic bodies is embedded, composing a multi-domain elastic system. For this type of problem, the frequency domain wave equation can be defined as

$$(\lambda + 2\mu)\nabla\nabla\cdot\mathbf{u} - \mu\,\nabla\times\nabla\times\mathbf{u} + \omega^2\rho\,\mathbf{u} = 0 \tag{1}$$

In this equation, $\mathbf{u}$ is the displacement vector, λ and μ are the Lamé's constants and ρ is the mass density of the elastic medium.

Consider now that in this physical system the solid inclusions have a constant geometry along the z direction, and the whole system is subjected to a harmonic point source, which is oscillating with an angular frequency ω. Since the system's geometry is infinite and does not change along the z direction, it is possible to apply a spatial Fourier transformation along that direction and to determine the 3D displacement field as a continuous integral of simpler bi-dimensional solutions, as

$$\mathbf{u}^{3D} = \int_{-\infty}^{+\infty} \mathbf{u}^{2.5D}(k)\mathrm{e}^{-ik(z-z_0)}dk \tag{2}$$

where k is the axial wavenumber. Discretising this integral and rewriting this equation as a discrete summation (assuming an infinite set of virtual sources equally spaced L_{vs} along the axial direction, such that $\Delta k_z = 2\pi/L_{vs}$), the 3D displacement field may be written as

$$\mathbf{u}^{3D} = \sum_{m=-\infty}^{+\infty} \mathbf{u}^{2.5D}(k_z)\mathrm{e}^{-\mathrm{i}k_z(z-z_0)}\Delta k_z \mathrm{e}^{-\mathrm{i}k_z(z-z_0)} \tag{3}$$

where $k_z = m\Delta k_z$. This summation converges, and can be approximated by a finite number of terms, ranging from $-M$ to $+M$. Note that, in order to avoid any spatial contamination from the virtual sources, the distance L_{vs} must be sufficiently large, and that the case with $k_z = 0$ corresponds to the 2D problem.

2.2 General formulation of the Method of Fundamental Solutions (MFS)

The MFS approximates the solution in a given elastic medium in terms of a linear combination of fundamental solutions of the governing differential equation. In fact, a set of NS fictitious sources is located outside the elastic domain (namely at points X_0^n, with $n = 1\ldots NS$), in order to circumvent singularities, and then the displacement field at any point of the host medium can be written as a summation of the contribution of those virtual loads, with a-priori unknown amplitudes A_j^n. Therefore, the displacement u_i, at a generic point X, along direction $i = x, y, z$ is represented by:

$$u_i(X, k_z, \omega) = \sum_{n=1}^{NS}\sum_{j=1}^{3} G_{ij}\left(X, X_0^n, k_z, \omega\right) A_j^n\left(k_z, \omega\right) \tag{4}$$

where G_{ij} corresponds to the displacement generated at point X, along direction i, by a unit load acting along direction j at point X_0^n, which can be determined by using the appropriate fundamental solutions.

This approach allows the incorporation of different Green's functions without introducing any significant changes to the method; thus it is quite simple to use Green's functions that account, for example, for the presence of a half-space or of a layered medium. For the simpler case of an infinite homogeneous elastic medium (full-space), the required Green's functions $G_{ij}\left(X, X_0^n, k_{zm}, \omega\right)$ for displacements in the different directions, with $i, j = x, y, z$, were presented by Tadeu and Kausel [10], and can be shortly defined as:

$$G_{xx}(X, X_0, k_z, \omega) = \frac{1}{4\mathrm{i}\,\rho\,\omega^2}\left[k_s^2\, H_{0\beta} - \frac{1}{r}B_1 + \gamma_x^2\, B_2\right] \tag{5a}$$

$$G_{yy}(X, X_0, k_z, \omega) = \frac{1}{4\mathrm{i}\,\rho\,\omega^2}\left[k_s^2\, H_{0\beta} - \frac{1}{r}B_1 + \gamma_y^2\, B_2\right] \tag{5b}$$

$$G_{zz}(X, X_0, k_z, \omega) = \frac{1}{4\mathrm{i}\,\rho\,\omega^2}\left[k_s^2\, H_{0\beta} - k_z^2\, B_0\right] \tag{5c}$$

$$G_{xy}(X, X_0, k_z, \omega) = G_{yx}(X, X_0, k_z, \omega) = \gamma_x\,\gamma_y\,\frac{1}{4\mathrm{i}\,\rho\,\omega^2}B_2 \tag{5d}$$

$$G_{xz}(X, X_0, k_z, \omega) = G_{zx}(X, X_0, k_z, \omega) = \mathrm{i}\,k_z\,\gamma_x\,\frac{1}{4\mathrm{i}\,\rho\,\omega^2}B_1 \tag{5e}$$

$$G_{yz}(X, X_0, k_z, \omega) = G_{zy}(X, X_0, k_z, \omega) = \mathrm{i}\, k_z\, \gamma_y \frac{1}{4\mathrm{i}\, \rho\, \omega^2} B_1 \tag{5f}$$

with $B_n = k_\beta^n H_{n\beta} - k_\alpha^n H_{n\alpha}$, $k_\alpha = \sqrt{k_p^2 - k_z^2}$ with $\operatorname{Im} k_\alpha < 0, k_p = \omega/\alpha$, $k_\beta = \sqrt{k_s^2 - k_z^2}$ with $\operatorname{Im} k_\beta < 0$, $k_s = \omega/\beta$, α and β are the P and S wave propagation velocities, $\gamma_l = \partial r / \partial x_l = x_l / r$ with $l = x, y$, $H_{n\alpha} = H_n^{(2)}(k_\alpha r)$, $H_{n\beta} = H_n^{(2)}(k_\beta r)$ ($H_n^{(2)}$ is the Hankel function of second type and n^{th} order) and r stands for the distance between the source and field points.

To define a solution methodology for the specific problem in which elastic inclusions are embedded in a host infinite medium, let us illustrate the formulation of the method for the case of a single inclusion (Fig. 1). For that case, it becomes necessary to define a number of discrete points along the solid-solid interface where the necessary boundary conditions need to be enforced (collocation points). Two sets of virtual sources need also to be defined, one of them located within the inclusion that allows simulation of the outer field, and the other surrounding the inclusion in order to simulate the wave field inside it. Using equn (4), the corresponding displacement fields can thus be calculated.

In the present case, boundary conditions imply continuity of displacements and equilibrium of tractions along the boundary between the two distinct domains. Imposing those conditions, a linear equation system with $2 \times 3 \times NS$ equations and $2 \times 3 \times NS$ unknowns is established, with the form

$$\begin{bmatrix} \mathbf{G}_{(1)}(k_z, \omega) - \mathbf{G}_{(2)}(k_z, \omega) \\ \mathbf{H}_{(1)}(k_z, \omega) - \mathbf{H}_{(2)}(k_z, \omega) \end{bmatrix} \mathbf{A}(k_z, \omega) = \begin{bmatrix} -\mathbf{u}_{inc}(k_z, \omega) \\ -\tilde{\ }_{inc}(k_z, \omega) \end{bmatrix} \tag{6}$$

where $\mathbf{G}_{(k)}$ and $\mathbf{H}_{(k)}$ stand for coefficients matrices related to medium k, assembled following the definition of terms G_{ij} and their combined derivatives together with the corresponding constitutive law, respectively; $\mathbf{A}$ stands for the unknown amplitude vector; and $\mathbf{u}_{inc}$ and $\tilde{\ }_{inc}$, incorporate the displacements and tractions originated by a real source at the collocation points, respectively.

Solution of this equation system allows obtaining the unknown amplitude coefficients of the virtual sources, which in turn allow computing the relevant wave fields inside and outside the inclusions. The process illustrated here for a single inclusion can easily be generalized for multiple inclusions.

2.3 Implementation details for periodic geometries

If a set of N inclusions is considered, equally spaced between them forming a regular matrix, the calculation of the MFS system matrix can be greatly accelerated. Indeed, for that case, a

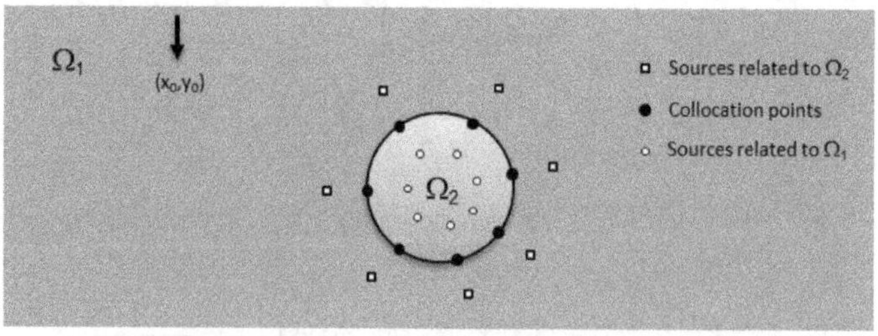

Figure 1: Schematic representation of the MFS model.

more general form of equn (6) must be used, in which each block itself is built as a sequence of smaller blocks. For example for $\mathbf{G}_{(1)}(k_z,\omega)$ and $\mathbf{G}_{(2)}(k_z,\omega)$, the corresponding form can be written as

$$\mathbf{G}_{(1)}(k_z,\omega)=\begin{bmatrix} \mathbf{G}_{(1)}^{1,1}(k_z,\omega) & \mathbf{G}_{(1)}^{1,2}(k_z,\omega) & \cdots & \mathbf{G}_{(1)}^{1,N}(k_z,\omega) \\ \mathbf{G}_{(1)}^{2,1}(k_z,\omega) & \mathbf{G}_{(1)}^{2,2}(k_z,\omega) & \cdots & \mathbf{G}_{(1)}^{2,N}(k_z,\omega) \\ \vdots & \vdots & \ddots & \vdots \\ \mathbf{G}_{(1)}^{N,1}(k_z,\omega) & \mathbf{G}_{(1)}^{N,2}(k_z,\omega) & \cdots & \mathbf{G}_{(1)}^{N,N}(k_z,\omega) \end{bmatrix}$$

$$\mathbf{G}_{(2)}(k_z,\omega)=\begin{bmatrix} \mathbf{G}_{(2)}^{1,1}(k_z,\omega) & \mathbf{0} & \cdots & \mathbf{0} \\ \mathbf{0} & \mathbf{G}_{(2)}^{2,2}(k_z,\omega) & \cdots & \mathbf{0} \\ \vdots & \vdots & \ddots & \vdots \\ \mathbf{0} & \mathbf{0} & \cdots & \mathbf{G}_{(2)}^{N,N}(k_z,\omega) \end{bmatrix} \quad (7)$$

In each block, the i,j superscript indicates the effect of the set of sources related to inclusion j on the set of collocation points around inclusion i; when the internal field of each inclusion is considered ($\mathbf{G}_{(2)}(k_z,\omega)$), a large number of null matrix blocks occurs, since the set of sources around the inclusion only influences its internal field.

It is important to notice that, if the inclusions are periodically placed, many of the blocks of $\mathbf{G}_{(1)}(k_z,\omega)$are repeated, and may only be calculated once; this can be easily realized, for example, for all diagonal blocks. Although this behaviour has been described for the .$\mathbf{G}_{(1)}(k_z,\omega)$. submatrix, it is equally applicable to $\mathbf{H}_{(1)}(k_z,\omega)$.

Given this observation, a practical implementation algorithm to compute $\mathbf{G}_{(1)}(k_z,\omega)$ and $\mathbf{H}_{(1)}(k_z,\omega)$can simply be written as:

- The calculation starts, looping *i* and *j* between 1 and *N*:
  - The vector connecting the centers of scatterer *i* and *j* is computed ($\mathbf{v}=\mathbf{x}_i^{SC}-\mathbf{x}_j^{SC}$), and compared to the ones calculated in previous loop steps.
    - If two blocks connected by $\mathbf{v}$ have been previously calculated, then there is no need to recalculate the block i,j, and the reference of the previously calculated block is associated to i,j;
    - Otherwise, block i,j is calculated and stored, and vector $\mathbf{v}$ is added to the list of already calculated blocks;
- At the end of this loop, all necessary blocks are calculated.

3 MODEL VERIFICATION AND PERFORMANCE

To first check the correctness of the proposed model and of its implementation, a homogeneous system is simulated, for which the analytical solution corresponds to the fundamental solutions described above. Using the proposed strategy, a matrix with 3x3 elastic inclusions, with radius 0.3 m, equally spaced 1.2 m is considered to be embedded in a host medium, and the same properties are ascribed to both the host medium and to the inclusions. These properties correspond to a Young modulus of 100 MPa, a density of 2000 kg/m^3 and a Poisson's ratio of 0.2. A source illuminates the system at x=–2.0 and y=0.0, vertically centred with the inclusions, and responses are computed at a receiver positioned at x=5.3 and y=1.8. Figure 2 illustrates the computed displacements at the receiver, originated by a vertical load, together with the analytical solution, and considering an apparent velocity of 1336 m/s (defined as $c_z=\omega/k_z$). The presented plots clearly evidence an excellent match between the two results.

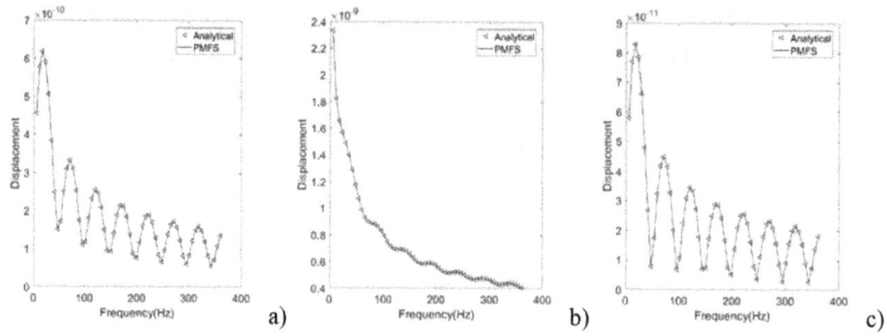

Figure 2: Verification of the implemented algorithm for x (a), y (b) and z(c) displacements.

To understand the performance gain between the proposed scheme (designated by PMFS) and a traditional MFS implementation, a sequence of tests has been performed, recording the CPU times required by each approach. Geometries with 4×4, 5×5, 6×6 and 7×7 inclusions are considered, with 30 collocation points describing each inclusion; a maximum MFS matrix size of 8820 × 8820 is thus obtained for the larger problem. The internal material of the inclusions is assumed to have a Young modulus of 30 GPa, a density of 2500 kg/m^3 and a Poisson's ratio of 0.3. Figure 3 illustrates the performance of the methods using a bar chart, in which each bar is divided in two parts: the lower (dark grey) part is related to the CPU time required for matrix formation; the upper (lighter grey) is related to the solution of the equation system. Since the same solver is used in both cases, the required CPU time for solution of the equation system is always the same for the two schemes. Analysing the plot, it is clear that a significant reduction of the CPU time occurs for the proposed implementation, which is related only to the matrix formation time. Comparing the dark grey parts of the plotted bars, this reduction is very large, and in general only 20% to 30% of the CPU time is required.

4 NUMERICAL APPLICATIONS

The presented model has been used to study the wave propagation patterns and the physical behaviour of a system composed by different numbers of inclusions. A number of results is shown for different geometries, composed of arrangements of circular cylinders. All results are presented in terms of normalized slowness and of normalized frequency (nondimensional

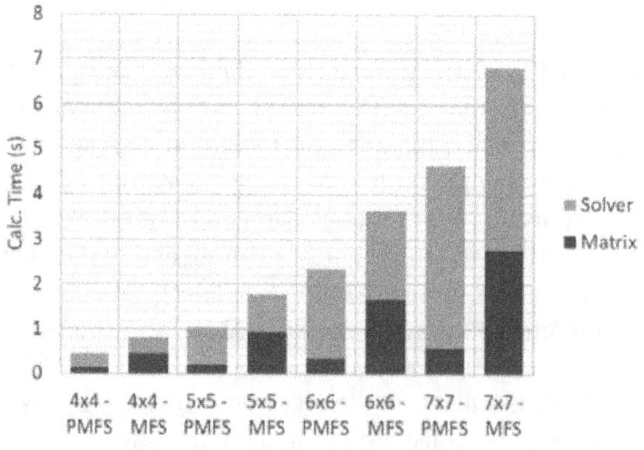

Figure 3: CPU times for the proposed (PMFS) and for the classical MFS implementations.

parameters). The normalized frequency is given in terms of the relation between the lattice constant of the arrangement of cylinders (spacing between consecutive centres) and the wavelength of the S waves in the solid host medium, allowing to define:

$$\overline{F} = a / \lambda_s = a \times f / c_s \tag{7}$$

The normalized slowness is defined as the relation of the axial wavenumber (k_z) to the S wave wavenumber, given as:

$$k = \frac{k_z}{\omega / c_s} \tag{8}$$

The following material properties are used in the present work:

In the presented results, different geometrical arrangements are analysed, including both square and triangular lattice distributions. The typical arrangement for each test case is represented in Fig. 4, for examples with 2x5 inclusions in a rectangular lattice, and for 2x3 inclusions in a triangular lattice.

The first set of examples presented here is related to the case in which perfectly rigid and unmovable inclusions are considered, embedded in a solid medium with the properties of M1, where S waves propagate at 144.3 m/s, and P waves at 235.7 m/s.

Figure 5 presents the amplitude level of the combined x, y and z velocities at R1. It is clear that for normalized slownesses above 1, very little energy propagates in the system, and, in

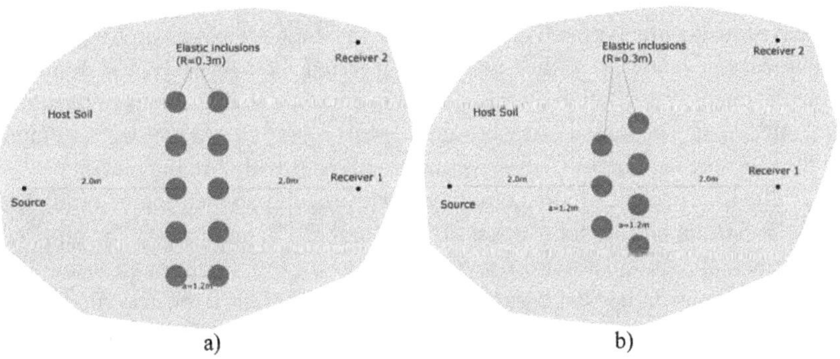

Figure 4: Illustrative scheme of the tested configurations for 2x5 square (a) and 2x3 triangular (b) lattices.

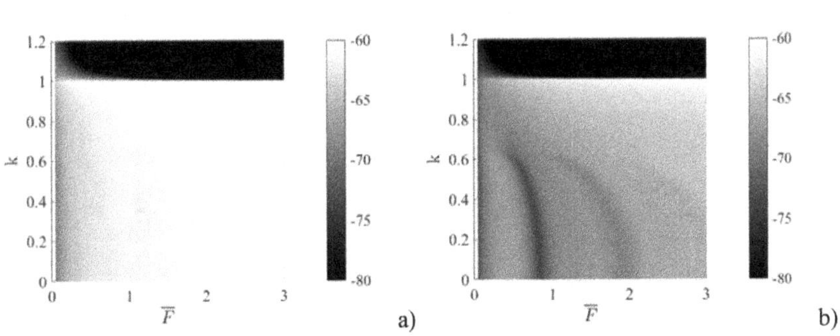

Figure 5: Absolute vibration velocity levels at R1 for an homogeneous medium (a) and in the presence of a single rigid inclusion (b).

fact, only evanescent waves exist. Bellow this slowness, since the elastic waves propagate freely, without obstacles, increasing vibration velocities are seen as the frequency increases. This scenario greatly changes in the presence of a rigid inclusion, as can be seen in Fig. 5b (where a rigid inclusion of radius 0.3 m is considered between the source and the receiver). Here, a significant decrease in the velocity levels is observed throughout the whole analysed $k-F$ domain, since the presence of the obstacle blocks the direct incidence of waves at the receiver.

To understand the effect of multiple inclusions, disposed in an organized pattern defined by a lattice constant *a*, in the wave propagation pattern, several tests were performed for different scenarios. To facilitate the interpretation of the effect of the multiple inclusions, the next plots illustrate the results in terms of "insertion loss" (IL), given by the difference of vibration levels registered at the receiver without and with the intermediate structures.

Figure 6 shows the IL computed for an array of 6 inclusions, disposed in a rectangular manner, considering a lattice constant a = 1.2 m and an inclusion radius of 0.3 m. The structure of the presented plots has now become much more complex, and includes several features. Indeed, multiple interactions occur between inclusions, originating constructive and destructive interferences. However, one particular feature merits special attention, particularly the presence of a high attenuation frequency range visible as a pattern of lighter shades in the presented plots. This zone occurs for the full sequence of *k* values and indicates that a strong attenuation is originated by the presence of the set of inclusions. This effect occurs mostly between normalized frequencies of 0.5 and 1.5, that is, between $f=0.5\,c_s/a$ and $f=1.5\,c_s/a$ (this zone is marked with 1 in the plot). Interestingly, this lower limit can be related to the theory of the acoustic attenuation by sonic crystal, in which the band gap frequency can usually be defined as $f=0.5\,c_s/a$; thus, it can be inferred that, possibly, the dominant effect controlling the energy attenuation by this system is the same as that observed in acoustics. An additional effect that is also seen in the presented plots is related to a constructive interference between the multiple inclusions (these regions are marked with 2 in the plot).

Figure 7 illustrates similar plots computed for triangular shaped inclusion distributions, considering 3x3 (total of 10) inclusions and 5x3 (total of 17) inclusions. The plots presented in this figure reveal a quite similar behaviour to those in Fig. 6, although some differences can be observed. Indeed, the "band-gap" identified above seems to be now more marked and narrower, even for the 3x3 case, and seems to become progressively more intense when more inclusions are added. For the case of 10 inclusions, the "band-gap" effect is indeed very well defined, and corresponds to a continuous bright patch in the Fig. 7b, occurring between normalized frequencies of 0.5 and 1.0. Additionally, it becomes clear that one of the constructive effects identified above (the leftmost one marked "2" in Fig. 6b) has now

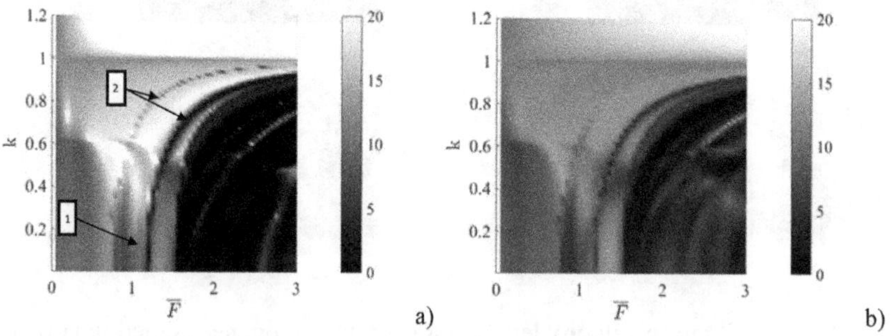

Figure 6: Insertion Loss at R1 (a) and R2 (b) for a 3x3 square lattice system.

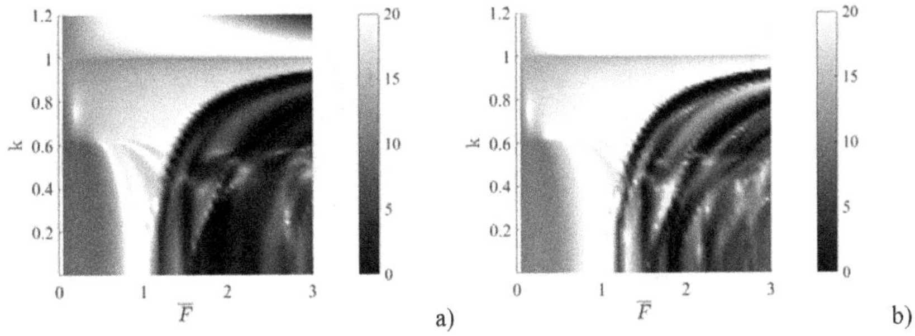

Figure 7: Insertion Loss at R1 for triangular lattices considering 3×3 (a) and 5×3 (b) groups.

disappeared; that effect can be related to the diagonal distance between inclusions in the rectangular distribution, which is eliminated when triangular lattices are considered.

To understand the effect of considering the inclusions as being made of an elastic material, simulations were performed for the geometrical configuration used in Fig. 7b (5×3 triangular lattice). Two different material properties were considered for the inclusions, namely those indicated for M2 and M3 in Table 1. Figure 8 illustrates the IL at receiver 1 for both materials.

Observing the plotted results, it is clear that new features appear due to the consideration of the elastic properties of the inclusions. Indeed, although the upper part of the plot remains essentially unchanged, there is a very distinct behaviour on the lower part. This behaviour may be associated with the longitudinal vibration modes of the inclusions, namely their flexural and torsional modes, which allow a significant part of the energy to pass through and

Table 1: Material properties

Material	Density (kg/m^3)	E (MPa)	Poisson
M1	2000	100	0.2
M2	7800	200000	0.3
M3	2500	30000	0.3

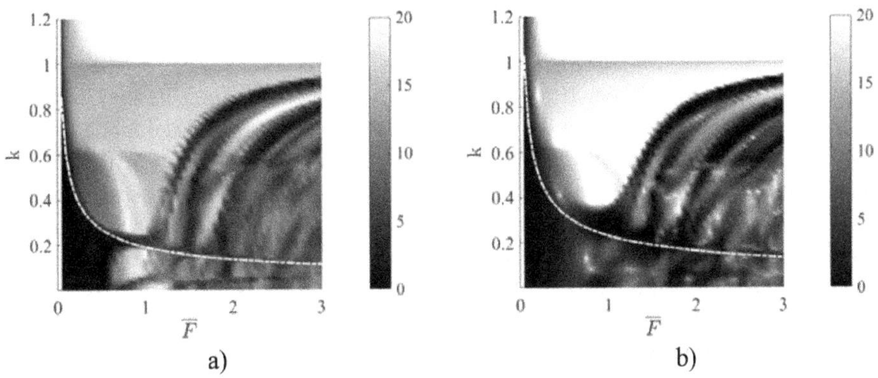

Figure 8: Insertion Loss at R1 when the inclusions are made of M2 (a) and M3 (b).

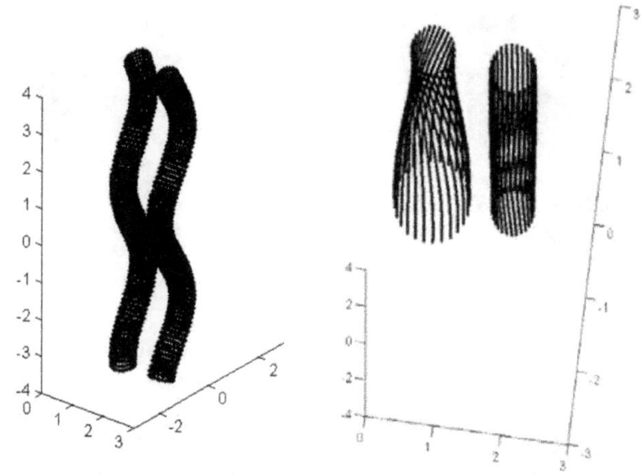

Figure 9: Deformed configuration of elastic inclusions computed for $k-\overline{F}$ pairs corresponding to the bending wave (left) and torsional wave (right) dispersion curves.

reach the receivers. To confirm this hypothesis, the dispersion curve for longitudinal bending waves has been computed for both material types, and is represented in the plots by a dashed line. Clearly, this new line divides the upper and lower parts of the plot, and below it, a completely different behaviour is seen when compared with the plots depicted in Figure 6 or 7. For lower values of k, the additional contribution of a torsional mode is seen, which also contributes to a reduced IL observed here. To confirm that these are indeed the referred modes, the deformed configuration of the inclusions has been calculated for $k-\overline{F}$ pairs located over the corresponding dispersion curves, and is depicted in Fig. 9. To facilitate the interpretation, only two inclusions are plotted. Observing those plots, it can be seen that a clear bending behaviour is observed on the left plot, while a torsional behaviour is seen on the right plot, thus confirming the described effects.

The observations made above for the case of elastic inclusions allow inferring that accounting for the correct elastic behaviour of the inclusions is essential, since the global behaviour is very different from that seen in the case of a rigid inclusion. The elastic behaviour of the inclusions seems to clearly decrease the attenuation efficiency of the periodic structures, mainly for lower values of the normalized slowness.

5 CONCLUSIONS

In the present paper, a study regarding the simulation of the dynamic behaviour of an elastic sonic-crystal-like structure, accounting for its 3D behaviour. The proposed model is based in a 2.5D elastodynamic formulation, making use of an improved version of the Method of Fundamental Solutions specifically tailored for periodic structures. It was shown that the proposed model is quite adequate and very efficient for the analysis of such structures, and allows simulations to be performed even for quite large systems.

Some application examples were presented regarding the analysis of that type of problems, and interesting conclusions could be drawn. It can be said that the simulated arrays of rigid cylinders essentially present the expected behaviour of a sonic-crystal, exhibiting a clear band-gap. The position of this band-gap can be approximately estimated considering the S wave velocity, in a similar manner to that used in acoustic problems. When square and triangular lattices were compared, the latter seem to exhibit a smoother and cleaner behaviour, with a very well defined band-gap region, although affecting a narrower frequency band. Finally, it was observed that when elastic inclusions are considered, a dramatic change occurs, and the vibrational behaviour of the inclusions introduces a very significant degradation of the IL, particularly for low values of the normalized slowness (k), below the dispersion curve associated with the bending waves of the inclusions. This behaviour represents a significant issue that must be addressed in order to efficiently use this concept for vibration mitigation.

ACKNOWLEDGEMENTS

The authors acknowledge the financial support of FCT – Foundation for Science and Technology and COMPETE, through research project PTDC/ECM-COM/1364/2014. This work is also financed by FEDER funds through the Competitivity Factors Operational Programme - COMPETE and by national funds through FCT – Foundation for Science and Technology within the scope of the projects POCI-01-0145-FEDER-007633 (ISISE) and POCI-01-0145-FEDER-007457 (CONSTRUCT).

REFERENCES

[1] Umnova, O., Attenborough, K. & Linton, C.M., Effects of porous covering on sound attenuation by periodic arrays of cylinders. The *Journal of the Acoustical Society of Amercia*, **119**(1), pp. 278–284, 2006.
https://doi.org/10.1121/1.2133715

[2] Martinez-Sala, R., Rubio, C., Garcia-Raffi, L., Sanchez-Perez, J.V., Sanchez-Perez, E.A. & Llinares, J., Control of noise by trees arranged like sonic crystals. *Journal Sound Vibration*, **291**(100), pp. 100–106, 2006.
https://doi.org/10.1016/j.jsv.2005.05.030

[3] Kim, S.H. & Das, M.P., Artificial seismic shadow zone by acoustic metamaterials. *Modern Physics Letters B*, **27**(20), pp. 1350140, 2013.
https://doi.org/10.1142/s0217984913501406

[4] Alagoz, B.B. & Alagoz, S., Towards earthquake shields: a numerical investigation of earthquake shielding with seismic crystals. *Open Journal of Acoustics*, **1**(03), p. 63, 2011.
https://doi.org/10.4236/oja.2011.13008

[5] Brûlé, S., Javelaud, E.H., Enoch, S. & Guenneau, S., Experiments on seismic metamaterials: molding surface waves. *Physical Review Letters*, **112**(13), pp. 133901, 2014.
https://doi.org/10.1103/physrevlett.112.133901

[6] Amado-Mendes, P. & Godinho, L., Reduction of vibrations transmitted through the soil by multiple buried inclusions–numerical analysis. *ICOVP – 11th International Conference on Vibration Problems* (in CD-Rom), 2013.

[7] Martins, M., Godinho, L. & Picado-Santos, L., Numerical evaluation of sound attenuation provided by periodic structures. *Archives of Acoustics*, **38**(4), pp. 503–516, 2013.
https://doi.org/10.2478/aoa-2013-0060

[8] Santos, P.G., Carbajo, J., Godinho, L. & Ramis, J., Sound propagation analysis on sonic crystal elastic structures using the Method of Fundamental Solutions (MFS). *Computers, Materials and Continua,* **43**(2), pp. 109–136, 2014.
https://doi.org/10.3970/cmc.2014.043.109

[9] Godinho, L. & Tadeu, A., The importance of a small wall deformation in the three-dimensional acoustic logging results. *Geophysical Journal International*, **151**(2), pp. 403–415, 2002.
https://doi.org/10.1046/j.1365-246X.2002.01792.x

[10] Tadeu, A.J. & Kausel, E., Green's functions for two-and-a-half-dimensional elastodynamic problems. *Journal of Engineering Mechanics*, **126**(10), pp. 1093–1097, 2000.
https://doi.org/10.1061/(ASCE)0733-9399(2000)126:10(1093)

BOUNDARY INTEGRAL EQUATIONS OF DYNAMICS PROBLEMS FOR MULTI-CONNECTED THERMOELASTIC SEMI-PLANE WITH A FREE BOUNDARY

LYUDMILA A. ALEXEYEVA[1] & BAKHYT N. ALIPOVA[2,3*]
[1]Institute of Mathematics and Mathematical Modeling of MES RK, Kazakhstan.
[2]International University of Information Technology, Kazakhstan, Samara University, Russia.
[3]Al-Farabi Kazakh National University, Kazakhstan.

ABSTRACT
The dynamics of multi-connected thermoelastic semiplane with the non-stationary power source and thermal effects by using of a model of coupled thermoelasticity is investigated. Green's tensor in the space of the Laplace transforms in time describes the displacements of medium under the effect of the impulse concentrated power and thermal sources. The generalized solution of the problem of the dynamics of thermoelastic semiplane with the free boundary under the effect of arbitrary mass forces and thermal sources in 2D-case is built.
Keywords: dynamics, green's tensor, semiplane, the laplace transformation, thermoelasticity.

1 INTRODUCTION

The various mathematical models of Mechanics of Deformable Solid are used to calculate the real properties of rock mass by the study of seismic processes in the Earth's crust. The most studied processes are wave propagation and diffraction in elastic media under actions of concentrated and distributed sources of different form. Theoretical studies of such problems based on classical methods of mathematical physics have sufficiently extensive bibliography [1–3].

Real solid massif, besides elastic properties, possesses the whole spectrum of other properties, which have an essential effect on processes of seismic wave propagation and stress-strained state of media. In this connection the complication of mathematical model is absolutely necessary taking into account the more complete operative factors during the study of seismic processes. The one of these factors is the temperature of a massif, which substantially influences on its stress-strained state by static and dynamic effects. Investigation of dynamics of thermoelastic medium under the actions of non-stationary power and thermal sources relates to the poorly studied problems of Mechanics and Mathematical Physics. The research of Novacki [4], Kupradze *et al.*, [5], Kech and Teodorescu [6] were coming to be regarded as a major contribution to the study of thermoelastic problems.

In connection with a complexity of constructing the solutions of the system of thermo-elastodynamics equations (its form relates to the systems of mixed hyperbolic-parabolic type), regularly these equations are simplified with disregarding the action of elastic deformations on the temperature field of medium. For this model, which are called *uncoupled thermoelasticity*, at first it is possible to define temperature field, solving the well studied heat conduction equation, then to define displacements and speeds of medium the classical equations of elastic body dynamics are used. The gradient of temperature field enters into as the mass force. But even for such mathematical model the class of solved problems is very limited.

In this article we consider the thermoelastic state of semi-plane with the cylindrical cavity under an action of non-stationary mass powers and thermal sources by using the model of

*ORCID: http://orcid.org/0000-0003-0915-2759

DOI: 10.2495/CMEM-V6-N6-1120-1126

coupled thermoelasticity. Thus we take into account the action of elastic deformation velocity on an temperature field. We develop boundary element method (BEM) for this problem based on generalized functions method (GFM) of construction of boundary value problem (BVP) solutions for differential equations.

2 MOTION EQUATIONS OF THERMOELASTIC MEDIA

The isotropic thermoelastic medium is characterized by the finite numbers of thermodynamic parameters: mass density ρ, elastic Lame constant λ and μ, thermoelastic and thermal constants γ, η, κ. All constants are positive.

Dynamics of thermoelastic medium in the Cartesian coordinates system is described by the system of differential equations of mixed hyperbolic-parabolic type [7]:

$$\begin{aligned} &\sigma_{ij,j} - \rho \ddot{u}_i + F_i = 0, \\ &\theta_{,jj} - \kappa^{-1}\dot{\theta} - \eta \dot{u}_{j,j} + Q = 0, \quad j = 1,...,N \end{aligned} \tag{1}$$

where $u_i(x, t)$, $\sigma_{ij}(x, t)$ are the components of the displacement vector and stress tensor, $u_3(x, t) = \theta(x,t)$ is temperature; $F=F_i e_i$ is the mass force; Q is the power of thermal source, λ, μ are elastic Lame parameters; $\gamma \equiv (3\lambda + 2\mu)\alpha_t$, α_t is coefficient of linear thermal expansion, $\kappa = \lambda_0 / b_\varepsilon$ is coefficient of thermal diffusivity, λ_0 is the coefficient of thermal conductivity, $\eta \equiv \dfrac{\gamma T_0}{\lambda_0}$,T_0 is the current absolute temperature of the medium in the initial state. The stress tensor σ_{ij} is connected with displacements and temperature by *the Duhamel – Neumann law:*

$$\sigma_{ij} = \lambda \delta_{ij} \delta_{kl} u_{k,l} + \mu(u_{i,j} + u_{j,i}) - \gamma \delta_{ij} \theta \tag{2}$$

We will use its general form

$$\sigma_{ij} = C_{ij}^{kl} u_{k,l} - \gamma \delta_{ij} \theta,$$

$$C_{ij}^{kl} = \lambda \delta_{ij} \delta_{kl} + \mu \left(\delta_{ik} \delta_{jl} + \delta_{il} \delta_{jk} \right)$$

Here δ_{ij} is Kronecker's symbol. The partial derivatives on the appropriate coordinates are designated by symbol after comma: $u_{i,j} \equiv \partial u_i / \partial x_j$, differentiation with respect to time t: $\mathring{u} = \partial u / \partial t$.

Here we consider the case of plane deformation: $x = (x_1, x_2)$, $i, j = 1,2$. Everywhere the tensor convolutions are taken on the similar indices.

3 THE STATEMENT OF THE BOUNDARY VALUE PROBLEM FOR A THERMOELASTIC SEMI-PLANE WITH CAVITY

Let consider the thermoelastic semi-plane $\Pi^+ = \{x_1 > 0, -\infty < x_2 < \infty\}$ (the boundary Π) with cylindrical cavity S^+, S is an boundary of S^+, S^- is region of elastic media without its boundary, $\Pi + S = B$, $n(x)$ is external normal to B. We suppose that acting forces $F(x,t)$ and thermal source $Q(x, t)$ are known and the initial state of medium is known also:

$$u_i(x,0) = 0,\ \theta(x,0) = \theta^0(x),\ x \in S^- + B;\ \dot{u}_i(x,0) = 0,\ x \in S^- \tag{3}$$

On the boundary of cavity the stresses and heat flow are given:

$$\sigma_{ij}(x,t) n_j(x) = q_i^S(x,t), \quad \frac{\partial \theta(x,t)}{\partial n} = \theta^S(x,t), \qquad x \in S \tag{4}$$

The boundary of semi-space is free from loads and heat fluxes:

$$\sigma_{ij}(x,t)=0,\quad j=1,2,\quad \frac{\partial\,\theta(x,t)}{\partial\, n}=0,\quad x_1=0 \tag{5}$$

The system (1) with use (2) can be written in the closed form

$$L_{ij}(\partial_x,\partial_t)u_j+F_i=0$$
$$L_{ij}=\left(\lambda+\mu\right)\partial_i\partial_j+(\mu\Delta-\rho\,\partial_t\partial_t)\delta_{ij}-\gamma\delta_{j(N+1)}\partial_i,\quad i=1,2 \tag{6}$$
$$L_{3j}=(\Delta-\kappa^{-1}\partial_t)\delta_{j(N+1)}-\eta\,(1-\delta_{j(N+1)})\partial_t\partial_j,\quad 1,2,3$$

It is the system of mixed hyperbolic-parabolic type. In this case there are thermoelastic shock waves with the jump of stresses, speeds and heat fluxes at their fronts appeared in the media [8, 9]. The solutions satisfy to the next conditions on the jumps at the wave fronts F_t :

$$\left[u\right]_{F_t}=0,\quad \left[\sigma_{ij}n_j+\rho c_k\dot{u}_i\right]_{F_t}=0,$$

$$\left[\theta\right]_{F_t}=0,\quad \left[\dot{\theta}\right]_{F_t}=\frac{\rho\kappa n_j}{c_k}\left[\dot{u}_j\right]_{F_t}\quad k=1,2 \tag{7}$$

Here $n=\{n_1,n_2\}$ is vector of unit normal to wave front F_t which propagates in a thermoelastic medium with one of the following velocities of thermoelastic waves:

$$c_1=\sqrt{\frac{\lambda+2\mu}{\rho}},\quad c_2=\sqrt{\frac{\mu}{\rho}},$$

These velocities coincide with velocity of propagation of dilatation and shear waves in elastic media.

It is necessary to define the displacements, temperature and stresses of thermoelastic semi-plane with cavity which must to satisfy the conditions of radiation on infinity.

4 GENERALIZED FUNCTION METHOD. DYNAMIC ANALOGUE OF SOMIGLIANA FORMULAE IN DISTRIBUTION SPACE

We'll consider BVP in the space of Laplace transformation of displacements in time t:

$$u(x,p)=\int_0^\infty u(x,t)e^{-pt}dt$$

Here we use the Generalized Function Method (GFM) to construct the generalized solution of this problem [10].

At first we introduce the general regular vector-functions:

$$\hat{u}(x,p)=u(x,p)H_B^-(x),\ \hat{F}=F(x,p)H_B^-(x),\ \hat{Q}=Q(x,p)H_B^-(x),$$

where $H_B^-(x)$ is characteristic function of the region S^-:

$$H_B^-(x)=\begin{cases}1, & x\in S^-\\ 0{,}5, & x\in B\\ 0, & x\notin S^-+B\end{cases}$$

Note, that generalized derivative of such functions is

$$\partial_j \hat{f} = H_B^-(x)\partial_j f - n_j f\,\delta_B(x),$$

where $n_j f\,\delta_S(x)$ is a singular function - *simple layer on surface B*.

By using generalized differentiation, we obtain singular distributions and their derivatives as simple layers in right part:

$$L_{ij}(\partial_x, p)\hat{u}_j(x,p) = -\bar{F}_i(x,p)H_B^-(x) - \hat{P}_i^B(x,p) \tag{8}$$

We obtain the densities of (8) if these simple layers depend on boundary displacements, temperatures, stresses and heat flows [11]. The shock waves are the generalized solutions of equations. The conditions on the wave front (7) give zero densities on the simple and double layers, which appear by generalized differentiation of shock thermoelastic waves. Therefore they disappeared in the eqn (8).

By using of the Laplace transformation of Green's tensor for the free thermoelastic semi-plane $\vec{V}_m^i(x,p)$, the convolution of right part (8) gives the expression of generalized solution of this problems:

Theorem 1. *If $u(x,p)$ is classic solutions of BVP, then generalized solution of BVP has the form of following convolution*

$$\hat{u}_m(x,p) = \vec{V}_m^i * (\rho\hat{F}_i + \hat{G}_i) + \vec{V}_m^i * \bar{q}_i^s \delta_S(x) + C_{ij}^{kl}\vec{V}_{m^{,}j}^i * \left(\bar{u}_k^B n_l \delta_S(x)\right) +$$

$$+\vec{V}_m^{N+1} * (\kappa^{-1}\hat{F}_{N+1} + \hat{G}_{N+1}) + \vec{V}_m^{N+1} * \left(\bar{q}^S - \eta p \bar{u}_j^B n_j\right)\delta_s(x) + \vec{V}_m^{N+1}{}_{,j} * \left(\bar{\theta}^B n_j \delta_S\right).$$

$$i,j,k,l = 1,2,\quad m = 1,\ldots,3 \tag{9}$$

Here we use the boundary conditions (5) on the boundary of semi-plane.

5 GREEN'S TENSOR FOR THE FREE THERMOELASTIC SEMI-PLANE AND FUNDAMENTAL TENSOR OF THERMO-STRESSES

The fundamental solutions $V_i^k(x,t)$ of thermoelasticity (1) for the free semi-space was constructed in the paper [11]. This matrix of solutions satisfies equations under the action of the impulse concentrated mass forces and thermal sources, described by the singular δ-functions:

$$L_{ij}\left(\partial_1,\partial_2,\partial_t\right)V_j^k(x,t) = \delta_i^k \delta(x)\delta(t) \tag{10}$$

$$L_{3j}\left(\partial_1,\partial_2,\partial_t\right)V_j^k(x,t) = \delta_i^3 \delta(x)\delta(t),\quad i=1,2,\quad j.k=1,2,3$$

There are the boundary conditions on the free surface:

$$\sum_{i1}^{m} = \left(\lambda V_{k,k}^m - \gamma V_3^m\right)\delta_{i1} + \mu\left(V_{i,1}^m + V_{i,1}^m\right) = 0,\quad x_1 = h \tag{11}$$

$$\frac{\partial V_3^m}{\partial x_1} = 0$$

here Σ_{ij}^m is the sum of the stresses, generated by the fundamental solutions of the V by fixed m:

$$\sum_{ij}^{m} = \lambda V_{k,k}^m \delta_{ij} - \gamma V_3^m \delta_{ij} + \mu\left(V_{i,j}^m + V_{j,i}^m\right)$$

The second condition (7) indicates the absence of heat flux on the border. It satisfies to the following conditions for radiation and damping at infinity also:

$$V_i^k(x,t) = 0 \;\; for\, \forall t < 0 \tag{12}$$
$$V_i^k(x,t) \to 0 \;\; for\, \forall t > 0, \;\; x \to \infty$$

Thus we specified some conditions for radiation, which describe reflected wave from the boundary of semi-plane. The analytic formulas for calculation have complex form [12, 13].

To express the formulae of Theorem 1 in integral form, we input the fundamental tensor of thermo-stresses $T_i^j(x,n,t)$: $i,j,k,l=\overline{1,N}, \quad m=\overline{1,N+1}$)

$$\overline{T}_m^k(x,n,p) = C_{ij}^{kl}\overline{V}_{m,j}^i n_l - \eta p \overline{V}_m^3 n_k, \quad \overline{T}_m^3 = \overline{V}_{m,j}^3 n_j \tag{13}$$

The following Theorems for this tensor are valid.

Theorem 2. *The tensor $\overline{T}_m^k$ for fixed k is fundamental solution of equations of thermoelasticity (6) by corresponding mass power source and thermal source of multipole type:*

for $k=\overline{1,N}$ $\qquad \overline{F}_s = \left(C_{sj}^{[k]l} n_l \partial_j - \eta p n_k \delta_s^{N+1}\right)\delta(x),$

for $k=N+1$ $\qquad \overline{F}_s = \delta_s^{[N+1]} n_j \delta_{,j} \quad (s=\overline{1,N+1}, \quad j,l=\overline{1,N}).$

Theorem 3. *(Dynamical analogue of Gauss Formulae). If* $D^- \in S^- + B$, and boundary *D* *is Lyapunov's surface [8], then*

$$V.P.\int_D \overline{T}_i^m(x-y,n(y),p)dS(y) + \rho p^2 \int_{D^-} \overline{V}_i^m(x-y,p)dV(y) = \delta_i^m H_D^-(x)$$
$$V.P.\int_D \overline{T}_{N+1}^m(x \quad y,n(y),p)dS(y) + p\kappa^{-1} \int_{D^-} \overline{V}_{N+1}^m(x-y,p)dV(y) = \delta_{N+1}^m H_D^-(x)$$

where in all integrals are regular for inner points of D, and are taken as Value Principal on boundary.

6 THE SOLUTION OF BVP. SINGULAR BOUNDARY INTEGRAL EQUATIONS

We can write the formula (9) in integral form using the fundamental tensors.

Theorem 4. *If classic solution $\bar{u}(x,p)$of BVP exists then it could be expressed in the following regular integral form:*

$$\begin{gathered}\bar{u}_m H_B^-(x) = \overline{V}_m^j * (\overline{F}_j + \overline{G}_j) + \\ + \int_S \left(\overline{V}_m^k(x-y,p)\overline{q}_k^S(y,p) + \overline{T}_m^k(x-y,p,n(y))\bar{u}_k^S(y,p)\right) dS(y)\end{gathered} \tag{14}$$

for $x \in S^- + \Pi$. *Here* $\overline{q}_3^S = \overline{q}^S(x,p), \quad u_3^S = \overline{\theta}^S(x,p)$.

We note that the equation's terms with displacements and temperature in formula (9) disappear on the boundary of the semi-plane, since the corresponding integral kernels are equal zero. Only the integrals on the cavity boundary remain.

The formula of the Theorem 4 expresses the displacements and the temperature inside the area through their boundary values, as well as the values of the load and the heat flux at the boundary. It is an analogue of the Somigliana Formulae of the static theory of elasticity, which expresses the displacement of an elastic medium through the boundary values of load and displacement [14]. Therefore, this formula can be called the *thermoelastic Analogue of the Formulae of Somigliana* [15, 16].

Since only a part of the boundary functions is known, the Formulae do not give a solution of the BVP. However, it is also valid for $x \in S$ in the sense of definition of $H_S^-(x)$ and gives the boundary integral equations for solving the BVPs. So, using Theorem 4, we can prove the following Theorem 5.

Theorem 5. *If classic solution of BVP $\vec{u}(x,p)$ satisfies Helder's condition on the boundary for any $p(\mathrm{Re}\, p > 0)$, then*

$$
\begin{aligned}
0,5\bar{u}_m &= \bar{V}_m^j * (\bar{F}_j + \bar{G}_j) + \int_S \left(\bar{V}_m^k (x-y,p) \bar{q}_k^S (y,p) \right) dS(y) + \\
&+ V.P. \int_S \left(\bar{T}_m^k (x-y,p,n(y)) \bar{u}_k^S (y,p) \right) dS(y), \quad m,k = 1,2,3
\end{aligned} \tag{15}
$$

Here $\bar{q}_3^S = \bar{q}^S(x,p)$, $u_3^S = \bar{\theta}^S(x,p)$. For $x \in S$ second integral is singular, it exists in Value Principle sense.

CONCLUSION

The boundary integral equations (BIE) allow finding unknown boundary functions, and obtained formulas of Theorem 5 are valid to find the needed solution. All received equations are amenable to a numerical solution based on the Boundary Element Method (BEM).

The algorithm of numerical realization of the Method of Boundary Integral Equations (BIEM) for the solution of plane BVPs ($R^2 \times t$) of coupled thermoelastodynamics is being developed. A numerical program for calculating the integral kernels of a system of BIEs in the space of Laplace transformation in time and a numerical program for calculating the first approximation of a system of singular BIEs for determining boundary displacements and temperature have been developed.

REFERENCES

[1] Aitaliev, Sh.M., Alexeyeva, L.A, Dildabaev, Sh.M. & Zhanbyrbaev, N.B., *Method of boundary integral equations in the problems of dynamics of elastic multicoupled solids* - Almaty: "Nauka"-1992.

[2] Guz, A.N., Kubenko, V.D. & Cherevko, M.A., *Diffraction of elastic waves*, Kiev: "Naukova Dumka", 1978.

[3] Erzhanov, Zh.S., Aitaliev, Sh.M. & Alexeyeva, L.A., *Dynamics of tonnels and underground pipelines*.- Almaty: "Nauka", 1989.

[4] Novacki, V., *Theory of elasticity.* Moscow: "Mir", 1975.

[5] Kupradze, V.D., Gegelia, T.G., Basheleishvili, E.O. & Burchuladze, T.V., *3D problems of mathematical theory of elasticity and thermoelasticity*. Moscow: "Nauka", 1976.

[6] Kech, V. & Teodorescu, P., *Introduction into theory of generalized functions with applications in technics*- Moscow: "Mir", 1978.

[7] Novacki, V., *Dynamical problems of thermoelasticity*. Moscow: "Mir", 1970.

[8] Alexeyeva, L.A. & Dadaeva, A.N., About uniqueness of solutions of boundary-valued problems of thermoelasticity with considering of thermal shoch waves. *Proceedings of KazNTU, Series of Mathematics, Mechanics and Informatics*, **28**, pp. 11–18, 2013.

[9] Alipova, B.N., Alexeyeva, L.A. & Dadayeva, A.N., Shock waves as generalized solutions of thermoelastodynamics equations. On the uniqueness of boundary value problems solutions. *AIP Conference Proceedings, ICNPAA 2016 World Congress 11th International Conference on Mathematical Problems in Engineering, Aerospace and Sciences*, July 4–8, 2016, La Rochelle, France, American Institute of Physics, 2017. https://doi.org/10.1063/1.4765466

[10] Vladimirov, V.S., *Generalized functions in mathematical physics*. Moscow: "Nauka", 1978.

[11] Alexeyeva, L.A. & Kupesova (Alipova), B.N., Method of generalized functions for boundary problems of coupled thermoelastodynamics. *Journal of Applied Mathematics and Mechanics (of Russian Academy of Sciences)*, **65**(2), pp. 334–345, 2001. https://doi.org/10.1016/s0021-8928(01)00037-5

[12] Alexeyeva, L.A. & Alipova B.N., Tensor of Green for thermoelastic semi-plane with free boundary. *Mathematical Journal of RK*, **17**(1), pp. 32–42, 2017.

[13] Alipova, B.N., Method of Boundary Integral Equations (BIEM) and generalized solutions of transient problems of thermoelastodynamics. *ICNPAA 2012, 9–15 July, AIP Conference Proceedings*, American Institute of Physics, Vien, Austria, 1493, **39**, pp. 39–46, 2012.
https://doi.org/10.1063/1.4765466

[14] Alipova, B., Dynamical analogue of formulae of Gauss of transient problems of thermoelastodynamics. AIP Conf. Proc. 1637, 26 (2014); 1 *10th International Conference On Mathematical Problems In Engineering, Aerospace And Sciences: ICNPAA 2014*, Narvik, Norway, 15–18 July 2014.
https://doi.org/10.1063/1.4904561

[15] Alipova, B., Fundamental tensor of tension of coupled thermoelastodynamics in 2-D and 3-D cases. *Journal | MESA*, [S.l.], **7**(3), pp. 513–524, 2016. Available at: http://nonlinearstudies.com/index.php/mesa/article/view/1179

[16] Alipova, B., Certificate of Copyright of intellectual property No 001494 dd 10/23/2013 *"Dynamical analogue of Somigliana's Formulae of coupled thermoelastodynamics"*. Available at: http://www.avtor.uz/index.php?option-com_desposition&task=display_desp_det&id=1600&itemid=88&lang=ru

AN INVESTIGATION OF EIGENFREQUENCIES OF BOUNDARY INTEGRAL EQUATIONS AND THE BURTON-MILLER FORMULATION IN TWO-DIMENSIONAL ELASTODYNAMICS

KEI MATSUSHIMA, HIROSHI ISAKARI, TORU TAKAHASHI & TOSHIRO MATSUMOTO
Nagoya University, Japan.

ABSTRACT

In this study, we investigate the distribution of eigenfrequencies of boundary integral equations (BIEs) of two-dimensional elastodynamics. The corresponding eigenvalue problem is classified as a nonlinear eigenvalue problem. We confirm that the Burton-Miller formulation can properly avoid fictitious eigenfrequencies. The boundary element method (BEM) is expected as a powerful numerical tool for designing sophisticated devices related to elastic waves such as acoustic metamaterials. However, the BEM is known that it loses its accuracy for certain frequencies, called as fictitious eigenfrequencies, for problems defined in the infinite domain. Recent researches It has also been revealed that not only the real-valued eigenfrequencies but also the complex-valued ones may affect the accuracy of the BEM results. We examine the distribution of complex eigenvalues obtained by BIEs for time-harmonic elastodynamic problems with the help of the Sakurai-Sugiura method which is applicable to nonlinear eigenvalue problems. We also examine its relation to the accuracy of the BEM numerical results. We also discuss an appropriate choice of the coupling parameter from a viewpoint of the distribution of fictitious eigenfrequencies.

Keywords: Boundary integral equation, Burton-Miller method, Elastodynamics, Fictitious eigenfrequency, Sakurai-Sugiura method, Transmission problem

1 INTRODUCTION

The boundary element method (BEM) is one of the main tools for numerical analyses of various boundary value problems together with the finite element method, finite difference method, and so forth. One of the most remarkable features of the BEM is that it can deal with unbounded domains rigorously without any approximation. This unique property enables us to analyse wave propagation problems, e.g. acoustic, elastic, and electromagnetic waves, defined in the unbounded domain with a high accuracy.

However, the BEM suffers from singularity treatment of the boundary integral equations (BIEs), and also it loses the uniqueness of the solution at certain frequencies, though the original boundary value problem has a unique solution. Those specific frequencies are called fictitious eigenfrequencies, and there existence is a serious drawback of the BEM, especially when the fictitious eigenvalues are real-valued. Thus, some modified BIEs have been proposed to avoid the real-valued fictitious eigenfrequencies, e.g. CHIEF method [1], the Burton-Miller formulation [2], the Müller formulation [3], and the PMCHWT formulation [4].

Recent studies have revealed that not only real-valued eigenfrequencies but also complex-valued ones with small imaginary parts affect the performance of the BEM [5, 6]. This implies that the fictitious eigenfrequency problem may not be avoided even though the modified BIEs are used. In addition, the exterior wave problems and transmission problems also have complex-valued resonance frequencies, which are called true eigenfrequencies in this study. They also affect the boundary element analysis when their imaginary parts are small. We will show later that they are rather small and have bad effects on the analysis in a certain case.

DOI: 10.2495/CMEM-V6-N6-1127-1137

In this study, we investigate the eigenfrequencies of the Burton-Miller-type BIEs in a two-dimensional elastodynamic transmission problem and its relation to the performance of the BEM. The Sakurai-Sugiura method (SSM) [7], which converts a nonlinear eigenvalue problem resulting from the discretized BIEs into a generalised eigenvalue problem, is employed for calculating the eigenfrequencies. Also, we discuss an appropriate choice of the coupling parameter of the Burton-Miller formulation for the elastodynamic problem.

2 BOUNDARY INTEGRAL EQUATIONS

2.1 Problem statement

In this study, we consider a two-dimensional elastodynamic transmission problem. As shown in Fig. 1, let $\Omega^{(1)}$ be an unbounded domain which is filled with a linear isotropic elastic medium whose mass density is $\rho^{(1)}$ and Lamé's constants are $\lambda^{(1)}$ and $\mu^{(1)}$. Similarly, let $\Omega^{(2)} = \mathbb{R}^2 \backslash \Omega^{(1)}$ be a bounded domain with the mass density $\rho^{(2)}$ and their Lamé's constants $\lambda^{(2)}$ and $\mu^{(2)}$. Assuming a plane-strain condition and time harmonic oscillations with time dependence $e^{-i\omega t}$, where ω is the angular frequency. The displacement $\boldsymbol{u}$ and the stress σ in $\Omega^{(1)} \cup \Omega^{(2)}$ are governed by the following transmission problem:

$$\sigma_{ji,j}(x) + \rho^{(1)}\omega^2 u_i(x) = 0 \qquad x \in \Omega^{(1)}, \tag{1}$$

$$\sigma_{ji,j}(x) + \rho^{(2)}\omega^2 u_i(x) = 0 \qquad x \in \Omega^{(2)}, \tag{2}$$

$$u_i^{(1)}(x) = u_i^{(2)}(x)(=: u_i(x)) \qquad x \in \Gamma, \tag{3}$$

$$\sigma_{ji}^{(1)}(x) n_j(x) = \sigma_{ji}^{(2)}(x) n_j(x)(=: t_i(x)) \qquad x \in \Gamma, \tag{4}$$

$$\text{Radiation condition for } u_i^{sc}(x) \text{ as } |x| \to \infty, \tag{5}$$

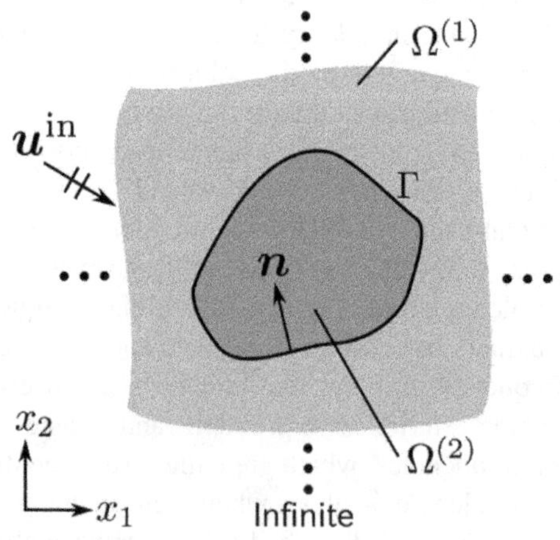

Figure 1: Problem statement.

where $\Gamma = \partial\Omega^{(1)} = \partial\Omega^{(2)}$ is the boundary, $\boldsymbol{n}$ is the unit normal vector on Γ outward to $\Omega^{(1)}$ and $\boldsymbol{u}^{\mathrm{in}}$ and $\boldsymbol{u}^{\mathrm{sc}} = \boldsymbol{u} - \boldsymbol{u}^{\mathrm{in}}$ are the incident and scattered waves, respectively. Also, $\boldsymbol{u}^{(\nu)}$ and $\sigma^{(\nu)}$ ($\nu = 1, 2$) denote the boundary traces of $\boldsymbol{u}$ and σ from $\Omega^{(\nu)}$ to Γ, respectively. The wavenumbers of the longitudinal wave $k_L^{(\nu)}$ and the transverse wave $k_T^{(\nu)}$ in $\Omega^{(\nu)}$ are given by

$$k_{\mathrm{L}}^{(\nu)} = \omega\sqrt{\frac{\rho^{(\nu)}}{\lambda^{(\nu)} + 2\mu^{(\nu)}}}, \tag{6}$$

$$k_T^{(\nu)} = \omega\sqrt{\frac{\rho^{(\nu)}}{\mu^{(\nu)}}}, \tag{7}$$

3 EIGENFREQUENCIES OF THE BIES

3.1 The Burton-Miller-type BIEs in the elastodynamic transmission problem

We obtain the following BIEs equivalent to the transmission problem (1)–(5) by the Burton-Miller formulation:

$$\frac{1}{2}u_i + \frac{\alpha}{2}t_i + \left(D_{ij}^{(1)} + \alpha N_{ij}^{(1)}\right)u_j - \left(S_{ij}^{(1)} + \alpha D_{ij}^{*(1)}\right)t_j = u_i^{\mathrm{in}} + \alpha C_{ijkl}^{(1)} u_{k,l}^{\mathrm{in}} n_j, \tag{8}$$

$$\frac{1}{2}u_i - D_{ij}^{(2)}u_j + S_{ij}^{(2)}t_j = 0, \tag{9}$$

where $\alpha \in \mathbb{C}$ is the coupling parameter, $S_{ij}^{(\nu)}$, $D_{ij}^{(\nu)}$, $D_{ij}^{*(\nu)}$ and $N_{ij}^{(\nu)}$ ($\nu = 1, 2$) are the integral operators defined for a scalar function w by

$$\left(S_{ij}^{(\nu)}w\right)(x) = \int_\Gamma G_{ij}^{(\nu)}(x, y)\, w(y)\mathrm{d}\Gamma_y, \tag{10}$$

$$\left(D_{ij}^{(\nu)}w\right)(x) = -\mathrm{v.p.}\int_\Gamma C_{kljm}^{(\nu)} G_{ki,l}^{(\nu)}(x, y) n_m(y) w(y)\mathrm{d}\Gamma_y, \tag{11}$$

$$\left(D_{ij}^{*(\nu)}w\right)(x) = \mathrm{v.p.}\int_\Gamma C_{klim}^{(\nu)} G_{kj,l}^{(\nu)}(x, y) n_m(x) w(y)\mathrm{d}\Gamma_y, \tag{12}$$

$$\left(N_{ij}^{(\nu)}w\right)(x) = -\mathrm{p.f.}\int_\Gamma C_{impq}^{(\nu)} C_{kljn}^{(\nu)} G_{kp,lq}^{(\nu)}(x, y) n_m(x) n_n(y) w(y)\mathrm{d}\Gamma_y, \tag{13}$$

where $\boldsymbol{C}^{(\nu)}$ is the elasticity tensor in $\Omega^{(\nu)}$, 'v.p.' and 'p.f.' denote Cauchy's principal value and the finite part of divergent integrals, respectively, and $G_{ij}^{(\nu)}$ is the fundamental solution of two-dimensional elastodynamics which is expressed by the Kronecker delta δ_{ij} and the Hankel function of first kind and order zero $H_0^{(1)}$ as

$$G_{ij}^{(\nu)}(x, y) = \frac{\mathrm{i}}{4\mu^{(\nu)}}\left\{H_0^{(1)}\left(k_T^{(\nu)}|x-y|\right)\delta_{ij} + \left(\frac{1}{k_T^{(\nu)}}\right)^2 \frac{\partial^2}{\partial y_i \partial y_j}\left(H_0^{(1)}\left(k_T^{(\nu)}|x-y|\right) - H_0^{(1)}\left(k_L^{(\nu)}|x-y|\right)\right)\right\}. \tag{14}$$

3.2 Classification of the eigenfrequencies of the Burton-Miller-type BIEs

Eigenfrequencies of the Burton-Miller-type BIEs (8) and (9) can be classified into the following three types:

- eigenfrequencies of the boundary value problem (1)–(5) (true eigenfrequencies)
- eigenfrequencies of the interior impedance problem (exterior fictitious eigenfrequencies):

$$C^{(1)}_{ijkl}u_{k,lj}(x)+\rho^{(1)}\omega^2 u_i(x)=0 \qquad x\in\Omega^{(2)}, \tag{15}$$

$$u_i(x)+\alpha C^{(1)}_{ijkl}u_{k,l}(x)n_j(x)=0 \qquad x\in\Gamma. \tag{16}$$

- eigenfrequencies of the exterior Dirichlet problem (interior fictitious eigenfrequencies):

$$C^{(2)}_{ijkl}u_{k,lj}(x)+\rho^{(2)}\omega^2 u_i(x)=0 \qquad x\in\Omega^{(1)}, \tag{17}$$

$$u_i(x)=0 \qquad x\in\Gamma, \tag{18}$$

$$\text{Radiation condition for } u_i(x) \text{ as } |x|\to\infty. \tag{19}$$

3.3 The block Sakurai-Sugiura method (SSM)

The block SSM is a contour integral method which solves a nonlinear eigenvalue problem which finds $\lambda\in\mathbb{C}$ such that there exists a non-trivial solution $\phi\in\mathbb{C}^k$ of

$$A(\lambda)\phi=0, \tag{20}$$

where $A\in\mathbb{C}^{k\times k}$ is an analytic matrix function.

In order to examine eigenfrequencies of the BIEs (8) and (9) numerically, one can set A to be a coefficient matrix which is obtained by discretising the BIEs and perform the block SSM. In this study, however, we limit the shape of the inclusion $\Omega^{(2)}$ to a cylinder of radius a centred at the origin and set A to be

$$\Delta^n_{\text{true}}(\omega)=\begin{bmatrix} U^n_1(a) & U^n_2(a) & -\tilde{U}^n_1(a) & -\tilde{U}^n_2(a) \\ V^n_1(a) & V^n_2(a) & -\tilde{V}^n_1(a) & -\tilde{V}^n_2(a) \\ \mu^{(1)}T^n_{11}(a) & \mu^{(1)}T^n_{12}(a) & -\mu^{(2)}\tilde{T}^n_{11}(a) & -\mu^{(2)}\tilde{T}^n_{12}(a) \\ \mu^{(1)}T^n_{41}(a) & \mu^{(1)}T^n_{42}(a) & -\mu^{(2)}\tilde{T}^n_{41}(a) & -\mu^{(2)}\tilde{T}^n_{42}(a) \end{bmatrix}, \tag{21}$$

$$\Delta^n_{\text{ext}}(\omega)=\begin{bmatrix} \tilde{U}^n_1(a)+\alpha\dfrac{2\mu^{(1)}}{a}\tilde{T}^n_{11}(a) & \tilde{U}^n_2(a)+\alpha\dfrac{2\mu^{(1)}}{a}\tilde{T}^n_{12}(a) \\ \tilde{V}^n_1(a)+\alpha\dfrac{2\mu^{(1)}}{a}\tilde{T}^n_{41}(a) & \tilde{V}^n_2(a)+\alpha\dfrac{2\mu^{(1)}}{a}\tilde{T}^n_{42}(a) \end{bmatrix}, \tag{22}$$

$$\Delta^n_{\text{int}}(\omega)=\begin{bmatrix} U^n_1(a) & U^n_2(a) \\ V^n_1(a) & V^n_2(a) \end{bmatrix}, \tag{23}$$

respectively (see Appendix). This enables us to reduce computational cost and separate the three types of the eigenfrequencies. For general configurations, Misawa *et al.* [6] have proposed a method to distinguish the fictitious eigenfrequencies from the true eigenfrequencies by modifying BIEs.

The nonlinear eigenproblems are solved by the following procedure:

1. Let γ be a closed Jordan curve in the complex plane. Determine a representative point z_0 of γ (e.g. centre of a circle) and compute $A(z_0) = [\boldsymbol{a}_0\ \boldsymbol{a}_1 \cdots \boldsymbol{a}_k]$.
2. Construct the Hankel matrices

$$H_{m'} = \begin{bmatrix} \mu_0 & \mu_1 & \cdots & \mu_{m'-1} \\ \mu_1 & \mu_2 & \cdots & \mu_{m'} \\ \vdots & \vdots & \ddots & \vdots \\ \mu_{m'-1} & \mu_{m'} & \cdots & \mu_{2m'-2} \end{bmatrix}, \tag{24}$$

$$H^{<}_{m'} = \begin{bmatrix} \mu_1 & \mu_2 & \cdots & \mu_{m'} \\ \mu_2 & \mu_3 & \cdots & \mu_{m'+1} \\ \vdots & \vdots & \ddots & \vdots \\ \mu_{m'} & \mu_{m'+1} & \cdots & \mu_{2m'-1} \end{bmatrix}, \tag{25}$$

where m' is a sufficiently large integer such that the number of eigenvalues in γ does not exceed m', and the moments μ_i ($i = 0, \cdots, 2m' - 1$) are defined as

$$\mu_i = \frac{1}{2\pi \mathrm{i}} \int_\gamma z^i f(\mathrm{z})\mathrm{dz}, \tag{26}$$

$$f(\mathrm{z}) = \mathrm{U}^H \tilde{A}^{-1}(\mathrm{z})\mathrm{V}, \tag{27}$$

with the scaled matrix $\tilde{A}(\mathrm{z}) = \mathrm{A}(\mathrm{z})\ \mathrm{diag}(1/\|a_0\|, 1/\|a_1\|, \ldots, 1/\|a_k\|)$ and random matrices $U, V \in \mathbb{C}^{k\times l}$. If μ_0 is a zero matrix, then we judge that there exists no eigenvalue in γ and terminate the procedure.
3. Compute eigenvalues of , $H^{<}_m - \lambda H_m$ where m is the number of non-zero singular values of $H_{m'}$. The obtained eigenvalues are identical to the eigenvalues of the nonlinear eigenvalue problem $A(\lambda)\phi = \mathbf{0}$ in γ.

4 NUMERICAL EXAMPLES

In this section, we investigate the effect of the eigenfrequencies on the accuracy of the BEM. The collocation method with constant elements is used for discretisation of the BIEs, and GMRES with tolerance 10^{-5} is utilised to solve linear algebraic equations. The boundary Γ is discretised into 500 boundary elements, and its radius is set as $a = 1$ [m]. The incident wave $\boldsymbol{u}^{\mathrm{in}}$ is set to be a plane S-wave propagating in x_2 direction. We first assume the host matrix $\Omega^{(1)}$ to be steel (mass density $\rho = 7.80 \times 10^3$ [kg], Young's modulus $E = 205$ [GPa], Poisson's ratio $\nu = 0.30$) and the inclusion $\Omega^{(2)}$ to be epoxy resin ($\rho = 1.85 \times 10^3$ [kg], $E = 3.00$ [GPa], $\nu = 0.34$).

Figure 2 shows the relative l_2 errors of the numerical solutions ($\boldsymbol{u}$, $\boldsymbol{t}$) on Γ against the analytical solutions, the numbers of iterations for convergence in the GMRES and the distribution of eigenfrequencies which is obtained by the block SSM. We confirm that the exterior fictitious eigenfrequencies (eigenfrequencies of the interior Dirichlet problem) are distributed only on the real axis but the true and interior fictitious eigenfrequencies are in the lower half-plane. The figure also points out that the accuracy of the BEM and the convergence property

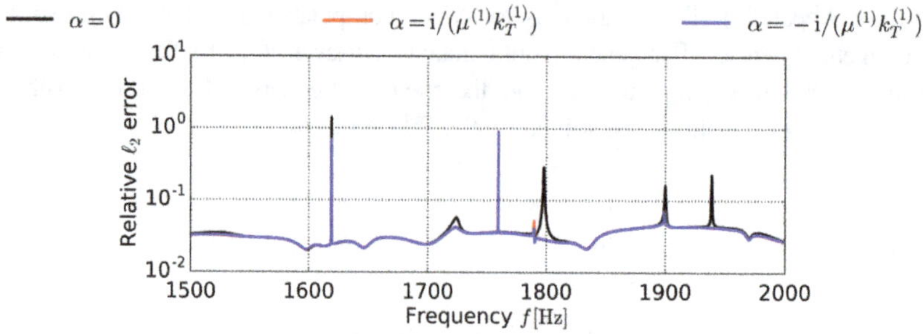

(a) Relative ℓ_2 error against the analytical solution

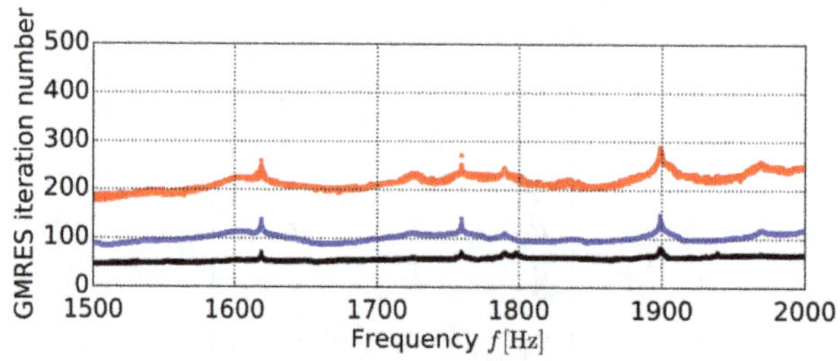

(b) Iteration numbers of the GMRES

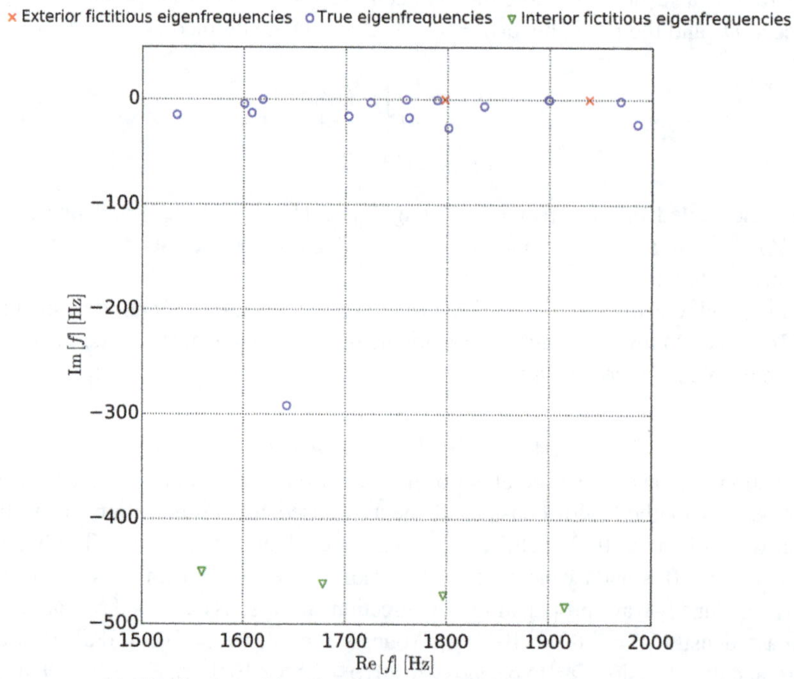

(c) Distribution of the eigenfrequencies in the case of $\alpha = 0$

Figure 2: Relation between the accuracy of the BEM, iteration numbers of the GMRES and distribution of the eigenfrequencies of the BIEs (8), (9) in the case that the host matrix $\Omega^{(1)}$ is the steel and the inclusion $\Omega^{(2)}$ is the epoxy resin.

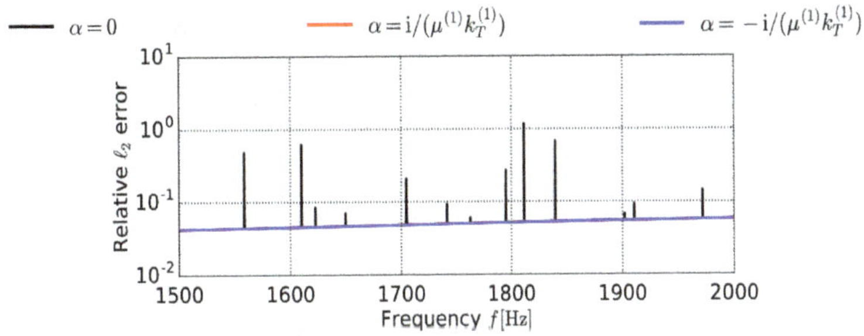

(a) Relative ℓ_2 error against the analytical solution

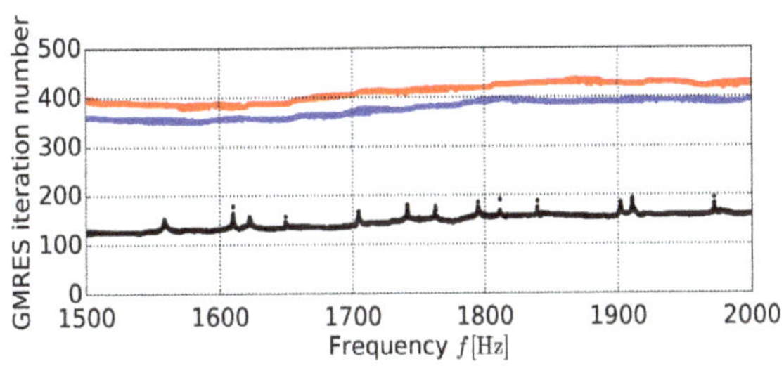

(b) Iteration numbers of the GMRES

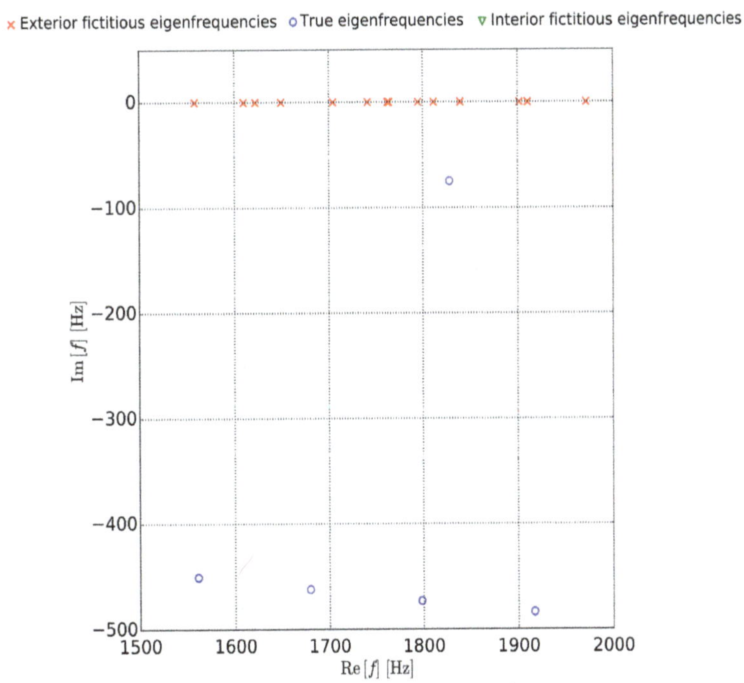

(c) Distribution of the eigenfrequencies

Figure 3: Relation between the accuracy of the BEM, iteration numbers of the GMRES and distribution of the eigenfrequencies of the BIEs (8), (9) in the case that the host matrix $\Omega^{(1)}$ is the epoxy resin and the inclusion $\Omega^{(2)}$ is the steel.

become worse when its frequency is a complex number near the real axis of the complex plane. It is remarkable that not only fictitious eigenfrequencies but also the true eigenfrequencies near the real axis affect the accuracy and convergence property. Note that the effects of the true eigenvalues are inevitable though the fictitious eigenfrequencies can be moved within the complex plane and their effects can be avoided, which will be confirmed below.

Next, we exchange the material parameters of the host matrix and inclusion (the host matrix is the epoxy resin and the inclusion is the steel in this case) and perform the same numerical example. The result is shown in Fig. 3. Different from the previous example, we see that the true eigenfrequencies are not distributed near the real axis and their effects are not observed. In fact, the wavelengths of both the longitudinal and transverse wave in the inclusion $\Omega^{(2)}$ are shorter than in the host matrix $\Omega^{(1)}$ in the previous case, and this would indicate

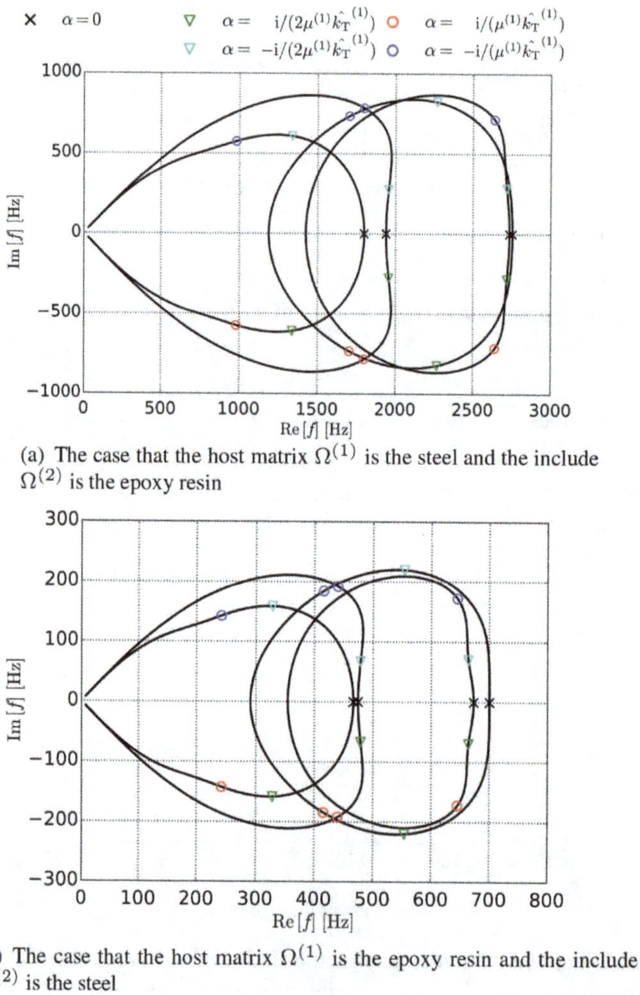

(a) The case that the host matrix $\Omega^{(1)}$ is the steel and the include $\Omega^{(2)}$ is the epoxy resin

(b) The case that the host matrix $\Omega^{(1)}$ is the epoxy resin and the include $\Omega^{(2)}$ is the steel

Figure 4: Loci of the exterior fictitious eigenfrequencies with variation of $\alpha \in \left\{ \mathrm{i}\alpha'/\left(\mu \hat{k}_T^{(1)}\right) \middle| \alpha' \in \mathbb{R} \right\}$.

that there exist some eigenmodes with small damping, which is represented as the negative number of the imaginary part of an eigenfrequency (note that the time dependence $e^{-i\omega t}$ is employed).

Finally, we investigate how the exterior fictitious eigenfrequencies are moved within the complex plane when the coupling parameter α is changed. Considering that some researchers have reported that $\alpha = \pm i/k$ are appropriate choices for Helmholtz' equations, e.g., [8–10], where k is the wavenumber, we examine α in the form of

$$\alpha = \frac{i\alpha'}{\mu^{(1)}\hat{k}_T^{(1)}}, \tag{28}$$

where $\alpha' \in \mathbb{R}$ and $\hat{k}_T^{(1)} \in \mathbb{R}$ are parameters which are independent of ω. Practically $\hat{k}_T^{(1)}$ is determined by the target frequency that is actually used in the boundary element analysis. Figure 4 shows loci of some exterior fictitious eigenfrequencies with variation of $\alpha' \in (-\infty, \infty)$. We employ the wavenumber of the transverse wave $\hat{k}_T^{(1)} \in \mathbb{R}$ at the intersection of the locus and real axis as $\hat{k}_T^{(1)}$ for each locus. We see that the loci are symmetric with respect to the real axis for $\alpha' \in (-\infty, 0]$ and $[0, \infty)$, which implies that the distances between the exterior fictitious eigenfrequencies and the real axis are independent of the sign of α' though it significantly affects the convergence property. Also, the loci indicate that $\alpha = \pm i/\left(\mu^{(1)}\hat{k}_T^{(1)}\right)$ are appropriate choices in the sense of distance from the real axis, and support the results in Figs 2 and 3 that those α can avoid the influence of the exterior fictitious eigenfrequencies. From these results, we conclude that $\alpha = -i/\left(\mu^{(1)}k_T^{(1)}\right)$ is an appropriate choice of the coupling parameter in the Burton-Miller formulation considering the accuracy and convergence properties.

5 CONCLUSION

In this study, we have investigated a distribution of eigenfrequencies of the Burton-Miller-type BIEs in two-dimensional elastodynamics. The Sakurai-Sugiura method (SSM) has been employed for the investigation. The numerical examples show that the Burton-Miller formulation can avoid the fictitious eigenfrequency problem in terms of the distribution and $\alpha = -i/\left(\mu^{(1)}k_T^{(1)}\right)$ is an appropriate choice of its coupling parameter.

ACKNOWLEDGEMENTS

This work was supported by JSPS KAKENHI Grant Numbers JP16H04255 and JP17K14146.

APPENDIX: SERIES REPRESENTATION OF WAVE PROPAGATIONS IN TWO-DIMENSIONAL ELASTODYNAMICS

A solution of the equation (1) which satisfies the radiation condition can be written in polar coordinates $x_1 = r\cos\theta$, $x_2 = r\sin\theta$ as follows [11]:

$$u_r(x) = \frac{1}{r}\sum_{n=-\infty}^{\infty}\left(a_n^{(1)}U_1^{n(1)}(r) + b_n^{(1)}U_2^{n(1)}(r)\right)e^{in\theta} \qquad x \in \Omega^{(1)}, \tag{29}$$

$$u_\theta(x) = \frac{1}{r}\sum_{n=-\infty}^{\infty}\left(a_n^{(1)}V_1^{n(1)}(r) + b_n^{(1)}V_2^{n(1)}(r)\right)e^{in\theta} \qquad x \in \Omega^{(1)}, \tag{30}$$

$$\sigma_{rr}(x) = \frac{2\mu^{(1)}}{r^2}\sum_{n=-\infty}^{\infty}\left(a_n^{(1)}T_{11}^{n(1)}(r) + b_n^{(1)}T_{12}^{n(1)}(r)\right)e^{in\theta} \qquad x \in \Omega^{(1)}, \tag{31}$$

$$\sigma_{r\theta}(x)=\frac{2\mu^{(1)}}{r^2}\sum_{n=-\infty}^{\infty}\left(a_n^{(1)}T_{41}^{n(1)}(r)+b_n^{(1)}T_{42}^{n(1)}(r)\right)\mathrm{e}^{in\theta}\qquad x\in\Omega^{(1)}, \tag{32}$$

$$\sigma_{\theta\theta}(x)=\frac{2\mu^{(1)}}{r^2}\sum_{n=-\infty}^{\infty}\left(a_n^{(1)}T_{21}^{n(1)}(r)+b_n^{(1)}T_{22}^{n(1)}(r)\right)\mathrm{e}^{in\theta}\qquad x\in\Omega^{(1)}, \tag{33}$$

where $a_n^{(\nu)}, b_n^{(\nu)}\in\mathbb{C}$ are constants and $U_i^{n(\nu)}, V_i^{n(\nu)}$ and $T_{ij}^{n(\nu)}(\nu=1,2)$ are defined as

$$U_1^{n(\nu)}(r)=-nH_n^{(1)}\left(k_L^{(\nu)}r\right)+k_L^{(\nu)}rH_{n-1}^{(1)}\left(k_L^{(\nu)}r\right), \tag{34}$$

$$U_2^{n(\nu)}(r)=\mathrm{i}nH_n^{(1)}\left(k_T^{(\nu)}r\right), \tag{35}$$

$$V_1^{n(\nu)}(r)=\mathrm{i}nH_n^{(1)}\left(k_L^{(\nu)}r\right), \tag{36}$$

$$V_2^{n(\nu)}(r)=nH_n^{(1)}\left(k_T^{(\nu)}r\right)-k_T^{(\nu)}rH_{n-1}^{(1)}\left(k_T^{(\nu)}r\right), \tag{37}$$

$$T_{11}^{n(\nu)}(r)=\left(n^2+n-\frac{1}{2}\left(k_T^{(\nu)}\right)^2r^2\right)H_n^{(1)}\left(k_L^{(\nu)}r\right)-k_L^{(\nu)}rH_{n-1}^{(1)}\left(k_L^{(\nu)}r\right), \tag{38}$$

$$T_{12}^{n(\nu)}(r)=-\mathrm{i}n\left((n+1)H_n^{(1)}\left(k_T^{(\nu)}r\right)-k_T^{(\nu)}rH_{n-1}^{(1)}\left(k_T^{(\nu)}r\right)\right) \tag{39}$$

$$T_{21}^{n(\nu)}(r)=-\left(n^2+n+\frac{\lambda^{(\nu)}}{2\mu^{(\nu)}}\left(k_L^{(\nu)}\right)^2r^2\right)H_n^{(1)}\left(k_L^{(\nu)}r\right)+k_L^{(\nu)}rH_{n-1}^{(1)}\left(k_L^{(\nu)}r\right), \tag{40}$$

$$T_{22}^{n(\nu)}(r)=\mathrm{i}n\left((n+1)H_n^{(1)}\left(k_T^{(\nu)}r\right)-k_T^{(\nu)}rH_{n-1}^{(1)}\left(k_T^{(\nu)}r\right)\right), \tag{41}$$

$$T_{41}^{n(\nu)}(r)=-\mathrm{i}n\left((n+1)H_n^{(1)}\left(k_L^{(\nu)}r\right)-k_L^{(\nu)}rH_{n-1}^{(1)}\left(k_L^{(\nu)}r\right)\right), \tag{42}$$

$$T_{42}^{n(\nu)}(r)=-\left(n^2+n-\frac{1}{2}\left(k_T^{(\nu)}\right)^2r^2\right)H_n^{(1)}\left(k_T^{(\nu)}r\right)+k_T^{(\nu)}rH_{n-1}^{(1)}\left(k_T^{(\nu)}r\right). \tag{43}$$

Similarly, we can obtain a solution of the equation (2) as follows:

$$u_r(x)=\frac{1}{r}\sum_{n=-\infty}^{\infty}\left(a_n^{(2)}\tilde{U}_1^{n(2)}(r)+b_n^{(2)}\tilde{U}_2^{n(2)}(r)\right)\mathrm{e}^{in\theta}\qquad x\in\Omega^{(2)}, \tag{44}$$

$$u_\theta(x)=\frac{1}{r}\sum_{n=-\infty}^{\infty}\left(a_n^{(2)}\tilde{V}_1^{n(2)}(r)+b_n^{(2)}\tilde{V}_2^{n(2)}(r)\right)\mathrm{e}^{in\theta}\qquad x\in\Omega^{(2)}, \tag{45}$$

$$\sigma_{rr}(x)=\frac{2\mu^{(2)}}{r^2}\sum_{n=-\infty}^{\infty}\left(a_n^{(2)}\tilde{T}_{11}^{n(2)}(r)+b_n^{(2)}\tilde{T}_{12}^{n(2)}(r)\right)\mathrm{e}^{in\theta}\qquad x\in\Omega^{(2)}, \tag{46}$$

$$\sigma_{r\theta}(x)=\frac{2\mu^{(2)}}{r^2}\sum_{n=-\infty}^{\infty}\left(a_n^{(2)}\tilde{T}_{41}^{n(2)}(r)+b_n^{(2)}\tilde{T}_{42}^{n(2)}(r)\right)\mathrm{e}^{in\theta}\qquad x\in\Omega^{(2)}, \tag{47}$$

$$\sigma_{\theta\theta}(x) = \frac{2\mu^{(2)}}{r^2} \sum_{n=-\infty}^{\infty} \left(a_n^{(2)} \tilde{T}_{21}^{n(2)}(r) + b_n^{(2)} \tilde{T}_{22}^{n(2)}(r) \right) \mathrm{e}^{\mathrm{i}n\theta} \qquad x \in \Omega^{(2)}, \tag{48}$$

where $\tilde{U}_i^{n(v)}, \tilde{V}_i^{n(v)}$ and $\tilde{T}_{ij}^{n(v)}$ are respectively functions which are obtained by replacing the Hankel functions $H_n^{(1)}$ in $U_i^{n(v)}, V_i^{n(v)}$ and $T_{ij}^{n(v)}$ with the Bessel function J_n.

REFERENCES

[1] Schenck, H.A., Improved integral formulation for acoustic radiation problems. *The Journal of the Acoustical Society of America*, **44**(1), pp. 41–58, 1968. https://doi.org/10.1121/1.1911085

[2] Burton, A.J. & Miller, G.F., The application of integral equation methods to the numerical solution of some exterior boundary-value problems. *Proceedings of the Royal Society of London A: Mathematical, Physical and Engineering Sciences*, **323**, pp. 201–210, 1971. https://doi.org/10.1098/rspa.1971.0097

[3] Müller, C., *Foundations of the Mathematical Theory of Electromagnetic Waves*. Springer Science & Business Media, **155**, 2013.

[4] Chew, W.C., Michielssen, E., Song, J.M. & Jin, J.M., *Fast and efficient algorithms in computational electromagnetics*. Artech House, Inc., 2001.

[5] Misawa, R., Niino, K. & Nishimura, N., An FMM for waveguide problems of 2-D Helmholtz' equation and its application to eigenvalue problems. *Wave Motion*, **63**, pp. 1–17, 2016. https://doi.org/10.1016/j.wavemoti.2015.12.006

[6] Misawa, R., Niino, K. & Nishimura, N., Boundary integral equations for calculating complex eigenvalues of transmission problems. *SIAM Journal on Applied Mathematics*, **77(2)**, pp. 770–788, 2017. https://doi.org/10.1137/16m1087436

[7] Asakura, J., Sakurai, T., Tadano, H., Ikegami, T. & Kimura, K., A numerical method for nonlinear eigenvalue problems using contour integrals. *JSIAM Letters*, **1**, pp. 52–55, 2009. https://doi.org/10.14495/jsiaml.1.52

[8] Kress, R., Minimizing the condition number of boundary integral operators in acoustic and electromagnetic scattering. *The Quarterly Journal of Mechanics and Applied Mathematics*, **38(2)**, pp. 323–341, 1985. https://doi.org/10.1093/qjmam/38.2.323

[9] Amini, S. & Harris, P.J., A comparison between various boundary integral formulations of the exterior acoustic problem. *Computer Methods in Applied Mechanics and Engineering*, **84(1)**, pp. 59–75, 1990. https://doi.org/10.1016/0045-7825(90)90089-5

[10] Zheng, C.J., Chen, H.B., Gao, H.F. & Du, L., Is the Burton-Miller formulation really free of fictitious eigenfrequencies? *Engineering Analysis with Boundary Elements*, **59**, pp. 43–51, 2015. https://doi.org/10.1016/j.enganabound.2015.04.014

[11] Eringen, A.C. & Suhubi, E.S., *Elastodynamics, Vol. II, Linear Theory*. Academic Press, 1975.

MULTI-SCALE CARDIOVASCULAR FLOW ANALYSIS BY AN INTEGRATED MESHLESS-LUMPED PARAMETER MODEL

LEONARDO A. BUENO[1], EDUARDO A. DIVO[1] & ALAIN J. KASSAB[2]
[1]Department of Mechanical Engineering, Embry-Riddle Aeronautical University, Daytona Beach, FL, USA.
[2]Department of Mechanical and Aerospace Engineering, University of Central Florida, Orlando, FL, USA.

ABSTRACT
A computational tool that integrates a Radial basis function (RBF)-based Meshless solver with a Lumped Parameter model (LPM) is developed to analyze the multi-scale and multi-physics interaction between the cardiovascular flow hemodynamics, the cardiac function, and the peripheral circulation. The Meshless solver is based on localized RBF collocations at scattered data points which allows for automation of the model generation via CAD integration. The time-accurate incompressible flow hemodynamics are addressed via a pressure-velocity correction scheme where the ensuing Poisson equations are accurately and efficiently solved at each time step by a Dual-Reciprocity Boundary Element method (DRBEM) formulation that takes advantage of the integrated surface discretization and automated point distribution used for the Meshless collocation. The local hemodynamics are integrated with the peripheral circulation via compartments that account for branch viscous resistance (R), flow inertia (L), and vessel compliance (C), namely RLC electric circuit analogies. The cardiac function is modeled via time-varying capacitors simulating the ventricles and constant capacitors simulating the atria, connected by diodes and resistors simulating the atrioventricular and ventricular-arterial valves. This multi-scale integration in an in-house developed computational tool opens the possibility for model automation of patient-specific anatomies from medical imaging, elastodynamics analysis of vessel wall deformation for fluid-structure interaction, automated model refinement, and inverse analysis for parameter estimation.
Keywords: Lumped-Parameter Model, Multi-scale CFD, Meshless Methods, DRBEM

1 INTRODUCTION

Multidimensional hemodynamics analysis of the cardiovascular system usually couples two types of computational models: a 2- or 3-dimensional section of interest where a full-detail hemodynamic analysis can be performed, and a 0-dimensional lumped-parameter model (LPM) of the peripheral circulation [1–4], as shown in Fig. 1. Current state-of-the-art couples, either loosely or strongly, the solution from in-house or commercial fluid solvers, for example, StarCCM+ or Fluent, with a LPM using in scripts to carry information about the boundary conditions back and forth. This is done at every iteration until the resultant parameters reach convergence [5–9]. Solutions such are typically sought for the pressure and flow dynamics, pulmonary artery pressure and flow, coronary artery flow and pressure, and the ratio of pulmonary blow flow and systemic blood flow [4].

The analysis of the full hemodynamics of the aortic arch is proposed using a meshless methods solver. A localized Radial-Basis Function Collocation Meshless Method (LRC-MM) addresses the high computational expense issue that would occur if a global RBF interpolation were to be used [10–12]. When decoupling of the incompressible governing equations of fluid flow, the resultant Poisson-like equations that arise from the solution of the pressure field at every time step are solved by a Dual Reciprocity Boundary Element Method (DR-BEM) which ensures high stability and accuracy. DR-BEM transfers the generation term of the resultant Poisson equation to the boundary via an RBF expansion [13, 14]. Since there is an existing point distribution from the LRC-MM, it can be used again when solving the boundary-only integration and determining the internal solution [15].

DOI: 10.2495/CMEM-V6-N6-1138-1148

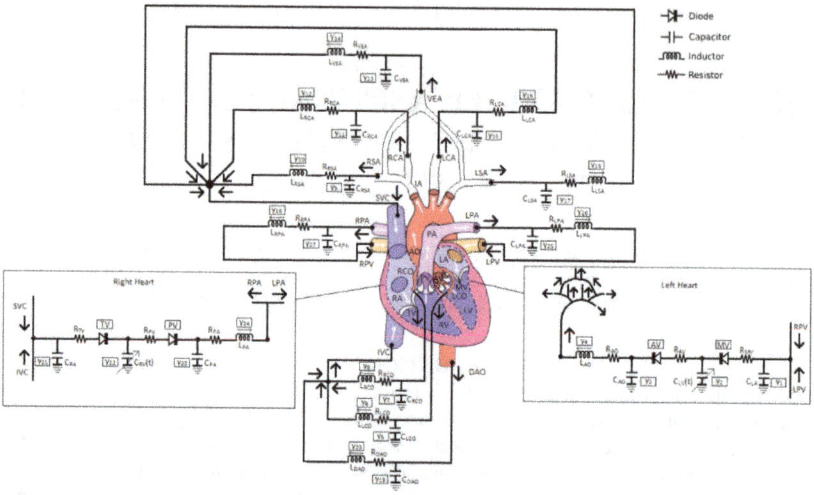

Figure 1: Visual representation of the integrated LRC-MM and LPM.

Beyond the region of interest where the details of the hemodynamics are to be resolved, the aortic bed for instance, the peripheral circulation can be simulated using a 0D lumped parameter model based on an electric analogy to fluid flow. Utilizing the Windkessel compartment model, the LPM is used to reproduce the impedance that the peripheral bed presents to the inlet and outlets of mutli-dimensional region of interest via a combination of effective viscous resistance to flow, compliance, and convective acceleration or inertance experienced by the flow in the peripheral vasculature [16]. In this instance, the multi-compartment model used is an RLC (resistance, inductance and capacitance) arrangement that can be used to flexibly model the entire network. Instead of treating the entire network as a single block, the different segments allow for the flow and pressure in different sections to be calculated [17]. The heart is also represented in this system though a combination of diodes that model heart valves, resitances and inductances, as well as a time-varying elastance function (inverse of the compliance) that drives the circuit. The latter serves as the analogue of the elastance of the myocardium [18].

2 3-D MESHLESS SOLVER APPROACH

The hemodynamics of the aorta will be solved using a 3-D meshless solver for the Navier-Stokes equations [10].

$$\begin{aligned} &\nabla \cdot \vec{V} = 0 \\ &\rho \frac{\partial \vec{V}}{\partial t} + \rho\left(\vec{V} \cdot \nabla\right)\vec{V} = \mu \nabla^2 \vec{V} - \nabla p + \vec{f} \end{aligned} \tag{1}$$

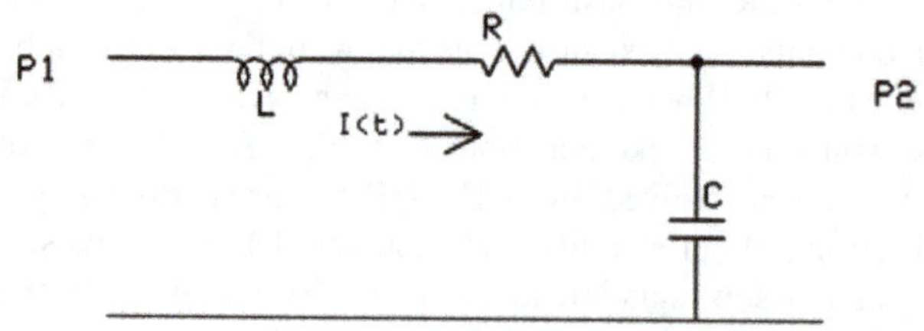

Figure 2: Single RLC compartment used in LPM.

For this incompressible case, the density, ρ, is constant [10]. The field variables of interest are the velocity, $\vec{V}(x,t)$ and the pressure $p(x,t)$. Therefore, the velocity advances as:

$$\vec{V}^{n+1}(x) = \vec{V}^{n}(x) + \Delta t\left[\frac{\mu}{\rho}\nabla^2\vec{V}^{n}(x) - \left(\vec{V}^{n}(x)\cdot\nabla\right)\vec{V}^{n}(x) - \frac{1}{\rho}\nabla p^{n+1}(x) + \frac{1}{\rho}\vec{f}^{n}(x)\right] \tag{2}$$

Given that there is no explicit pressure equation, a pressure correction field is introduced:

$$p^{n+1}(x) = p^{n}(x) + \phi^{n}(x) \tag{3}$$

Substituting (3) into (2), eventually leads to a Poisson equation whose solution results in the pressure correction, ϕ, from (3) [10]:

$$\nabla^2\phi^{n}(x) = \frac{\rho}{\Delta t}\left(\nabla\cdot\vec{V}^{*}(x)\right) \tag{4}$$

2.1 Dual Reciprocity Boundary Element Method (DR-BEM)

Given the need to solve (4) for the field variable, the pressure correction $\phi(x)$, a boundary element method (BEM) approach is proposed through the formulation of a boundary integral equation (BIE). The purpose of the BIE, obtained through an RBF expansion, is to obtain a BEM solution to the necessary domain integration that results from Green's free-space solution [10]. For the Poisson equation in (4), the right hand side is known and non-homogeneous, $B(x)$. Considering a weight function, $G(x,\zeta)$, applying Green's 2nd identity to the resultant equation while letting $G(x,\zeta)$ be Green's free-space solution of the Laplace equation that solved the adjoint equation, $\nabla^2 G(x,\zeta) = -\delta(x,\zeta)$, where $\delta(x,\zeta)$ is the Dirac Delta function. For example, in 2D the free-space solution is $G(x,\zeta) = -\ln(r)/2\pi s$, and there results an integral equation in terms of the field variable:

$$c(\zeta)\phi^{n}(\zeta) + \oint_{\Gamma}\phi^{n}(\zeta)H(x,\zeta)d\Gamma = \oint_{\Gamma}q^{n}(x)G(x,\zeta)d\Gamma + b(\zeta) \tag{5}$$

This solution still contains terms that require a domain integral:

$$q^{n}(x) = \frac{\partial\phi^{n}(x)}{\partial n} \qquad b(\zeta) = \int_{\Omega} B(x)G(x,\zeta)d\Omega$$
$$H(x,\zeta) = \frac{\partial G(x,\zeta)}{\partial n} \qquad c(\zeta) = \int_{\Omega}\delta(x,\zeta)d\Omega \tag{6}$$

However, the parameter $c(\zeta)$ is 1 inside the domain and 0.5 at the boundary. $b(\zeta)$ still contains a domain integral but with an RBF expansion of $B(x)$, using polyharmonic RBFs, eventually leads to its boundary form of:

$$b_i = \sum_{k=1}^{N}\alpha_k\left[\oint_{\Gamma}p_k(x)G(x,\zeta_i)d\Gamma - \oint_{\Gamma}u_k(x)H(x,\zeta_i)d\Gamma - c(\zeta_i)u_k(\zeta_i)\right]$$
$$p_k(x) = \frac{\partial u_k(x)}{\partial n} \tag{7}$$
$$H(x,\zeta) = \frac{\partial G(x,\zeta)}{\partial n}$$

And the BIE can be fully written in terms of boundary integrals:

$$c_i\phi_i^n + \sum_{j=1}^{N_B} \hat{H}_{ij}\phi_j^n = \sum_{j=1}^{N_B} G_{ij}q_j^n + \sum_{k=1}^{N} \alpha_k \left[\sum_{j=1}^{N_B} G_{ij}p_j^k - \sum_{j=1}^{N_B} H_{ij}u_j^k \right] \tag{8}$$

From the boundary solution, the field variable can be calculated at point inside the domain without the use of domain integrals. In addition, any derivative of the field variable can now be determined through the discretized form of the BIE.

3 LUMPED-PARAMETER MODEL

In the RLC compartment used to simulate the state of a set of vessels each component represents a physical characteristic: a resistance (R), accounts for viscous drag, an inductance (L), for flow inertia, and a capacitance (C), for vascular wall compliance. In the case of relevant variables, pressure (P) is voltage (V) and flow rate (Q) is current (I) [19], [20]. A time-varying capacitance, $C(t)$, represents the contractile behavior of the ventricle. A diode acts as the corresponding cardiac valve and insures unidirectional flow over the relaxation and contraction phases of the cardiac cycle [18, 21].

Given the incompressible Navier-Stokes equation from (1) and assuming fully developed axial flow and neglecting the external body force, the pressure gradient results in:

$$\frac{\partial p(z,t)}{\partial z} = \frac{\mu}{r}\frac{\partial}{\partial r}\left[r\frac{\partial u(r,t)}{\partial r} \right] - \rho\frac{\partial u(r,t)}{\partial t} \tag{9}$$

The axial velocity profile in the circular cross-section of radius $\Re$, assuming laminar flow, is:

$$u(r,t) = 2\bar{u}(t)\left[1 - \left(\frac{r}{\Re} \right)^2 \right] \tag{10}$$

Where $\bar{u}(t)$ is the time-varying cross-section-averaged mean axial velocity. After evaluating and integrating over the cross-section:

$$\frac{\partial p(z,t)}{\partial z} A_c = -\frac{8\mu}{\Re^2} Q(t) - \rho\frac{dQ(t)}{dt} \tag{11}$$

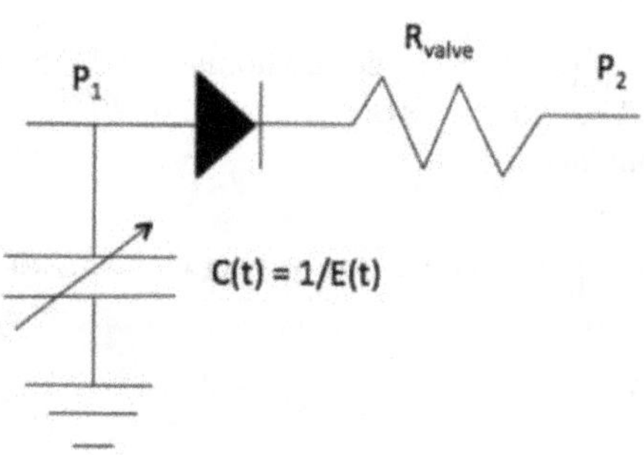

Figure 3: Electrical compartment containing the time-varying compliance.

Where A_c is the cross-section and $Q(t)$ is the volumetric flow rate. The resultant pressure drop, Δp, along a distance, Δp, is:

$$\Delta p(t) = \left(\frac{8\mu l}{\Re^2 A_c}\right) Q(t) + \left(\frac{\rho l}{A_c}\right) \frac{dQ(t)}{dt} \tag{12}$$

The analogy of the cardiovascular behavior and the electrical components is given in Table 1 [20]. The heaviside step function is used to represent the behavior of the heart valves, where the pressure drop controls the valve opening or closing depending on a pressure difference ΔP thus allowing the blood to flow in only one direction.

Since the total voltage drop through a resistor-inductor is:

$$\Delta v(t) = R\,i(t) + L\frac{di(t)}{dt} \tag{13}$$

Yielding the actual resistance and inductance in hydraulic flow terms from (12):

$$R = \frac{8\mu l}{\Re^2 A_c} \qquad L = \frac{\rho l}{A_c} \tag{14}$$

3.1 The Heart in LPM

Elastance is related to the ventricle pressure and volume according to [18]:

$$E(t) = \frac{P(t)}{V(t) - V_0} \tag{15}$$

Where V_0 is the volume of the ventricle at zero pressure. Mathematically, the time varying elastance is given by:

$$E(t) = (E_{max} - E_{min}) E_n(t_n) + E_{min} \tag{16}$$

The ventricular function can be simulated using the "double hill function" that represents the normalized elastance $E_n(t_n)$ [21]:

$$E_n(t_n) = 1.55\left[\frac{\left(\frac{t_n}{0.7}\right)^{1.9}}{1 + \left(\frac{t_n}{0.7}\right)^{1.9}}\right]\left[\frac{1}{1 + \left(\frac{t_n}{1.17}\right)^{21.9}}\right] \tag{17}$$

Table 1. Cardiovascular behavior and corresponding electrical analogy.

Cardiovascular Behavior	Electrical Component	Relationship
Vessel Resistance	Resistor	$\Delta P = QR$
Blood Vessel Compliance	Capacitor	$Q = CdP/dt$
Flow Inertia	Inductor	$\Delta P = LdQ/dt$
Heart valve	Diode	$Q = (\Delta P / R_{valve}) H(\Delta P)$

Where: $t_n = t/T_{\max}$ and $T_{\max} = 0.2 + 0.15t_c$ such that t_c is the cardiac cycle interval in seconds.

For a healthy adult left ventricle: $E_{\max} = 2\,mmHg/ml$ and $E_{\min} = 0.06\,mmHg/ml$, [21]. The time-dependant elastance is the reciprocal of the time-varying capacitance, $E(t) = 1/C(t)$ and are shown in Fig. 4 for a heart rate, HR, of 60 beats per minute (bpm).

4 COUPLING OF LRC-MM AND LPM

The multi-scale analysis requires the full system to be integrated such that the LPM is connected to the LRC-MM in order to have a closed-loop system. At points of coupling, relevant flow information will have to be converted from 3D- to 0D-analysis and back to 3D in order to complete the close-loop as in Fig. 1. The concept behind this coupling process is illustrated in Fig. 5.

The proposed approach is different on whether coupling is from the meshless solver to the LPM or vice versa. The two relevant parameters used are the local pressure and the volumetric flow rate.

- 3D-0D connection: the average values from the nodes at the cross-sectional surface used as input for the 0D compartment(s).
- 0D-3D connection: due to the assumption of Poiseuille flow, as mentioned earlier, the velocity from the LPM, u_{mean}, can be used to calculate the parabolic shape of flow field across the nodes on the surface. The local pressure from the LPM is transferred uniformly across the surface.

5 PRELIMINARY RESULTS

RBF-based Meshless solvers have been able to provide reliable simulations for several applications [11, 12, 15]. As a preliminary test to the multi-scale algorithms, a simplified model of

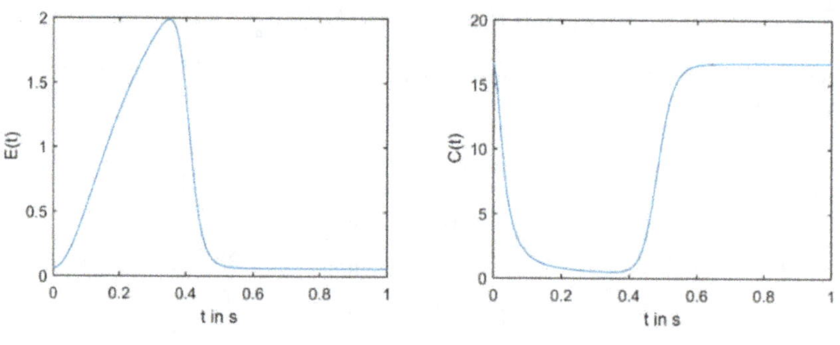

Figure 4. Elastance, $E(t)$ and ventricular compliance, $C(t)$, for a healthy heart at 60 bpm.

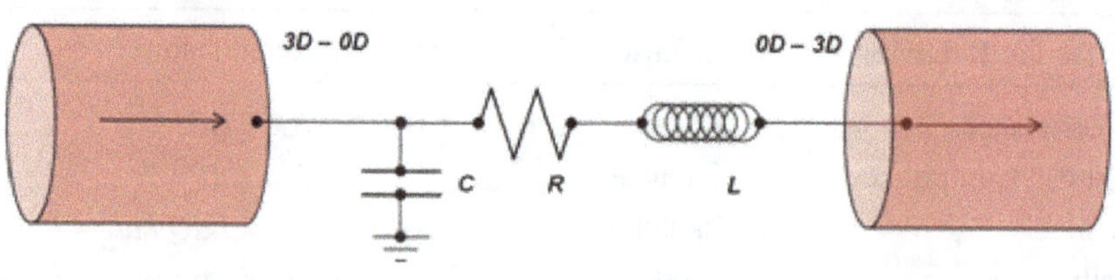

Figure 5: Coupling between an RLC compartment and 3D LRC-MM analysis.

a bulging and bifurcating vessel is configured to be loosely coupled with the LPM. The volumetric flow rate resulting from the LPM is passed at very time step as the inlet condition to the meshless model. The fluid properties of blood are used with the density $\rho = 1{,}060\ Kg/m^3$ and the viscosity $\mu = 3.5 \cdot 10^{-3}\, Kg/m \cdot s$.

The LPM is tested using a simplified model of the cardiovascular system, accounting for the compliance of the left atrium (C_A), mitral valve (D_M, R_M), left ventricle, driving the system with the time-varying capacitance ($C(t)$), aortic valve (D_A, R_A), aorta compliance and characteristic resistance and inductance (C_{AO}, R_{AO}, L_{AO}), and systemic circulation compliance and resistance (C_S, R_S) [18, 21]. The values associated with each of the resistances, compliances, or inductances, are well documented in the literature [21–23]. The arrangement using RLC compartments and a variable capacitance for the left ventricular compliance is shown in Fig. 6.

The LPM is solved for a healthy adult at 60 bpm over a period of 5 seconds. The resultant pressures at different locations: the left atrium (LA), P_{LA}, the left ventricle (LV), P_{LV}, the aorta (AO), P_{AO}, and the systemic pressure, P_{SYS} are shown in Fig. 7.

The meshless model consists of a bulging and bifurcating vessel with an inlet hydraulic diameter of 3cm bulging to 5cm after 10cm of entry length. Bifurcation occurs after 30cm and the outlets have hydraulic diameters of 1.5cm and 2cm. The overall length of the model

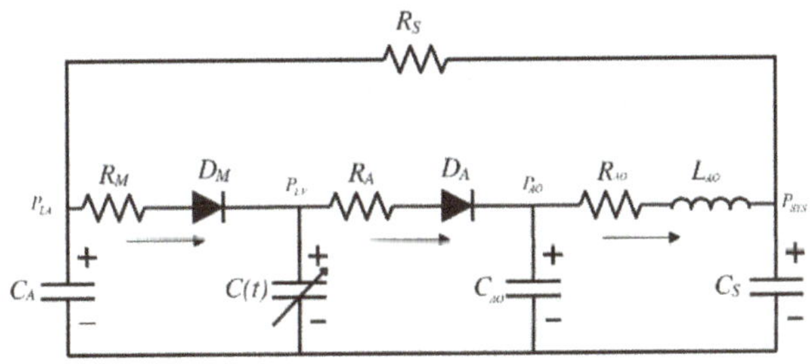

Figure 6: Simplified Cardiovascular circuit model [18].

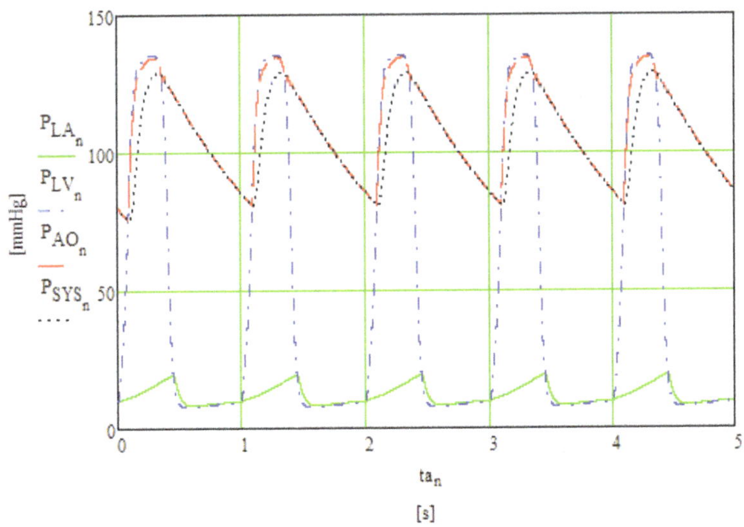

Figure 7: Pressure and Volumetric Flow after 5 cycles in LPM.

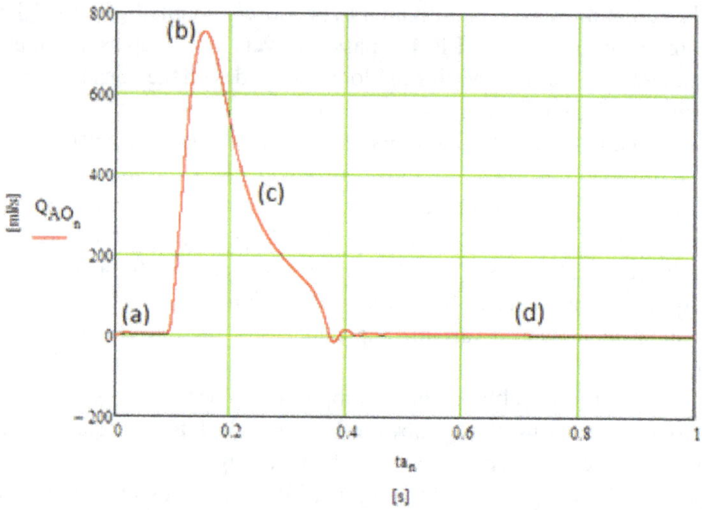

Figure 8: Volumetric flow rate for a single cycle of LPM.

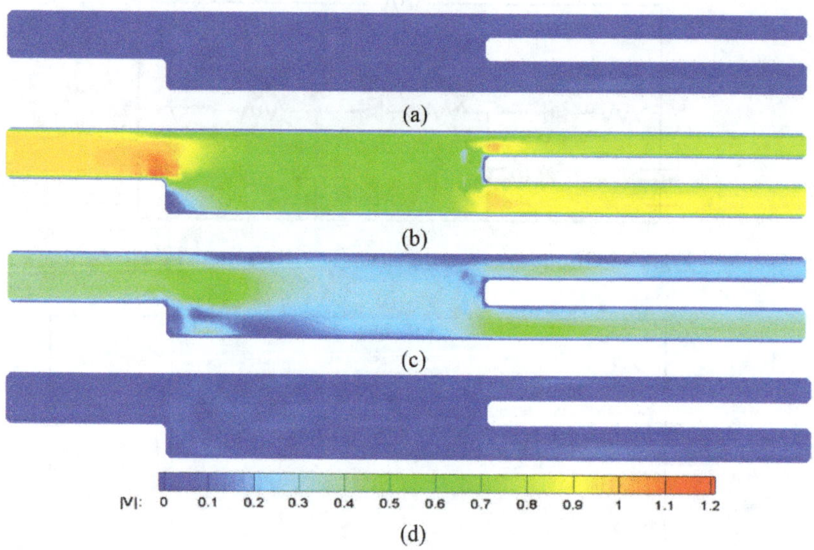

Figure 9: Velocity contour plots at representative locations throughout heart cycle: (a) onset of systole, (b) peak systole, (c) end of systole, and (d) diastole.

is 50cm. The resulting LPM flow rates yield an average representative Reynolds number around 1,200. Representative locations throughout the cardiac cycle, as shown in Fig. 8 for the LPM-generated flow rates, are selected to plot the corresponding velocity contours as shown in Fig. 9, for the last of the 5 generated cardiac cycles.

6 CONCLUSION

The proposed approach to address the multi-scale analysis of cardiovascular flow utilizes a Localized Radial-Basis Function Collocation Meshless Method (LRC-MM) solver aided by a stable and robust Dual-Reciprocity Boundary Element Method (DR-BEM) solver to address the pressure field at every time step of the solution for the local hemodynamics at a 3D scale. The coupled LRC-MM/DR-BEM approach is implemented over the same boundary discretization and internal point distribution which is automatically generated and allows for direct CAD integration. The peripheral circulation and heart function is represented in a 0D scale by electrical circuit analogies of the Lumped Parameter Models (LPM) where the viscous drag is represented by a resistor (R), the flow inertia is represented by an inductor (L), the wall compliance is represented by a capacitor (C), and the heart valve is represented by a diode (D). This electrical elements are lumped together in compartments representing the different branches of the cardiovascular circuit. The LRC-MM/DR-BEM and 0D LPM are effectively coupled to form a multi-scale closed circuit of the cardiovascular system. The in-house computational tool will further enable analysis of patient-specific geometry, fluid-structure interaction analysis of vessel wall deformation, automated model refinement. LPM model parameters can be obtained on a patient specific basis given measured pressure and flow waveforms on the basis for inverse analysis.

REFERENCES

[1] Migliavacca, F., Pennati, G., Dubini, G., Fumero, R., Pietrabissa, R., Urcelay, G., Bove, E.L., Hsia, T-Y. & de Leval, M.R., Modeling of the Norwood circulation: effects of shunt size, vascular resistances, and heart rate. *American Journal of Physiology - Heart and Circulatory Physiology*, **280**(5), pp. H2076–H2086, 2001.

[2] Migliavacca, F., Balossino, R., Pennati, G., Dubini, G., Hsia, T-Y., de Leval, M.R. & Bove, E.L., Multiscale modelling in biofluidynamics: application to reconstructive paediatric cardiac surgery. *Journal of Biomechanics*, **39**(6), pp. 1010–1020, 2006. https://doi.org/10.1016/j.jbiomech.2005.02.021

[3] DeCampli, W.M., Argueta-Morales, I.R., Divo, E. & Kassab, A.J., Computational fluid dynamics in congenital heart disease. *Cardiology in the Young*, **22**(06), pp. 800–808, 2012.
https://doi.org/10.1017/s1047951112002028

[4] Hsia, T-Y., Cosentino, D., Corsini, C., Pennati, G., Dubini, G., Migliavacca, F. & Modeling of Congenital Hearts Alliance (MOCHA) Investigators. Use of mathematical modeling to compare and predict hemodynamic effects between hybrid and surgical Norwood palliations for hypoplastic left heart syndrome. *Circulation*, **124**(11 Suppl), pp. S204–S210, 2011.
https://doi.org/10.1161/circulationaha.110.010769

[5] Ceballos, A., *A Coupled CFD-Lumped Parameter Model of the Human Circulation: Elucidating the Hemodynamics of the Hybrid Norwood Palliative Treatment and Effects of the Reverse Blalock-Taussic Shunt Placement and Diameter*. UCF, Orlando, Florida, 2015.

[6] Ceballos, A., Argueta-Morales, I.R., Divo, E., Osorio, R., Caldarone, C.A., Kassab, A.J. & DeCampli, W.M., Computational analysis of hybrid norwood circulation with distal aortic arch obstruction and reverse Blalock-Taussig Shunt. *The Annals of Thoracic Surgery*, **94**(5), pp. 1540–1550, 2012.
https://doi.org/10.1016/j.athoracsur.2012.06.043

[7] Prather, R., *A Multi-Scale CFD Analysis of Patient-Specific Geometries to Tailor LVAD Cannula Implantation Under Pulsatile Flow Conditions: an investigation aimed at reducing stroke incidence in LVADs*. University of Central Florida, Orlando, 2015.

[8] Prather, R.O., Kassab, A., Ni, M.W., Divo, E., Argueta, I.R. & DeCampli, W.M., Multi-scale pulsatile CFD modeling of thrombus transport in a patient-specific LVAD implantation. *International Journal of Numerical Methods for Heat & Fluid Flow*, **27**(5), pp. 1022–1039, 2017.
https://doi.org/10.1108/hff-10-2016-0378

[9] Quarteroni, A., Ragni, S. & Veneziani, A., Coupling between lumped and distributed models for blood flow problems. *Computing and Visualization in Science*, **4**(2), pp. 111–124, 2001.
https://doi.org/10.1007/s007910100063

[10] Pepper, D.W., Kassab, A. & Divo, E., *An Introduction to Finite Element, Boundary Element, and Meshless Methods with Applications to Heat Transfer and Fluid Flow*, ASME Press, New York, 2014.

[11] Divo, E. & Kassab, A.J., Localized meshless modeling of natural-convective viscous flows. *Numerical Heat Transfer, Part B: Fundamentals*, **53**(6), pp. 487–509, 2008.
https://doi.org/10.1080/10407790802083190

[12] Divo, E. & Kassab, A.J., An efficient localized radial basis function meshless method for fluid flow and conjugate heat transfer. *Journal of Heat Transfer*, **129**(2), pp. 124–136, 2006.
https://doi.org/10.1115/1.2402181

[13] Wrobel, L.C. & Brebbia, C.A., The dual reciprocity boundary element formulation for nonlinear diffusion problems. *Computer Methods in Applied Mechanics and Engineering*, **65**(2), pp. 147–164, 1987.
https://doi.org/10.1016/0045-7825(87)90010-7

[14] Divo, E. & Kassab, A.J., Transient non-linear heat conduction solution by a dual reciprocity boundary element method with an effective posteriori error estimator. *Heat Transfer,* **1**, pp. 77–86, 2004.
https://doi.org/IMECE2004-59262

[15] Bueno, L.A., Divo, E.A. & Kassab, A.J., A coupled localized Rbf Meshless/DRBEM formulation for accurate modeling of incompressible fluid flows. *International Journal of Computational Methods and Experimental Measurements*, **5**(3), pp. 359–368, 2017.
https://doi.org/10.2495/cmem-v5-n3-359-368

[16] Kind, T., Faes, T.J.C., Lankhaar, J.W., Vonk-Noordegraaf, A. & Verhaegen, M., Estimation of three- and four-element windkessel parameters using subspace model identification. *IEEE Transactions on Biomedical Engineering*, **57**(7), pp. 1531–1538, 2010.
https://doi.org/10.1109/tbme.2010.2041351

[17] Kokalari, I., Karaja, T. & Guerrisi, M., Review on lumped parameter method for modeling the blood flow in systemic arteries. *Journal of Biomedical Science and Engineering*, **06**(01), pp. 92–99, 2013.
https://doi.org/10.4236/jbise.2013.61012

[18] Faragallah, G., Wang, Y., Divo, E. & Simaan, M., A new control system for left ventricular assist devices based on patient-specific physiological demand. *Inverse Problems in Science and Engineering*, **20**(5), pp. 721–734, 2012.
https://doi.org/10.1080/17415977.2012.667092

[19] Formaggia, L. & Veneziani, A., Reduced and multiscale models for the human cardio-vascular system. *Lecture Notes VKI Lecture Series*, **7**, 2003.

[20] Creigen, V., Ferracina, L., Hlod, A., van Mourik, S., Sjauw, K., Rottschäfer, V., Vellekoop, M. & Zegeling, P., Modeling a heart pump. *European Study Group Mathematics with Industry*, Utrecht, p. 7, 2007.

[21] Simaan, M.A., Ferreira, A., Chen, S., Antaki, J.F. & Galati, D.G., A dynamical state space representation and performance analysis of a feedback-controlled rotary left ventricular assist device. *IEEE Transactions on Control Systems Technology*, **17**(1), pp. 15–28, 2009.
https://doi.org/10.1109/tcst.2007.912123

[22] Yu, Y-C., Boston, J.R., Simaan, M.A. & Antaki, J.F., Minimally invasive estimation of systemic vascular parameters. *Annals of Biomedical Engineering*, **29**(7), pp. 595–606, 2001.
https://doi.org/10.1114/1.1380420

[23] Yu, Y-C., Boston, J.R., Simaan, M.A. & Antaki, J.F., Estimation of systemic vascular bed parameters for artificial heart control. *IEEE Transactions on Automatic Control*, **43**(6), pp. 765–778, 1998.
https://doi.org/10.1109/9.679017

INVERSE HEAT CONDUCTION IN ANISOTROPIC AND FUNCTIONALLY GRADED MEDIA BY FINITE INTEGRATION METHOD

J. JIN[1], J.L. ZHENG[1], T. HUANG[1], J.J. YANG[1], H.S. WANG[2], P.H.WEN[2] & J.M. LI[3]
[1] School of Communication and Transportation Engineering, Changsha University of Science and Technology, China.
[2]School of Engineering and Materials Science, Queen Mary University of London, UK.
[3]Department of Thermal Engineering, Tsinghua University, Beijing, China.

ABSTRACT

Based on the recently developed finite integration method (FIM) for solving one and two dimensional ordinary and partial differential equations, this paper extends FIM to both stationary and transient heat conduction inverse problems for anisotropic and functionally graded materials with high degree of accuracy. Lagrange series approximation is applied to determine the first order of integral and differential matrices, which are used to form the system equation matrix for two and three dimensional problems. Singular Value Decomposition (SVD) is applied to solve the ill-conditioned system of algebraic equations obtained from the integral equation, boundary conditions and scattered temperature measurements. The convergence and accuracy of this method are investigated with two examples for anisotropic media and functionally graded materials.

Key words: finite integral method, integration matrix, inverse heat conduction, Lagrange series.

1 INTRODUCTION

Inverse heat conduction problems (IHCP) are used to estimate unknown quantities on the unreachable boundaries based on temperature and heat flux measurements in the analysis of physical problems in thermal engineering. As an example, the inverse heat conduction problem is used to compute the unknown temperature and heat flux at an unreachable boundary from scattered temperature measurements either at reachable interior points or boundary of the domain. In order to obtain the maximum information regarding the physical problem under study, IHCP enables a much closer collaboration between experimental and theoretical researchers. Since the solutions may become unstable as a result of errors inherent the measurements the difficulties encountered in the solution of IHTP should be recognized.

IHCP is mathematically classified as ill-posed in a general sense and is main reason that inverse problems were initially taken as not of physical interest. Traditional computational methods including the finite element method, boundary element method and meshless method for well-posed direct problems fail to produce acceptable solutions to these kinds of inverse problems. Several techniques have been proposed for solving a one-dimensional IHCP [1–5]. Applications with difference techniques can be found by the finite difference [6, 7], boundary element [8, 9] and finite elements [10, 11] for two-dimensional IHCP. More recently, the method of fundamental solution [12, 13] and meshless methods have been developed to IHCP in two and three dimensions such as meshless local Petrov-Galerkin method [14, 15], which have recently attracted great attention in science and engineering communities for numerically solving time-dependent problems.

Despite the great success of the finite element method, boundary element method and meshless methods, there is still a need for developing numerical schemes for multi-dimensional IHCP. Recently, Wen and his colleagues [16–18] developed the Finite Integration Method (FIM) for solving one and two-dimensional, differential equation problems and

DOI: 10.2495/CMEM-V6-N6-1149-1160

demonstrated its applications to non-local elasticity problems. It has been shown that the FIM gives much higher degree of accuracy than the Finite Difference Method (FDM) and the Point Collocation Method (PCM). In this paper, the FIM is further extended to solve IHCP for anisotropic media and functionally graded materials.

2 FORMULATION OF INVERSE HEAT CONDUCTION PROBLEMS

We assume that in engineering, the thermal conductivities $\lambda_{ik}(\mathbf{x})$are directionally dependent properties of an anisotropic material. All the properties of the material could be dependent on the spatial coordinates in a non-homogeneous material. The governing equation for steady heat conduction in anisotropic and continuously non-homogeneous media is given by the following partial differential equations with spatially variable properties:

$$\begin{aligned} &q_{i,i}(\mathbf{x}) = w(\mathbf{x}), && \mathbf{x}(x,y) \in \Omega \\ &q_i(\mathbf{x}) = -\lambda_{ik}(\mathbf{x}) T_{,k}(\mathbf{x}) && \\ &\alpha \overline{T}(\mathbf{x}) + \beta \overline{q}_i(\mathbf{x}) n_i = g(\mathbf{x}), && \mathbf{x}(x,y) \in \Gamma \end{aligned} \tag{1}$$

where T denotes the temperature distribution and$w(\mathbf{x})$ represents the heat source term, the first term is the rate of internal energy with time, the second term$q_i(\mathbf{x})$ is the heat flux vector, $\lambda_{ik}(\mathbf{x})$ is the thermal conductivity tensor, coefficients α and β are constants, n_i is the unit outward normal to the boundary. In addition, $T_0(\mathbf{x})$ is a given initial temperature in the domain and $g(\mathbf{x},t)$ gives the boundary value. For inverse heat conduction problem, it is convenient to assume that the boundary Γ consists of three parts $\Gamma = \left(\Gamma_u \cup \Gamma_q \cup \Gamma_\sigma\right)$ and the boundary conditions are given as

$$T(\mathbf{x}) = \overline{T}(\mathbf{x}) \qquad \mathbf{x} \in \Gamma_u, P_T \qquad \text{(Dirichlet b.c. and Pint temperature)} \tag{2a}$$

$$q(\mathbf{x}) = q_i(\mathbf{x}) n_i(\mathbf{x}) = \overline{q}(\mathbf{x}) \qquad \mathbf{x} \in \Gamma_q \qquad \text{(Neumann b.c.)} \tag{2b}$$

where $\overline{T}(\mathbf{x})$ and $\overline{q}(\mathbf{x})$are temperature and heat flux specified on the boundary. The IHCP to be investigated in this paper is to determine the temperature on the unreachable boundary Γ_ω from given Dirichlet data on Γ_u, Neumann data on Γ_q and scattered temperature measurements $T_P(\mathbf{x})$ at some points P_T either in the domain Ω or on the Neumann data boundary Γ_q.

3 FIM USING LAGRANGE SERIES

The simplest computation scheme for integration was introduced in Refs. [16, 17] and named as the Ordinary Linear Approach (OLA). For convenience of analysis in the following sections, coordinates x and y are used to replace x_1 and x_2, respectively, and subscripts i or k presents the nodal number. Consider an integral and apply trapezoidal integral rule

$$U(x_k) = \int_0^{x_k} u(\xi) d\xi = \sum_{i=1}^{k} a_{ki} u(x_i) \tag{3}$$

For trapezoidal rule, the coefficients are

$$a_{1i} = 0,$$
$$a_{ki} = \begin{cases} h, & i = 1, \\ 0.5h, & i = 2,3,...,k-1, \\ 0.5h, & i = k, \\ 0, & i > k, \end{cases} \tag{4}$$

where $x_i = (i-1)h,\ h = a/(N-1),\ i = 1,2,...,N$, are nodal points in [0,a], and $x_1 = 0, x_N = a$. Note that (3) can be written in a matrix form as

$$\mathbf{U} = \mathbf{A}\mathbf{u} \tag{5}$$

where $\mathbf{U} = \left[U_1, U_2, ..., U_N\right]^T$, $\mathbf{u} = \left[u_1, u_2, ..., u_N\right]^T$, and the first order of integration matrix

$$\mathbf{A} = \left(a_{ki}\right) = \Delta \begin{pmatrix} 0 & 0 & 0 & 0 & 0 & 0 \\ 1/2 & 1/2 & 0 & 0 & 0 & 0 \\ 1/2 & 1 & 1/2 & 0 & 0 & 0 \\ 1/2 & 1 & 1 & 1/2 & 0 & 0 \\ ... & ... & ... & ... & ... & ... \\ 1/2 & 1 & 1 & 1 & 1 & 1/2 \end{pmatrix}_{N\times N}, \tag{6}$$

and $U_i = U(x_i),\ u_i = u(x_i)$ are the values of integration and the integral function respectively at each nodes.

The first order of integral matrix A in eqn (6) can be evaluated using the Lagrange interpolation series. The integral function is approximated as Ref. [17]

$$u(x) = \sum_{k=1}^{N} c_k x^{k-1} \tag{7}$$

where the coefficients $\{c_k\}_{k=1}^{N}$ are defined by

$$\begin{bmatrix} 1 & x_1 & x_1^2 & ... & x_1^{N-1} \\ 1 & x_2 & x_2^2 & ... & x_2^{N-1} \\ ... & ... & ... & ... & ... \\ 1 & x_N & x_N^2 & ... & x_N^{N-1} \end{bmatrix} \begin{pmatrix} c_1 \\ c_2 \\ ... \\ c_N \end{pmatrix} = \begin{pmatrix} u_1 \\ u_2 \\ ... \\ u_N \end{pmatrix} \tag{8}$$

or the integral function can be written as

$$u(x) = \sum_{j=1}^{N} \prod_{\substack{k=1 \\ k\neq j}}^{N} \frac{(x - x_k)}{(x_j - x_k)} u_j \tag{9}$$

Therefore, we have

$$U(x_i) = \int_0^{(i-1)h} u(\xi)d\xi = \sum_{k=1}^{N} \frac{c_k}{k} x_i^k = \sum_{k=1}^{N} \frac{c_k}{k} (i-1)^2 h^2 \quad i = 1,2,...,N \tag{10}$$

For different boundary conditions, the first and second order of derivatives should be considered. Then for any points including two ends, from eqn (9), we have

$$\frac{du}{dx} = \sum_{k=2}^{N} (k-1) c_k x^{k-2} \text{ and } \frac{d^2u}{dx^2} = \sum_{k=3}^{N} (k-1)(k-2) c_k x^{k-3}. \tag{11}$$

From eqn (9), the first order derivative can also be written as

$$\frac{du}{dx} = u'(x) = \sum_{j=1}^{N} u_j \prod_{k=1, k\neq j}^{N} (x_j - x_k)^{-1} \sum_{i=1}^{N} \prod_{k=1, k\neq j, k\neq i}^{N} (x - x_k) = \sum_{j=1}^{N} d_j(x) u_j \tag{12}$$

Similar to the integration matrix, the first order derivative can be written, in matrix form, as

$$\mathbf{D} = \mathbf{B}\mathbf{u} \tag{13}$$

where $\mathbf{D} = [u'_1, u'_2, \ldots u'_N]^T$, $u'_k = u'(x_k)$ and matrix B is determined from eqn (12).

Thereafter, consider a multi-integral for one-dimensional problem

$$U^{(2)}(x) = \int_0^x \int_0^{\xi_2} u(\xi_1) d\xi_1 d\xi_2, \qquad x \in [0,a] \tag{14}$$

Applying the above OLA technique again for integral function $U^{(2)}(x)$, we have

$$U^{(2)}(x_k) = \int_0^{x_k} \int_0^{\xi_2} u(\xi_1) d\xi_1 d\xi_2 = \sum_{i=0}^{k} \sum_{j=0}^{i} a_{ki} a_{ij} u(x_i) = \sum_{i=0}^{k} a_{ki}^{(2)} u(x_i). \tag{15}$$

In the same way for the first order integration Eq.(5), the multi-integral above can also be written in a matrix form as

$$\mathbf{U}^{(2)} = \mathbf{A}^{(2)}\mathbf{u} = \mathbf{A}^2\mathbf{u} \tag{16}$$

Therefore, we can extend this conclusion to any multi-layer integrals as

$$\mathbf{U}^{(m)} = \mathbf{A}^{(m)}\mathbf{u} = \mathbf{A}^m\mathbf{u} \tag{17}$$

In the same way, for higher order derivative to the function u, we have

$$\mathbf{D}^{(m)} = \mathbf{B}^{(m)}\mathbf{u} = \mathbf{B}^m\mathbf{u} \tag{18}$$

4 FIM FOR TWO DIMENSIONAL PROBLEMS

For two-dimensional problems, consider a uniform distribution of collocation points as shown in Fig. 1, the integration matrix is defined as

$$U_x(x,y) = \int_0^x u(\xi, y) d\xi, \; U_{xk}(x_k, y_k) = \int_0^{x_k} u(\xi, y_k) d\xi \tag{19}$$

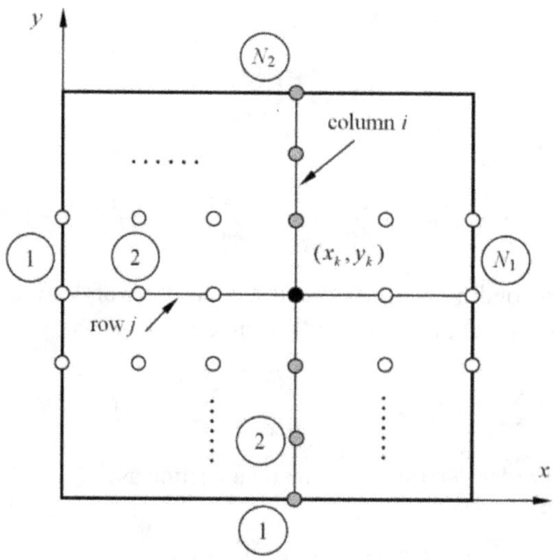

Figure 1. Uniform distribution of collocation points and the number of nodes.

and the total number of points is $k = N_1(j-1)+i$, where *i* and *j* denote the number of column and the number of row, respectively. This numbering system is called the global number system. We can also express each nodal value of integration Eq. (19) in a matrix form as

$$\mathbf{U}_x = \mathbf{A}_x\mathbf{u} \tag{20}$$

where integral value $\mathbf{U}_x = [U_{x1}, U_{x2}, \dots U_{xN}]^T$, nodal value function $\mathbf{u} = \left[u_1, u_2, \dots, u_M\right]^T$ and M is the total number of collocation points ($M = N_1 \times N_2$) for grid shown in Fig. 1. For a rectangular domain, the first order integration matrix is

$$\mathbf{A}_x = \underbrace{\begin{pmatrix} \mathbf{A} & 0 & \dots & 0 \\ 0 & \mathbf{A} & 0 & 0 \\ \dots & \dots & \dots & \dots \\ 0 & 0 & 0 & \mathbf{A} \end{pmatrix}}_{N_2} \tag{21}$$

in which, $\mathbf{A}$ is the integration matrix for one-dimensional problem given in eqn (5) with dimension $N_1 \times N_1$. Following the same procedure, the integration along y axis

$$U_y(x,y) = \int_0^y u(x,\eta)d\eta,\; U_{yk}(x_k, y_k) = \int_0^{y_k} u(x_k,\eta)d\eta \tag{22}$$

can be written in the matrix form as

$$\mathbf{U}_y = \mathbf{A}'\mathbf{u} \tag{23}$$

in the local system for the collocation points, where $k = N_2(i-1)+j$. The first order integration matrix in the local system is

$$\mathbf{A}' - \underbrace{\begin{pmatrix} \mathbf{A} & 0 & \dots & 0 \\ 0 & \mathbf{A} & 0 & 0 \\ \dots & \dots & \dots & \dots \\ 0 & 0 & 0 & \mathbf{A} \end{pmatrix}}_{N_1} \tag{24}$$

in which $\mathbf{A}$ is the integration matrix for one-dimensional problem given in eqn (5) with dimension $N_2 \times N_2$. By a simple re-arrangement of the number of the nodes from the local number system to global number system, eqn (23) can be rewritten, in the global system, as

$$\mathbf{U}_y = \mathbf{A}_y\mathbf{u}. \tag{25}$$

For multi-integration in the two-dimensional problem in a rectangular domain, we consider the following integral with respect to coordinate x

$$U_x^{(2)}(x_i, y) = \int_0^{x_i}\int_0^{\xi_2} u(\xi_1, y)d\xi_1 d\xi_2, \qquad x_i \in [0,a],\; y \in [0,b] \tag{26}$$

where$a\times b$is integral domain. Using the same procedure for one-dimension, we have

$$U_x^{(2)}(x_k,y_k)=\int_0^{x_k}\int_0^{\xi_2}u(\xi_1,y_k)d\xi_1 d\xi_2=\sum_{i=0}^{k}\sum_{j=0}^{i}(a_{kj})_x(a_{ji})_x u_i \tag{27}$$

or in matrix form

$$\mathbf{U}_x^{(2)}=\mathbf{A}_x^2\mathbf{u} \tag{28}$$

where

$$\mathbf{A}_x^2=\mathbf{A}_x\mathbf{A}_x=\underbrace{\begin{pmatrix}\mathbf{A}^2 & 0 & \dots & 0\\ 0 & \mathbf{A}^2 & 0 & 0\\ \dots & \dots & \dots & \dots\\ 0 & 0 & 0 & \mathbf{A}^2\end{pmatrix}}_{N_2} \tag{29}$$

Similarly, we have multi-integration$U_y^{(2)}(x,y)$ with respect to coordinate y

$$U_y^{(2)}(x_k,y_k)=\int_0^{y_k}\int_0^{\eta_2}u(x_k,\eta_1)d\eta_1 d\eta_2=\sum_{i=0}^{k}\sum_{j=0}^{i}(a_{kj})_y(a_{ji})_y u_i \tag{30}$$

and

$$\mathbf{U}_y^{(2)}=\mathbf{A}_y^2\mathbf{u}. \tag{31}$$

In addition, this method can be extended to multi-layers integration with two coordinates x and y as follow:

$$U^{(mn)}(x_k,y_k)=\int_0^{x_k}\dots\int_0^{\xi_2}\int_0^{y_k}\dots\int_0^{\eta_2}u(\xi,\eta)\underbrace{d\xi_1\dots d\xi_m}_{m-\text{layer}}\ \underbrace{d\eta_1\dots d\eta_n}_{n\text{-layer}} \qquad x_k\in[0,a],\ y_k\in[0,b] \tag{32}$$

and the nodal values of the above integration are obtained in the matrix form as

$$\mathbf{U}^{(mn)}=\mathbf{A}_x^m\mathbf{A}_y^n\mathbf{u}. \tag{33}$$

This method can be easily extended to multi-layer differentials with two coordinates x and y as

$$\mathbf{D}^{(mn)}=\mathbf{B}_x^m\mathbf{B}_y^n\mathbf{u}. \tag{34}$$

5 APPLICATION OF FIM TO TWO DIMENSION

Example 1. Steady state heat conduction in an anisotropic square plate

Consider a square plate of width $a=b=1$, as shown in Fig. 2(a), with thermal conductivity tensor: $\lambda_{11}=\lambda_{22}=\lambda_0$ and $\lambda_{12}=\lambda_{21}=0.2\lambda_0$. No heat source is considered, i.e. $w=0$. The analytical solution [15] is given as

$$T^*(x,y)=x^2+x-5xy \tag{35}$$

which satisfies the stationary counterpart of governing eqn (1) for the considered thermal conductivity tensor. Then the boundary conditions are given as

$$\bar{T}(x,0)=x^2+x,\ \ \bar{T}(x,1)=x^2+x-5x,\ \ \bar{q}(1,y)=4-5y \tag{36}$$

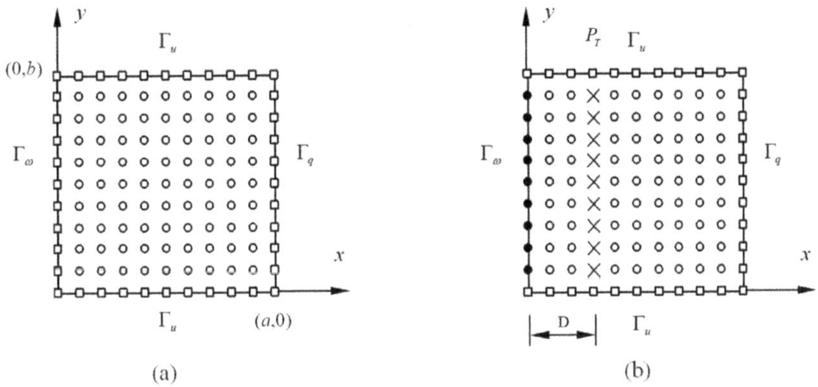

Figure 2. Distribution of collocation points: (a) internal and boundary points; (b) boundary conditions for inverse heat conduction problems, • unreachable boundary; □ Neumann and Dirichlet boundary; × measurement points P_T in domain, where D/a indicates the gap.

and the scattered temperature measurement points are collected either on boundaries or in the domain, as shown in Fig. 2(b). The unreachable boundary is$x=0$ (Γ_ω). Using integration matrix introduced in sections 3 or 4 , we have following governing equation in matrix form in terms of nodal values

$$\left[\lambda_{11}\mathbf{A}_y^2+2\lambda_{12}\mathbf{A}_y\mathbf{A}_x+\lambda_{22}\mathbf{A}_x^2\right]\ \mathbf{T}=\mathbf{X}\Psi_y\mathbf{f}_1+\Psi_y\mathbf{f}_0+\mathbf{Y}\Psi_x\mathbf{g}_1+\Psi_x\mathbf{g}_0 \tag{37}$$

where $\mathbf{T}=\{T_1,T_2,...,T_M\}^T$, $\mathbf{X}=\{x_1,x_2,...,x_M\}$, $\mathbf{Y}=\{y_1,y_2,...,y_M\}$, $\mathbf{f}_k=\{f_1^k,f_2^k,...,f_q^k\}^{\mathrm{T}}$, $\mathbf{g}_k=\{g_1^k,g_2^k,...,g_p^k\}^{\mathrm{T}}$, ($k$=0,1), q and p are numbers of points to be used for interpolation of functions $f(y)$ and $g(x)$ collocated on y and x axes, respectively, Ψ_x and Ψ_y are shape function matrices of one-dimensional shape functions with respect to coordinates x and y, respectively. Integral functions $f_0(y)$, $f_1(y)$, $g_0(x)$ and $g_1(x)$can be interpolated in terms of the nodal values described in Ref. [17] with following two steps:

1. Determine the regions of functions$f(y)$ and $g(x)$, i.e. $[\bar{y}_1,\bar{y}_q]$, $[\bar{x}_1,\bar{x}_p]$, and uniformly distributed points in these regions;
2. Determine one-dimensional shape function matrices Ψ_y and Ψ_x

$$\begin{aligned}f(y)&=0 \quad \text{if } \bar{y}_1<y<\bar{y}_{n-1} \text{ or } \bar{y}_n<y<\bar{y}_q\\ g(x)&=0 \quad \text{if } \bar{x}_1<x<\bar{x}_{m-1} \text{ or } \bar{x}_m<x<\bar{x}_p\end{aligned} \tag{38}$$

otherwise

$$\begin{aligned}f(y)&=\frac{\bar{y}_n-y}{\bar{y}_n-\bar{y}_{n-1}}f_{n-1}+\frac{y-\bar{y}_{n-1}}{\bar{y}_n-\bar{y}_{n-1}}f_n \quad \text{if } \bar{y}_{n-1}<y<\bar{y}_n\\ g(x)&=\frac{\bar{x}_m-x}{\bar{x}_m-\bar{x}_{m-1}}g_{m-1}+\frac{x-\bar{x}_{m-1}}{\bar{x}_m-\bar{x}_{m-1}}g_m \quad \text{if } \bar{x}_{m-1}<x<\bar{x}_m\end{aligned} \tag{39}$$

For a regular distribution of nodes, we select $N_1 = N_2 = N$, $M = N^2$, $q = p = N$. Then the shape function matrices $\Psi_x = \Psi_y = \mathbf{I}$. It is obvious that the number of equations from boundary conditions is $L = 4 \times N - 4$. Thus the numbers of unknown nodal values T_i is M and the numbers of nodal values of integral functions $f_k(y)$ and $g_k(x)$ $2\times(p+q) = 4\times N$. SVD scheme is used to solve a set of linear algebraic equations [17]. The average error is defined as

$$\varepsilon = \frac{1}{M_\omega}\sum_{i=1}^{M_\omega}\left|T_i - T_i^*\right| \tag{40}$$

where M_ω is the number of nodes on the unreachable boundary, as shown in Fig. 2(b). The interior and boundary points are selected, as shown in Fig. 2(a) with $N_1 = N_2 = M_\omega = 11$ and number of point measurements $M_P = N - 2$. The average errors with different gaps between the unreachable boundary and measurement points for different noise levels are presented in Fig. 3. Apparently in the cases when $\sigma = 0$, where σ indicates the magnitude of noise, the numerical solutions are all close to the analytical solutions with maximum error of $10^{-3}(\%)$. For the cases of without noise, we take the measurements of temperature by $T(x_P, y_P) = T^*(x_P, y_P) + \sigma\sin(\pi y_P)$. It is seen that the solutions are stable in the region of $D/a \le 0.4$. When $D/a > 0.5$, as expected, the error increases significantly since the measurement points are far away from the unreachable boundary. From Fig. 3, it can be observed that more accurate solutions could be obtained with smaller gaps. In addition, the convergence is observed with respect to the distribution density of nodes in the domain and on the boundary. When the interior and boundary points are selected as $N = 17$, the differences between the numerical results for the two node distributions are seen to be very small.

Example 2. Two-dimensional steady state heat conduction problem

Consider steady state heat conduction in a quarter of disk (2D) with isotropic media, as shown in Fig. 4(a). The governing equation (1) in polar coordinate system is

$$\frac{\partial^2 T}{\partial r^2} + \frac{1}{r}\frac{\partial T}{\partial r} + \frac{1}{r^2}\frac{\partial^2 T}{\partial \theta^2} = 2r^2 \qquad 1 \le r \le 2,\ 0 \le \theta \le \pi/2 \tag{41}$$

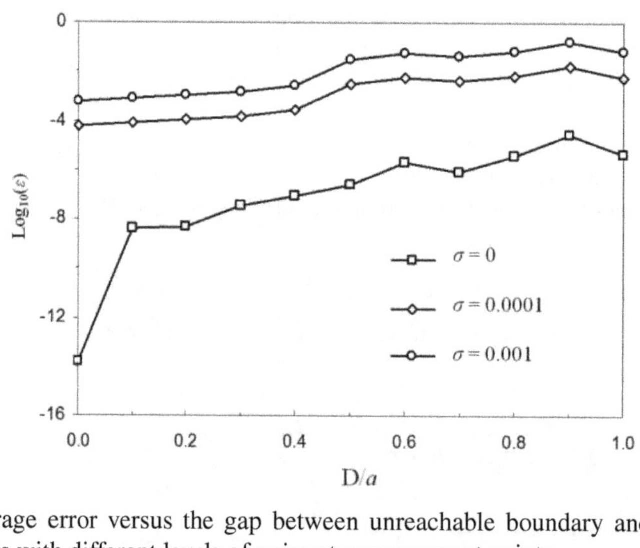

Figure 3. Average error versus the gap between unreachable boundary and measurement points with different levels of noise at measurement points.

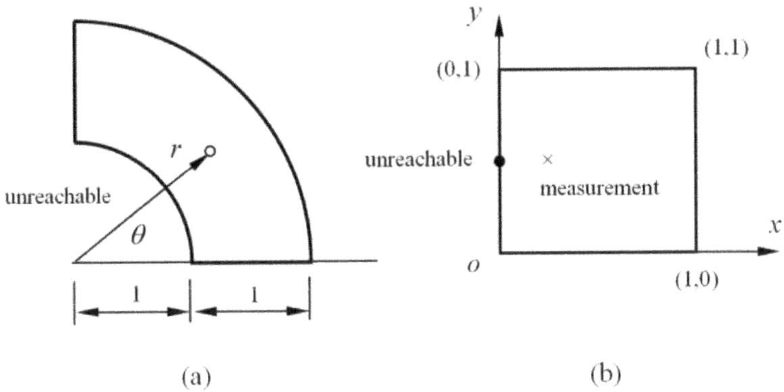

Figure 4. (a) Quarter of ring; (b) transformation of coordinates.

with the boundary conditions $T(r,0)=T(r,\pi/2)=0$; $T(1,\theta)=1/4\sin^2 2\theta$; $T(2,\theta)=4\sin^2 2\theta$. The exact solution is obtained as $T^*(r,\theta)=r^4/4\sin^2 2\theta$. By coordinate transform, Eq. (41) can be normalised as

$$\frac{\partial^2 T}{\partial x^2}+\frac{\partial}{\partial x}\left(\frac{T}{1+x}\right)+\frac{T}{(1+x)^2}+\frac{4}{(1+x)^2\pi^2}\frac{\partial^2 T}{\partial y^2}=2(1+x)^2 \tag{42}$$

where transformed coordinates $x=r-1$ and $y=\pi\theta/2$ and the domain is transformed into a square as shown in Fig. 4(b). Using the integration matrix in section 4, we have following governing equation, in matrix form, in terms of nodal values

$$\left[\mathbf{A}_y^2+\mathbf{A}_y^2\mathbf{A}_x+\mathbf{A}_x^2\mathbf{A}_y^2\mathbf{R}_1+\frac{4}{\pi^2}\mathbf{A}_x^2\mathbf{R}_2\right]\mathbf{T}=\mathbf{A}_y^2\mathbf{A}_x^2\mathbf{w}+\mathbf{X}\Psi_y\mathbf{f}_1+\Psi_y\mathbf{f}_0+\mathbf{Y}\Psi_x\mathbf{g}_1+\Psi_x\mathbf{g}_0 \tag{43}$$

where

$$\mathbf{R}_1=\mathbf{diag}\{1/(1+x_i\},\mathbf{R}_2=\mathbf{R}_1^2,\mathbf{w}=\mathbf{diag}\{2(1+x_i)^2\} \tag{44}$$

in which the integral functions on the right hand side are defined in eqn (37). Same nodes distribution in Example 1 are used. The average relative error is defined as

$$\varepsilon_{\text{re}}=\frac{1}{M}\sum_{i=1}^{M}\left|T_i-T_i^*\right|/T_{\max}^*,\quad T_{\max}^*=T^*(2,\pi/4)=4 \tag{45}$$

Variation of the logarithms of the average relative error ε_{re} with different numbers of grid N is shown in Fig. 5. For comparison the numerical solutions obtained by the finite difference method (FDM) is also given in Fig. 5. It is seen that the accuracy of FDM solutions increases linearly with increasing the number of nodes. As expected, the accuracy of the numerical solutions obtained by by FIM is much higher than that by FDM even for a small number of grid N. Secondly, we consider that the unreachable boundary Γ_ω is collocated at inner surface ($x=0$). By changing M to M_ω in eqn (45) and selecting the number of grid $N=11$, Fig. 6 shows the average errors over the unreachable boundary for different gaps D and different levels of noise at the points of measurement. To observe the effect of measurement noise, the

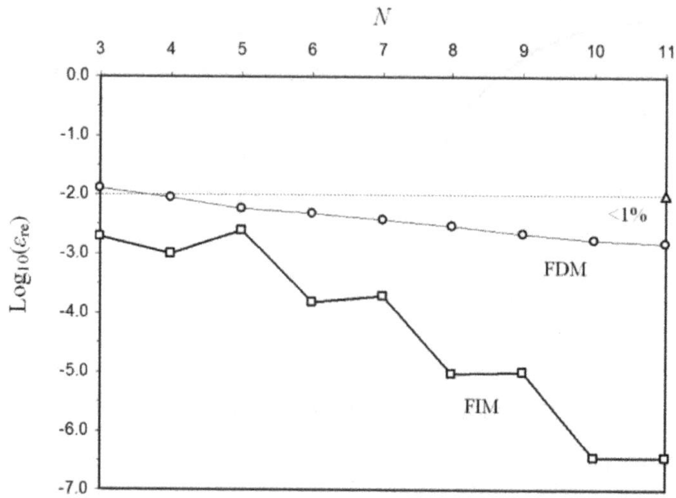

Figure 5. The logarithms of the average relative error versus the segment number N.

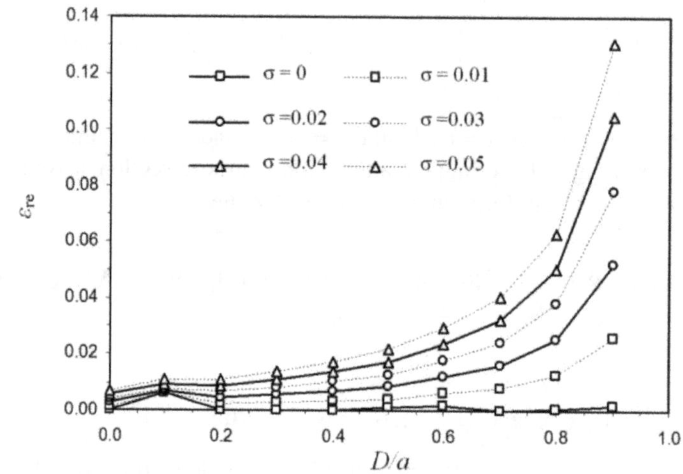

Figure 6. The average errors in the domain versus the gap between unreachable boundary and measurement points for different levels of noise.

measurements of temperature are taken by $T(x_P, y_P) = T^*(x_P, y_P) + \sigma \sin(\pi y_P)$. Apparently in the region of $D/a \leq 0.6$, it is seen that the temperature obtained on unreachable boundary is stable. However, when $D/a > 0.7$, the error increases significantly with noise.

6 CONCLUSIONS

The finite integration method with Lagrange series expansion has been demonstrated for two and three dimensional inverse heat conduction problems for the first time. Anisotropic media and functionally graded materials are considered. Approximation by Lagrange series is used to determine the first order integration and derivative matrices. The Singular Value

Decomposition (SVD) is applied to solve the ill-posed linear system of algebraic equations, derived from the governing equation, boundary conditions and scattered temperature measurements in the present work. The stability and accuracy were observed for inverse heat conduction problems in three examples. It has been shown that the finite integration method provides an efficient algorithm for solving inverse heat conduction problems. However, if the Dirichlet condition (measurements) and Neumann condition are specified at the same points of the boundary, the prediction of the temperature on the unreachable boundary becomes unstable with certain noises from measurement.

REFERENCES

[1] Jonas, P. & Louis, A.K., Approximate inverse for a one-dimensional inverse heat conduction problem. *Inverse Problems*, **16**, pp. 175–185, 2000. https://doi.org/10.1088/0266-5611/16/1/314

[2] Lesnic, D. & Elliott, L., The decomposition approach to inverse heat conduction. *Journal of Mathematical Analysis and Applications*, **232**, pp. 82–98, 1999. https://doi.org/10.1006/jmaa.1998.6243

[3] Liu, J., A stability analysis on Beck's procedure for inverse heat conduction problem. *Journal of Computational Physics*, **123**, pp. 65–73, 1996. https://doi.org/10.1006/jcph.1996.0005

[4] Shen, S.Y., A numerical study of inverse heat conduction problems. *Computers & Mathematics with Applications*, **38**, pp. 173–188, 1999. https://doi.org/10.1016/s0898-1221(99)00248-5

[5] Hon, Y.C. & Wei, T., A fundamental solution method for inverse heat conduction problem. *Engineering Analysis with Boundary Elements*, **28**, pp. 489–495, 2004. https://doi.org/10.1016/s0955 7997(03)00102-4

[6] Guo, L. & Murio, D.A., A modified space-marching finite-difference algorithm for the two-dimensional inverse heat conduction problem with slab symmetry. *Inverse Problems*, **7**, pp. 247–259, 1991. https://doi.org/10.1088/0266-5611/7/2/008

[7] Khalidy, N.A., A general space marching algorithm for the solution of two-dimensional boundary inverse heat conduction problems. *Numerical Heat Transfer Part B*, **34**, pp. 339–360, 1998. https://doi.org/10.1080/10407799808915062

[8] Chantasiriwan, S., An algorithm for solving multidimensional inverse heat conduction problem. *International Journal of Heat and Mass Transfer*, **44**, pp. 3823–3832, 2001. https://doi.org/10.1016/s0017-9310(01)00037-0

[9] Kurpisz, K. & Nowak, A.J., BEM approach to inverse heat conduction problems. *Engineering Analysis with Boundary Elements*, **10**, pp. 291–729, 1992. https://doi.org/10.1016/0955-7997(92)90142-t

[10] Hsu, T.R., Sun, N.S., Chen, G.G. & Gong, Z.L., Finite element formulation for two-dimensional inverse heat conduction analysis. *ASME Journal of Heat Transfer*, **114**, pp. 553–557, 1992. https://doi.org/10.1115/1.2911317

[11] Reinhardt, H.J., A numerical method for the solution of two-dimensional inverse heat conduction problems. *International Journal for Numerical Methods in Engineering*, **32**, pp. 363–283, 1991. https://doi.org/10.1002/nme.1620320209

[12] Hon, Y.C. & Wei, T., The method of fundamental solutions for solving multidimensional inverse heat conduction problems. *Computer Modeling in Engineering & Sciences*, **7**, pp. 19–32, 2005.

[13] Yu, G.X., Sun, J., Wang, H.S., Wen, P.H. & Rose, J.W., Meshless inverse method to determine temperature and heat flux at boundaries for 2D steady-state heat conduction problems. *Experimental Thermal and Fluid Science,* **52**, pp. 156–163, 2004. https://doi.org/10.1016/j.expthermflusci.2013.09.006

[14] Atluri, S.N., *The Meshless Method (MLPG) for Domain and BIE Discretizations,* Tech Science Press Forsyth GA, 2004.

[15] Sladek, J., Sladek, V. & Hon, Y.C., Inverse heat conduction problems by meshless local Petrov–Galerkin method. *Engineering Analysis with Boundary Elements*, **30**, pp. 650–661, 2006.
https://doi.org/10.1016/j.enganabound.2006.03.003

[16] Li, M., Hon, Y.C., Korakianitis, T. & Wen, P.H., Finite integration method for nonlocal elastic bar under static and dynamic loads. *Engineering Analysis with Boundary Elements*, **37**(5), pp. 842–849, 2013.
https://doi.org/10.1016/j.enganabound.2013.01.018

[17] Wen, P.H., Hon, Y.C., Li, M. & Korakianitis, T., Finite integration method for partial differential equations. *Applied Mathematical Modelling*, **37**, pp. 10092–10106, 2013. https://doi.org/10.1016/j.apm.2013.05.054

[18] Li, M., Tian, Z.L., Hon, Y.C., Chen, C.S. & Wen, P.H., Improved finite integration method for partial differential equations. *Engineering Analysis with Boundary Elements*, **64**, pp. 230–236, 2016.
https://doi.org/10.1016/j.enganabound.2015.12.012

THERMOELASTIC ANALYSIS OF BENDING PROBLEMS IN FGM PLATES

V. SLADEK, L. SATOR & J. SLADEK
Institute of Construction and Architecture, Slovak Academy of Sciences, Bratislava, Slovakia.

ABSTRACT

It is well known that the original 3D elasticity problem in plate structures subjected to transversal loading can be converted to a 2D problem. In addition to in-plane displacements, we need to introduce the deflection and/or rotation field variables in the plate mid-plane, in order to describe displacements and deformations within the plate structure. Thus, one can develop unified formulation for bending and in-plane deformation modes within the classical Kirchhoff-Love theory for bending of thin elastic plates and the shear deformation plate theories (the first order – FSDPT, and the third order - TSDPT). In this paper, we extend the derivation of the 2D formulation for coupled problems of thermoelasticity in plate structures. Three material coefficients play the role in stationary problems, namely the Young modulus, coefficient of linear thermal extension and the heat conduction coefficient. The influence of continuous gradation of these coefficients on the response of the plate subjected to thermal loadings is investigated in numerical simulations. The element-free strong formulation with using meshless approximations for spatial variation of field variables is developed.

Keywords: functional gradations of material coefficients, MLS approximations, Plate bending theories, strong formulation, thermal loading, unified formulation of 2D coupled problems

1 INTRODUCTION

Functionally graded materials (FGM) refers to the new class of advanced composite materials utilized in modern engineering structures and frequently studied [1, 2] because of their superior properties. It is a well-known elimination of delamination effect in plate-like structures if the sandwich composite structures are replaced by FGM composite ones. From the physical point of view as well as from the view of engineering applications the transversal gradation of material properties is more interesting than the in-plane gradation in plate problems. The transverse gradation of Young modulus (YM) gives rise to coupling between the deflections and in-plane deformations [3, 4] even in the case of thin plate theory and static problems. However, it is interesting to study also the plate structures with combined gradation of material coefficients in transversal and in-plane directions because new physical effects can arise [5] in multi-graded plates.

In this paper, we shall deal with the plate problems considered within classical thermoelasticity. The unified 2D formulation is developed for FGM plates with taking into account the assumption of classical Kirchhoff-Love theory for bending of thin elastic plates as well as assumptions of the shear deformation plate theory. The material coefficients are allowed to be functionally dependent in both the transversal as well as in-plane directions. An efficient formulation is developed for numerical solution of coupled multi-field problems and some interesting effects are illustrated in numerical simulations.

2 UNIFIED FORMULATION FOR FGM COMPOSITE PLATES UNDER THERMAL LOADING

In the unified formulation for all three plate bending theories (KLT, FSDPT and TSDPT), the displacements can be expressed in terms of the in-plane displacements $u_\alpha(\mathbf{x})$, transversal displacements (deflections) $w(\mathbf{x})$and rotations of the normal to the mid-surface $\varphi_\alpha(\mathbf{x})$ by

DOI: 10.2495/CMEM-V6-N6-1161-1172

$$v_i(\mathbf{x},x_3) = \delta_{i\alpha}\left\{u_\alpha(\mathbf{x}) + \left[c_1\phi(x_3) - x_3\right]w_{,\alpha}(\mathbf{x}) + c_1\phi(x_3)\varphi_\alpha(\mathbf{x})\right\} + \delta_{i3}w(\mathbf{x}) \tag{1}$$

in $\mathbf{x} \in \Omega$, $x_3 \in \left[-h/2,\, h/2\right]$ with $h \ll L$ where L is the characteristic length in Ω. Note that $i \in \{1,2,3\}$, $\alpha \in \{1,2\}$,

$\phi(x_3) := x_3 - c_2\psi(x_3)$, $\psi(x_3) := 4h\left(x_3/h\right)^3/3$ with

$$c_1 = \begin{cases} 0, & \text{KLT} \\ 1, & \text{SDPT} \end{cases},\quad c_2 = \begin{cases} 0, & \text{FSDPT, KLT} \\ 1, & \text{TSDPT} \end{cases}$$

being two key factors for switching among the three theories (KLT, FSDPT, TSDPT). Having known the dependence of displacements on x_3, we also know the dependence of strains and elastic stresses on x_3 and the original 3D thermo-elastic problem can be recast into 2D thermo-elastic plate bending problem. The total and thermal strains are given as

$$e_{\alpha\beta}(\mathbf{x},z) := \left(v_{\alpha,\beta} + v_{\beta,\alpha}\right)/2 = \varepsilon_{\alpha\beta}(\mathbf{x}) + \left[c_1\phi(x_3) - x_3\right]w_{,\alpha\beta}(\mathbf{x}) + c_1\phi(x_3)\eta_{\alpha\beta}(\mathbf{x}) \tag{2}$$

$$\varepsilon_{\alpha\beta} = \left(u_{\alpha,\beta} + u_{\beta,\alpha}\right)/2,\ \eta_{\alpha\beta} = \left(\varphi_{\alpha,\beta} + \varphi_{\beta,\alpha}\right)/2$$

$$e_{kl}^{\theta}(\mathbf{x},z) = \alpha\left(\theta - \theta_0\right)\delta_{kl} \tag{3}$$

and the elastic stresses in isotropic elastic plate are given as

$$\begin{aligned}\sigma_{\alpha\beta}(\mathbf{x},x_3) = \frac{E}{1-\nu^2}\frac{1-\nu}{H}\Big[&\tau_{\alpha\beta}^{(u)}(\mathbf{x}) + c_1\phi(x_3)\tau_{\alpha\beta}^{(\varphi)}(\mathbf{x}) + \left(c_1\phi(x_3) - x_3\right)\tau_{\alpha\beta}^{(w)}(\mathbf{x}) - \\ &-\alpha\tau_{\alpha\beta}^{(\theta)}(\mathbf{x},x_3)\Big]\end{aligned} \tag{4}$$

$$\sigma_{\alpha 3}(\mathbf{x},x_3) = \frac{E}{1+\nu}\frac{c_1}{2}\phi_{,3}(x_3)\left[w_{,\alpha}(\mathbf{x}) + \varphi_\alpha(\mathbf{x})\right]$$

$$\begin{aligned}\sigma_{33}(\mathbf{x},x_3) = \frac{E\nu}{1-\nu^2}\frac{1-\nu}{H}\Big[&u_{\gamma,\gamma}(\mathbf{x}) + c_1\phi(x_3)\varphi_{\gamma,\gamma}(\mathbf{x}) + \left(c_1\phi(x_3) - x_3\right)w_{,\gamma\gamma}(\mathbf{x}) - \\ &-\alpha\frac{1+\nu}{\nu}\theta(\mathbf{x},x_3)\Big]\end{aligned}$$

in which α is the linear thermal extension coefficient, $H = 1-2\nu$, with ν and E being the Poisson ratio and Young modulus (YM), respectively, and $\tau_{\alpha\beta}^{(\circ)}$ stand for particular strain contributions

$$\tau_{\alpha\beta}^{(u)}(\mathbf{x}) := H\varepsilon_{\alpha\beta}(\mathbf{x}) + \nu\delta_{\alpha\beta}\varepsilon_{\gamma\gamma}(\mathbf{x}),\ \tau_{\alpha\beta}^{(\varphi)}(\mathbf{x}) := H\eta_{\alpha\beta}(\mathbf{x}) + \nu\delta_{\alpha\beta}\eta_{\gamma\gamma}(\mathbf{x}), \tag{5}$$

$$\tau_{\alpha\beta}^{(w)}(\mathbf{x}) := Hw_{,\alpha\beta}(\mathbf{x}) + \nu\delta_{\alpha\beta}w_{,\gamma\gamma}(\mathbf{x}),\ \tau_{\alpha\beta}^{(\theta)}(\mathbf{x},x_3) := (1+\nu)\theta(\mathbf{x},x_3)\delta_{\alpha\beta}.$$

In the case of FGM plate, we assume the material coefficients to be continuous functions of Cartesian coordinates as

$$E(\mathbf{x},x_3) = E_0 E_H(\mathbf{x}) E_V(x_3),\ E_V(x_3) = 1 + \zeta\left(\frac{1}{2} \pm \frac{x_3}{h}\right)^p,\ E_H(\mathbf{x}) = 1 + \kappa_0 (x_1)^{b_0}$$

$$\alpha(\mathbf{x},x_3)=\alpha_0\alpha_H(\mathbf{x})\alpha_V(x_3),\ \alpha_V(x_3)=1+\xi\left(\frac{1}{2}\pm\frac{x_3}{h}\right)^r,\ \alpha_H(\mathbf{x})=1+\kappa_1(x_1)^{b_1} \tag{6}$$

$$k(\mathbf{x},x_3)=k_0k_H(\mathbf{x})k_V(x_3),\ k_V(x_3)=1+\omega\left(\frac{1}{2}\pm\frac{x_3}{h}\right)^s,\ k_H(\mathbf{x})=1+\kappa_2(x_1)^{b_2}$$

with k being the heat conduction coefficient.

It is well known that the governing equations as well as the boundary conditions in unified formulation for bending problems can be derived from the principle of virtual work

$$\delta U-\delta W_e=0,\ \delta U=\int_\Omega\left(\int_{-h/2}^{h/2}\sigma_{ij}(\mathbf{x},x_3)\delta e_{ij}(\mathbf{x},x_3)dx_3\right)d\Omega,\ \delta W_e=\int_\Omega p(\mathbf{x})\delta w(\mathbf{x})d\Omega \tag{7}$$

provided that the plate is subjected to transversal loading. In order to derive the semi-integral 2D formulation, we need to perform the integration with respect to x_3 in closed form. Therefore, we need to know the dependence of the whole intengrand in (7) on x_3. Up to now, we presented the dependence of elastic fields on x_3 and in what follows we assume the expansion of temperature on x_3 as

$$\theta(\mathbf{x},x_3)\approx\theta_0+\vartheta_0(\mathbf{x})+z\vartheta_1(\mathbf{x})+z^2\vartheta_2(\mathbf{x}),\ z=x_3/h\in[-0.5,\ 0.5] \tag{8}$$

with $\vartheta_a(\mathbf{x})$ for $(a=0,1,2)$, being new field variables. Such an expansion is consistent with assumptions made in plate structure problems formulations when $h\ll L$. Then,

$$\tau_{\alpha\beta}^{(\theta)}(\mathbf{x},x_3)=\sum_{a=0}^{2}z^a\tau_{\alpha\beta}^{(\vartheta_a)}(\mathbf{x}),\ \tau_{\alpha\beta}^{(\vartheta_a)}(\mathbf{x}):=(1+\nu)\vartheta_a(\mathbf{x})\delta_{\alpha\beta}\ \text{for}\ (a=0,1,2) \tag{9}$$

If we inserted (8) into the governing heat conduction equation, we would obtain the PDE

$$\left[k_H(\mathbf{x})k_V(x_3)\left(\sum_{a=0}^{2}(x_3/h)^a\vartheta_a(\mathbf{x})\right)_{,j}\right]_{,j}=0 \tag{10}$$

which is still dependent on x_3. Therefore, Eq. (10) will be considered in the averaged sense over the plate thickness. Introducing dimensionless coordinates and temperature fields

$$x_\beta^*:=x_\beta/L\ ,\ x_3^*:=x_3/h_0=h^*(\mathbf{x})z\,,\ \vartheta_a^*(\mathbf{x}):=\vartheta_a(\mathbf{x})/\theta_0 \tag{11}$$

the averaged heat conduction equation becomes

$$\sum_{a=1}^{2}C^{(\theta\vartheta_a)}(\mathbf{x})\vartheta_a^*(\mathbf{x})+\sum_{a=0}^{2}G_\beta^{(\theta\vartheta_a)}(\mathbf{x})\vartheta_{a,\beta}^*(\mathbf{x})+\sum_{a=0}^{2}G^{(\theta\vartheta_a)}(\mathbf{x})\vartheta_{a,\beta\beta}^*(\mathbf{x})=0, \tag{12}$$

where

$$C^{(\theta\vartheta_a)}(\mathbf{x}):=\left(\frac{L}{h_0}\right)^2\frac{k_H(\mathbf{x})}{\left(h^*(\mathbf{x})\right)^2}\left[k_V(1/2)+(-1)^ak_V(-1/2)\right],\ a=1,2$$

$$G_\beta^{(\theta\vartheta_a)}(\mathbf{x}):=\left(d_{(0)a}+\omega d_{(s)a}\right)k_{H,\beta}(\mathbf{x})\ ,\ a=0,1,2 \tag{13}$$

$$G^{(\theta\vartheta_a)}(\mathbf{x}):=\left(d_{(0)a}+\omega d_{(s)a}\right)k_H(\mathbf{x})\ ,\ a=0,1,2$$

in which $d_{(0)a} := \int_{-1/2}^{1/2} z^a dz = \frac{1+(-1)^a}{(a+1)2^{a+1}}$,

$$d_{(p)a} := \int_{-1/2}^{1/2} z^a \left(\frac{1}{2} \pm z\right)^p dz = (\pm 1)^a \int_0^1 \left(y - \frac{1}{2}\right)^a y^p dy = (\pm 1)^a \sum_{k=0}^{a} \binom{a}{k} \left(-\frac{1}{2}\right)^k \frac{1}{p+1+a-k}$$

Eq. (12) should be supplemented with thermal boundary conditions on the top and bottom surface of the plate. Usually, we distinguish three kinds of boundary conditions:

Dirichlet b.c.: $\vartheta_0^*(\mathbf{x}) \pm \frac{1}{2}\vartheta_1^*(\mathbf{x}) + \frac{1}{4}\vartheta_2^*(\mathbf{x}) = \frac{\bar{\theta}(\mathbf{x}, \pm 1/2)}{\theta_0} - 1$; $\bar{\theta}(\mathbf{x}, \pm 1/2)$-prescribed

$$\text{Neuman b.c.: } \pm k_0 k_H(\mathbf{x}) k_V(\pm 1/2)\left[\vartheta_1^*(\mathbf{x}) \pm \vartheta_2^*(\mathbf{x})\right] = \frac{\bar{q}(\mathbf{x}, \pm 1/2)}{\theta_0}; \tag{14}$$

$\bar{q}(\mathbf{x}, \pm 1/2)$- prescribed heat fluxes

Robin b.c.: $A\left[1 + \vartheta_0^*(\mathbf{x}) \pm \frac{1}{2}\vartheta_1^*(\mathbf{x}) + \frac{1}{4}\vartheta_2^*(\mathbf{x})\right] \pm B k_0 k_H(\mathbf{x}) k_V(\pm 1/2)\left[\vartheta_1^*(\mathbf{x}) \pm \vartheta_2^*(\mathbf{x})\right] = 0.$

In view of prescribed boundary conditions, one can eliminate ϑ_1 and ϑ_2 form (12) and $\vartheta_0^*(\mathbf{x}) = \theta(\mathbf{x}, x_3 = 0)/\theta_0$ is the dimensionless temperature at the mid-plane of the plate. For ϑ_0^*, we need to prescribe also boundary conditions on the edge of the plate.

In order to proceed in the derivation of the variational formulation, we define certain semi-integral fields

$$M_{\alpha\beta}^{(\varphi)}(\mathbf{x}) := c_1 \int_{-h/2}^{h/2} \phi(x_3) \sigma_{\alpha\beta}(\mathbf{x}, x_3) dx_3 ,$$

$$M_{\alpha\beta}^{(w)}(\mathbf{x}) := \int_{-h/2}^{h/2} \left(x_3 - c_1 \phi(x_3)\right) \sigma_{\alpha\beta}(\mathbf{x}, x_3) dx_3, \tag{15}$$

$$T_{3\beta}^{(w\varphi)}(\mathbf{x}) := c_1 \left[\tilde{T}_{3\beta}(\mathbf{x}) - c_2 Q_{3\beta}(\mathbf{x})\right],$$

$$T_{3\beta}(\mathbf{x}) := \int_{-h/2}^{h/2} \sigma_{\alpha\beta}(\mathbf{x}, x_3) dx_3, \; Q_{3\beta}(\mathbf{x}) := \int_{-h/2}^{h/2} \psi_{,3}(x_3) \sigma_{\alpha\beta}(\mathbf{x}, x_3) dx_3$$

where $\tilde{T}_{3\beta}(\mathbf{x})$ is the Reissner modification of the shear stresses $T_{3\beta}(\mathbf{x})$ by the shear correction factor as

$$\tilde{T}_{3\beta}(\mathbf{x}) := \left[(1 - c_2)\kappa + c_2\right] T_{3\beta}(\mathbf{x}). \tag{16}$$

Now, in view of (2)-(4), (7) and (15), (16), we may write

$$\delta U - \delta W = -\int_\Omega \Big\{ T_{\alpha\beta,\beta}(\mathbf{x}) \delta u_\alpha(\mathbf{x}) + \left[M_{\alpha\beta,\alpha\beta}^{(w)}(\mathbf{x}) + T_{3\beta,\beta}^{(w\varphi)}(\mathbf{x}) + q(\mathbf{x}) \right] \delta w(\mathbf{x}) +$$

$$+ \left[M_{\alpha\beta,\beta}^{(\varphi)}(\mathbf{x}) - T_{3\alpha}^{(w\varphi)}(\mathbf{x}) \right] \delta\varphi_\alpha(\mathbf{x}) \Big\} d\Omega + \tag{17}$$

$$+\int_{\partial\Omega}\left\{n_\beta(\mathbf{x})T_{\alpha\beta}(\mathbf{x})\delta u_\alpha(\mathbf{x})-n_\alpha(\mathbf{x})n_\beta(\mathbf{x})M^{(w)}_{\alpha\beta}(\mathbf{x})\delta\left(\frac{\partial w}{\partial\mathbf{n}}(\mathbf{x})\right)+\right.$$

$$\left.+n_\beta(\mathbf{x})M^{(\varphi)}_{\alpha\beta}(\mathbf{x})\delta\varphi_\alpha(\mathbf{x})+V(\mathbf{x})\delta w(\mathbf{x})\right\}d\Gamma=0,$$

where

$$V(\mathbf{x}):=n_\alpha(\mathbf{x})\left(M^{(w)}_{\alpha\beta,\beta}(\mathbf{x})+T^{(w\varphi)}_{3\alpha}(\mathbf{x})\right)+\frac{\partial}{\partial\mathbf{t}}T^{(w)}(\mathbf{x})-\sum_c\delta(\mathbf{x}-\mathbf{x}^c)\left[\!\left[T^{(w)}(\mathbf{x}^c)\right]\!\right] \tag{18}$$

$$T^{(w)}(\mathbf{x}):=t_\alpha(\mathbf{x})n_\beta(\mathbf{x})M^{(w)}_{\alpha\beta}(\mathbf{x})$$

being the generalized shear force and the twisting moment defined on the plate edge. Furthermore, the jump at a corner on the oriented boundary edge $\partial\Omega$ is defined as

$$\left[\!\left[A(\mathbf{x}^c)\right]\!\right]:=A(\mathbf{x}^c+0)-A(\mathbf{x}^c-0). \tag{19}$$

Since $\{\delta u_\alpha,\delta w,\delta\varphi_\alpha\}$ are arbitrary in Ω and variation of particular fields are arbitrary on that part of the boundary where the field variable is not prescribed by the boundary conditions, one can deduce the governing equations and boundary conditions from the variational formulation Eq. (17). In the linear theory, it is convenient to utilize the formulation for dimensionless fields

$$u^*_\beta(\mathbf{x}):=u_\beta(\mathbf{x})/h_0,\ \varphi^*_\beta(\mathbf{x},t):=\varphi_\beta(\mathbf{x},t),\ w^*(\mathbf{x}):=w(\mathbf{x})/h_0\ .$$

Then, we obtain

governing equations:

$$T^*_{\alpha\beta,\beta}(\mathbf{x})=0 \tag{20}$$

$$M^{*(w)}_{\alpha\beta,\alpha\beta}(\mathbf{x})+\left(L/h_0\right)^2T^{*(w\varphi)}_{3\beta,\beta}(\mathbf{x})=-p^*(\mathbf{x}) \tag{21}$$

$$M^{*(\varphi)}_{\alpha\beta,\beta}(\mathbf{x})-\left(L/h_0\right)^2T^{*(w\varphi)}_{3\alpha}(\mathbf{x})=0 \tag{22}$$

boundary conditions:

$$u^*_\alpha(\mathbf{x})\big|_{\partial\Omega_u}\text{ is prescribed and/or }n_\beta(\mathbf{x})T^*_{\alpha\beta}(\mathbf{x})\big|_{\partial\Omega_t}=0$$

$$\frac{\partial w^*}{\partial\mathbf{n}}(\mathbf{x})\bigg|_{\partial\Omega_{wn}}\text{ is prescribed and/or }n_\alpha(\mathbf{x})n_\beta(\mathbf{x})M^{*(w)}_{\alpha\beta}(\mathbf{x})\big|_{\partial\Omega_M}=0 \tag{23}$$

$$\varphi^*_\alpha(\mathbf{x})\big|_{\partial\Omega_\varphi}\text{ is prescribed and/or }n_\beta(\mathbf{x})M^{*(\varphi)}_{\alpha\beta}(\mathbf{x})\big|_{\partial\Omega_{M\varphi}}=0$$

$$w^*(\mathbf{x})\big|_{\partial\Omega_w}\text{ is prescribed and/or }V^*(\mathbf{x})\big|_{\partial\Omega_V}=0.$$

The explicit expressions for the semi-integral fields are given as

$$T^*_{\alpha\beta}(\mathbf{x})=C^{(uu)}(\mathbf{x})\tau^{*(u)}_{\alpha\beta}(\mathbf{x})+C^{(u\varphi)}(\mathbf{x})\tau^{*(\varphi)}_{\alpha\beta}(\mathbf{x})+C^{(uw)}(\mathbf{x})\tau^{*(w)}_{\alpha\beta}(\mathbf{x})+$$

$$+\sum_{a=0}^{2}C^{(u\vartheta_a)}(\mathbf{x})\tau^{*(\vartheta_a)}_{\alpha\beta}(\mathbf{x})=\frac{L^2}{D_0}T_{\alpha\beta}(\mathbf{x}) \tag{24}$$

$$C^{(uu)}(\mathbf{x}) := 12\frac{1-\nu}{H}\frac{L}{h_0}\left(d_{(0)0} + \zeta d_{(p)0}\right)D_{1H}(\mathbf{x})$$

$$C^{(u\varphi)}(\mathbf{x}) := 12\frac{1-\nu}{H}\frac{L}{h_0}c_1\zeta\left(d_{(p)1} - \frac{4}{3}c_2 d_{(p)3}\right)D_{2H}(\mathbf{x})$$

$$C^{(uw)}(\mathbf{x}) := 12\frac{1-\nu}{H}\zeta\left[(c_1-1)d_{(p)1} - \frac{4}{3}c_1 c_2 d_{(p)3}\right]D_{2H}(\mathbf{x})$$

$$C^{(u\vartheta_a)}(\mathbf{x}) := -12\frac{1-\nu}{H}\left(\frac{L}{h_0}\right)^2 \alpha_0\theta_0\left(d_{(0)a} + \xi d_{(r)a} + \zeta d_{(p)a} + \xi\zeta d_{(p+r)a}\right)B_{1H}(\mathbf{x})$$

$$T_{3\alpha}^{*(w\varphi)}(\mathbf{x},t) = C^s(\mathbf{x})\left[w_{,\alpha}^*(\mathbf{x}) + \frac{L}{h_0}\varphi_\alpha^*(\mathbf{x})\right] = \frac{Lh_0}{D_0}T_{3\alpha}(\mathbf{x})$$

$$C^s(\mathbf{x}) := c_1 6(1-\nu)\left\{\left[(1-c_2)\kappa + c_2\right]\left(d_{(0)0} + \zeta d_{(p)0}\right) - 8c_2\left(d_{(0)2} + \zeta d_{(p)2}\right) + \right.$$
$$\left. + 16c_2\left(d_{(0)4} + \zeta d_{(p)4}\right)\right\}D_{1H}(\mathbf{x})$$

$$M_{\alpha\beta}^{*(\varphi)}(\mathbf{x}) = C^{(\varphi u)}(\mathbf{x})\tau_{\alpha\beta}^{*(u)}(\mathbf{x}) + C^{(\varphi\varphi)}(\mathbf{x})\tau_{\alpha\beta}^{*(\varphi)}(\mathbf{x}) + C^{(\varphi w)}(\mathbf{x})\tau_{\alpha\beta}^{*(w)}(\mathbf{x}) +$$
$$+ \sum_{a=0}^{2} C^{(\varphi\vartheta_a)}(\mathbf{x})\tau_{\alpha\beta}^{*(\vartheta_a)}(\mathbf{x}) = \frac{L^2}{D_0 h_0}M_{\alpha\beta}^{(\varphi)}(\mathbf{x})$$

$$C^{(\varphi u)}(\mathbf{x}) = C^{(u\varphi)}(\mathbf{x})$$

$$C^{(\varphi\varphi)}(\mathbf{x}) := 12\frac{1-\nu}{H}\frac{L}{h_0}c_1\left[d_{(0)2} - \frac{8}{3}c_2 d_{(0)4} + \frac{16}{9}c_2 d_{(0)6} + \right.$$
$$\left. + \zeta\left(d_{(p)2} - \frac{8}{3}c_2 d_{(p)4} + \frac{16}{9}c_2 d_{(p)6}\right)\right]D_{3H}(\mathbf{x})$$

$$C^{(\varphi w)}(\mathbf{x}) := -16\frac{1-\nu}{H}c_1 c_2\left[d_{(0)4} - \frac{4}{3}d_{(0)6} + \zeta\left(d_{(p)4} - \frac{4}{3}d_{(p)6}\right)\right]D_{3H}(\mathbf{x})$$

$$C^{(\varphi\vartheta_a)}(\mathbf{x}) := -12\frac{1-\nu}{H}\left(\frac{L}{h_0}\right)^2 \alpha_0\theta_0 c_1\left[d_{(0)a+1} - \frac{4}{3}c_2 d_{(0)a+3} + \xi\left(d_{(r)a+1} - \frac{4}{3}c_2 d_{(r)a+3}\right) + \right.$$
$$\left. + \zeta\left(d_{(p)a+1} - \frac{4}{3}c_2 d_{(p)a+3}\right) + \xi\zeta\left(d_{(p+r)a+1} - \frac{4}{3}c_2 d_{(p+r)a+3}\right)\right]B_{2H}(\mathbf{x})$$

$$M_{\alpha\beta}^{*(w)}(\mathbf{x}) = C^{(wu)}(\mathbf{x})\tau_{\alpha\beta}^{*(u)}(\mathbf{x}) + C^{(w\varphi)}(\mathbf{x})\tau_{\alpha\beta}^{*(\varphi)}(\mathbf{x}) +$$
$$+ C^{(ww)}(\mathbf{x})\tau_{\alpha\beta}^{*(w)}(\mathbf{x}) + \sum_{a=0}^{2} C^{(w\vartheta_a)}(\mathbf{x})\tau_{\alpha\beta}^{*(\vartheta_a)}(\mathbf{x}) = \frac{L^2}{D_0 h_0}M_{\alpha\beta}^{(w)}(\mathbf{x})$$

$$C^{(wu)}(\mathbf{x}) := 12\frac{1-\nu}{H}\frac{L}{h_0}\zeta\left[(1-c_1)d_{(p)1} + \frac{4}{3}c_1c_2d_{(p)3}\right]D_{2H}(\mathbf{x}) = -\frac{L}{h_0}C^{(uw)}(\mathbf{x})$$

$$C^{(w\varphi)}(\mathbf{x}) := 16\frac{1-\nu}{H}\frac{L}{h_0}c_1c_2\left[d_{(0)4} - \frac{4}{3}d_{(0)6} + \zeta\left(d_{(p)4} - \frac{4}{3}d_{(p)6}\right)\right]D_{3H}(\mathbf{x}) = -\frac{L}{h_0}C^{(\varphi w)}(\mathbf{x})$$

$$C^{(ww)}(\mathbf{x}) := 12\frac{1-\nu}{H}\left\{(c_1-1)d_{(0)2} - \frac{16}{9}c_1c_2d_{(0)6} + \right.$$

$$\left. + \zeta\left[(c_1-1)d_{(p)2} - \frac{16}{9}c_1c_2d_{(p)6}\right]\right\}D_{3H}(\mathbf{x})$$

$$C^{(w\vartheta_a)}(\mathbf{x}) := 12\frac{1-\nu}{H}\left(\frac{L}{h_0}\right)^2\alpha_0\theta_0\left\{(c_1-1)d_{(0)a+1} - \frac{4}{3}c_1c_2d_{(0)a+3} + \right.$$

$$+ \xi\left[(c_1-1)d_{(r)a+1} - \frac{4}{3}c_1c_2d_{(r)a+3}\right] + \zeta\left[(c_1-1)d_{(p)a+1} - \frac{4}{3}c_1c_2d_{(p)a+3}\right] +$$

$$\left. + \xi\zeta\left[(c_1-1)d_{(p+r)a+1} - \frac{4}{3}c_1c_2d_{(p+r)a+3}\right]\right\}B_{2H}(\mathbf{x}),$$

where

$$\begin{aligned}
\tau_{\alpha\beta}^{*(u)}(\mathbf{x}) &= \frac{H}{2}\left(u^*_{\alpha,\beta}(\mathbf{x}) + u^*_{\beta,\alpha}(\mathbf{x})\right) + \nu\delta_{\alpha\beta}u^*_{\gamma,\gamma}(\mathbf{x}) - \frac{L}{h_0}\tau_{\alpha\beta}^{(u)} \\
\tau_{\alpha\beta}^{*(\varphi)}(\mathbf{x}) &= \frac{H}{2}\left(\varphi^*_{\alpha,\beta}(\mathbf{x}) + \varphi^*_{\beta,\alpha}(\mathbf{x})\right) + \nu\delta_{\alpha\beta}\varphi^*_{\gamma,\gamma}(\mathbf{x}) = L\tau_{\alpha\beta}^{(\varphi)} \\
\tau_{\alpha\beta}^{*(w)}(\mathbf{x}) &= Hw^*_{,\alpha\beta}(\mathbf{x}) + \nu\delta_{\alpha\beta}\nabla^2 w^*(\mathbf{x}) = \frac{L^2}{h_0}\tau_{\alpha\beta}^{(w)} \\
\tau_{\alpha\beta}^{*(\vartheta_a)}(\mathbf{x}) &= (1+\nu)\vartheta_a^*(\mathbf{x})\delta_{\alpha\beta} = \tau_{\alpha\beta}^{(\vartheta_a)}/\theta_0 \\
D_{jH}(\mathbf{x}) &:= \left(h^*(\mathbf{x})\right)^j E_H(\mathbf{x}),\ B_{jH}(\mathbf{x}) := \left(h^*(\mathbf{x})\right)^j E_H(\mathbf{x})\alpha_H(\mathbf{x}), \\
D_0 &= \frac{E_0(h_0)^3}{12(1-\nu^2)},\ p^*(\mathbf{x}) := \frac{L^4}{D_0 h_0}p(\mathbf{x}).
\end{aligned} \tag{25}$$

3 NUMERICAL IMPLEMENTATION

The MLS-approximation belongs to mesh-free approximations, since no predefined connectivity among nodal points is required [6]. The nodal points are freely distributed inside the analyzed domain and on its boundary. Beside the standard MLS-approximation, one can utilize also the Central Approximation Node (CAN) concept [7, 8] which enables to utilize the local support character of the approximation efficiently. In the case of strong formulations, however, the advantages of the local support approximation are utilized immediately

[9, 10]. The MLS-approximation of a field variable $g(\mathbf{x})$ is expressed in terms of the shape functions $\psi^{\bar{a}}(\mathbf{x})$ and certain nodal unknowns $\hat{g}^{\bar{a}}$ as

$$g(\mathbf{x})\big|_{\Omega^q} = \sum_{a=1}^{N^q} \hat{g}^{\bar{a}} \psi^{\bar{a}}(\mathbf{x}) \,,\, \Omega \supset \Omega^q \ni \mathbf{x}^q, \tag{26}$$

where the central approximation nodal point $\mathbf{x}^q$ can be selected as the nearest nodal point to the point of approximation $\mathbf{x}$, and $\bar{a} := n(q,a)$ is the global number while a is the local number of the node from the support domain of the nodal point $\mathbf{x}^q$.

In order to eliminate the 3rd and 4th order derivatives of field variables in the governing equations, we propose to utilize the decomposition of the derived system of the PDE into a set of PDE with derivatives not higher than second order [3] by introducing new field variables

$$m^*(\mathbf{x}) := \nabla^2 w^*(\mathbf{x}),\ s^*_\alpha(\mathbf{x}) := \nabla^2 u^*_\alpha(\mathbf{x}),\ f^*_\alpha(\mathbf{x}) := \nabla^2 \varphi^*_\alpha(\mathbf{x}) \text{ in } \Omega \cup \partial\Omega \tag{27}$$

4 NUMERICAL EXAMPLES

Let us consider the square plate $L \times L$ $(L = 1)$ with clamped edges and constant thickness $h = h_0 = L/50$. The plates are either homogeneous or FGM with power-law gradation of material coefficients according to (6) and constant Poisson ratio $\nu = 0.3$. In presentation of numerical results, we focus on the study of the deflection response to thermal loading with prescribed temperatures on the bottom and top of the plate, $\theta^*(\mathbf{x},-h/2) = 0$, $\theta^*(\mathbf{x},h/2) = 20$, and vanishing heat flux on the lateral sides $kn_\alpha(\mathbf{x})\vartheta^*_{0,\alpha}(\mathbf{x})\big|_{\partial\Omega} = 0$.

The first conclusion is that there is no deflection response to the prescribed thermal loading neither in the homogeneous nor in FGM plates even if any one of the coefficients $\{E,\alpha,k\}$ is graded non-linearly in the transversal direction. Neither the combination of gradations of material coefficients in transversal direction gives rise to finite deflections in both the KLT and TSDPT.

Another observation is that the deflections are independent on the heat conduction coefficient gradation.

It can be seen from the analysis of governing equations that there is a coupling between deflections and thermal fields in the KLT only if $\nabla^2\left(E_H(\mathbf{x})\alpha_H(\mathbf{x})\right) \neq 0$. The numerical results (Fig. 1) confirm this finding. Really, finite deflections appeared when at least one of the coefficients $\left(E_H(\mathbf{x}),\ \alpha_H(\mathbf{x})\right)$ was graded non-linearly, or both of them were graded at least linearly.

In contrast to the KLT, finite deflections occur in the TSDPT even if $\nabla\left(E_H(\mathbf{x})\alpha_H(\mathbf{x})\right) \neq 0$ because of coupling between rotations φ_α and temperature and another coupling between deflections and rotations in TSDPT. This indirect coupling between deflections and temperature is significantly weaker than the direct $(w-\vartheta_0)$ -coupling occurring in FGM plates with such an in-plane gradation of material coefficients that $\nabla^2\left(E_H(\mathbf{x})\alpha_H(\mathbf{x})\right) \neq 0$ (see Fig. 2).

The additional transversal gradation of the YM and/or linear heat extension coefficient has an influence on finite deflections in FGM plates with combined in-plane gradations of $E_H(\mathbf{x})$ and $\alpha_H(\mathbf{x})$. The additional transversal gradation of the heat extension coefficient increases the deflections in both the KLT and TSDPT, while the gradation of the YM has an opposite effect on deflections. Note that in the case of TSDPT, the deflections change the sign when the transversal gradation of YM is applied additionally (Fig. 3).

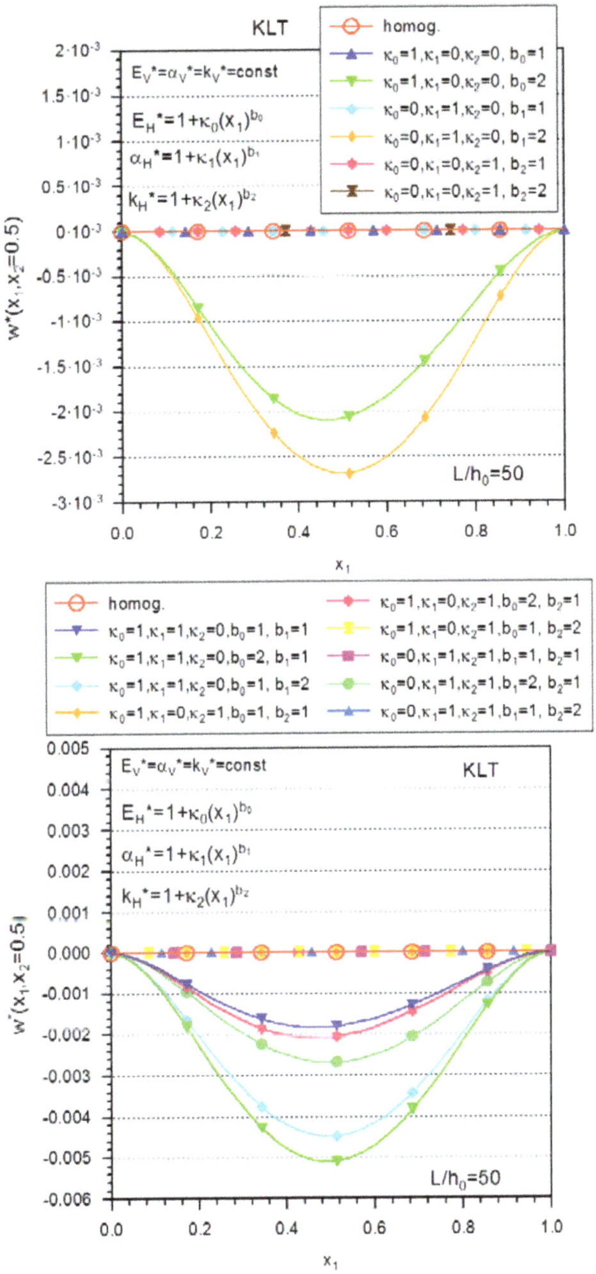

Figure 1: Deflections in FGM plates with in-plane gradation of material coefficients; results by KLT.

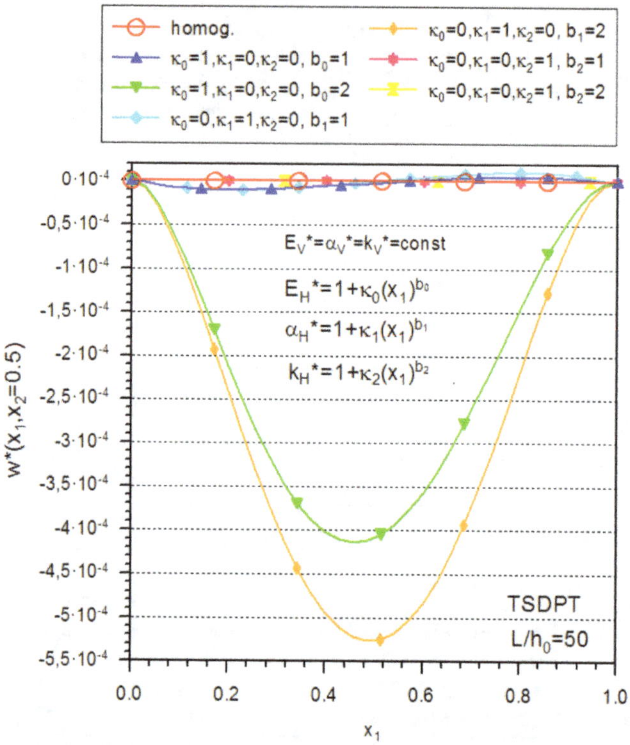

Figure 2: Deflections in FGM plates with in-plane gradation of material coefficients; results by TSDPT.

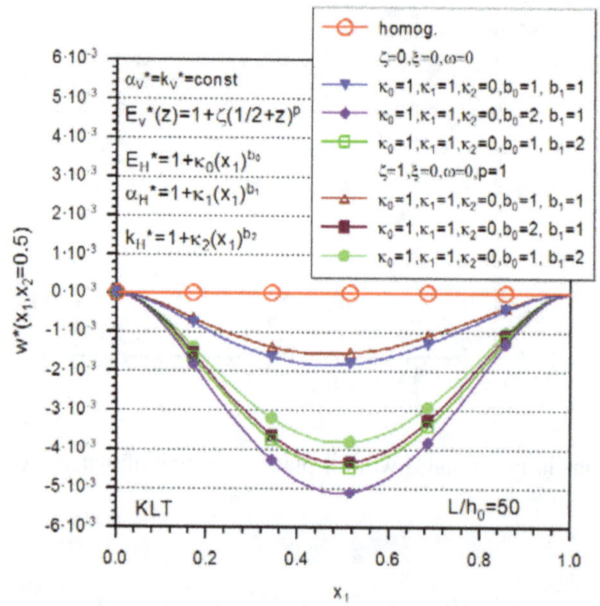

Figure 3: Deflections in FGM plates with combined in-plane gradation of material coefficients and transversal gradation of YM; results by KLT and TSDPT.

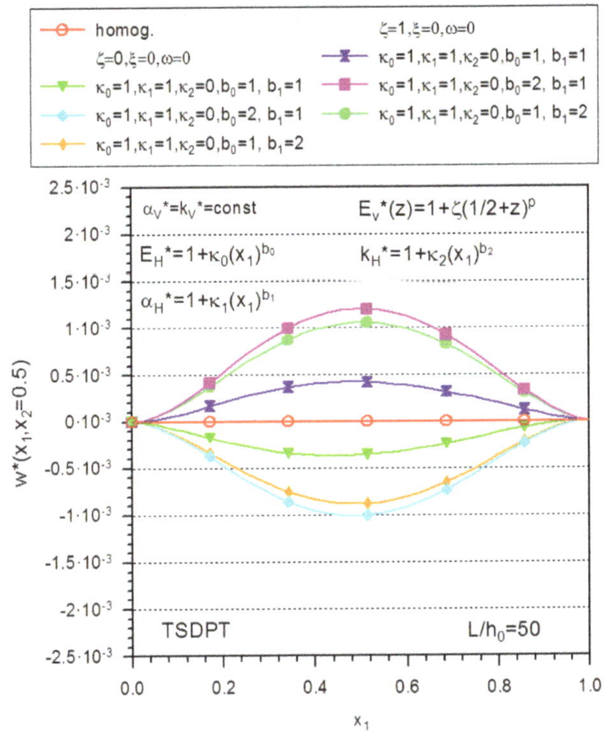

Figure 3: (*Continued*)

ACKNOWLEDGEMENTS

This work was partially supported by the Slovak Research and Development Agency under the contract No. APVV-14-0440.

REFERENCES

[1] Birman, V. & Byrd, L.W., Modeling and analysis of functionally graded materials and structures. *Applied Mechanics Reviews*, **60**, pp. 195–216, 2007. https://doi.org/10.1115/1.2777164

[2] Daniel, I.M. & Ishai, O., *Engineering Mechanics of Composite Materials*. Oxford University Press: New York, 2006.

[3] Sladek, V., Sladek, J. & Sator, L., Physical decomposition of thin plate bending problems and their solution by mesh-free methods. *Engineering Analysis with Boundary Elements*, **37**, pp. 348–365, 2013. https://doi.org/10.1016/j.enganabound.2012.11.005

[4] Sator, L., Sladek V. & Sladek J., Coupling effects in elastic analysis of FGM composite plates by mesh-free methods. *Composite Structures*, **115**, pp. 100–110, 2014. https://doi.org/10.1016/j.compstruct.2014.04.016

[5] Sator, L., Sladek, V. & Sladek, J., Multi-gradation coupling effects in FGM plates. *Composite Structures*, **171**, pp. 515–527, 2017. https://doi.org/10.1016/j.compstruct.2017.03.063

[6] Lancaster, P. & Salkauskas, K., Surfaces generated by moving least square method. *Mathematics of Computation*, **37**, pp. 141–158, 1981. https://doi.org/10.1090/s0025-5718-1981-0616367-1

[7] Sladek, V., Sladek, J. & Zhang, Ch., Computation of stresses in non-homogeneous elastic solids by local integral equation method: a comparative study. *Computational Mechanics*, **41**, pp. 827–845, 2008.
https://doi.org/10.1007/s00466-006-0155-5
[8] Sladek V., Sladek J. & Zhang Ch., Local integral equation formulation for axially symmetric problems involving elastic FGM. *Engineering Analysis with Boundary Elements,* **32**, pp. 1012–1024, 2008.
https://doi.org/10.1016/j.enganabound.2007.09.006
[9] Sladek V., Sladek J. & Zhang Ch., On increasing computational efficiency of local integral equation method combined with meshless implementations. *CMES- Computer Modeling in Engineering & Sciences,* **63**, pp. 243–263, 2010.
[10] Sladek, V. & Sladek, J., Local integral equations implemented by MLS-approximation and analytical integrations. *Engineering Analysis with Boundary Elements*, **34**, pp. 904–913, 2010.
https://doi.org/10.1016/j.enganabound.2010.03.015

BOUNDARY VALUE PROBLEMS FOR ELASTIC HALF-PLANES POSED IN TERMS OF STRESS AND DISPLACEMENT ORIENTATIONS

A. N. GALYBIN
The Schmidt Institute of Physics of the Earth (IPE RAS), Moscow, Russia.

ABSTRACT

This study investigates solvability of boundary value problems of plane elasticity formulated in terms of principal directions of the stress tensor and the orientations of the displacement vector. The analysis of solvability is performed by using the following approach. Firstly, boundary values of the complex potentials are represented by the Cauchy-type integrals with unknown density. Then a system of singular integral equations is obtained by satisfying particular boundary conditions. This system is further reduced to the system of the Riemann boundary value problems for the determination of sectionally holomorphic functions. Solvability of the Riemann problems is investigated by calculating their indexes. This allows one to determine the number of linearly independent solutions and hence the number of arbitrary parameters entering into the general solution.

Two novel formulations have been investigated for the case of elastic half-planes. In both cases the initial system of equations has been reduced to the form that allow for successive solution of its equations.

Keywords: boundary value problems, complex potentials, plane elasticity, singular integral equations

1 INTRODUCTION

In geomechanics the boundary values of stresses and/or displacements are often unknown on the boundary, which makes it impossible to apply classical formulations accepted in solid mechanics. However even incomplete information, such as orientations of principal stresses or displacements can serve as possible boundary conditions, BC. Accordingly, several non-classical formulations of elastic boundary value problems, BVP, employing BC in terms of orientations of principal stresses, displacements or tractions have been investigated in Refs. [1–6]. The mentioned studies are mostly dedicated to the investigation of solvability of the BVP with incomplete BC, which leads to non-unique solutions. The latter is proved by using the following approach. The complex potentials are presented by the Cauchy integrals with unknown densities. Their boundary values are found by employing the Sokhotski-Plemelj formulas. These are used to express the orientations of tractions, principal stresses and displacements on the contour in the form that contains two unknown complex valued functions. Further on, one of these functions is excluded by imposing the condition of continuity of the stress vector across the contour. Then by using two scalar BC in terms of the orientations one can derive a system of two real singular equations, SIE. These are combined into one linear complex SIE of the general form that contains the unknown function, its conjugate, the singular and regular integrals and their conjugates. An approach suggested by Gakhov [7] is used for analysing the solvability of the complex SIE. It assumes separation of the dominant part of the complex SIE followed by its conversion into the equivalent Riemann problem. The solvability of the latter depends on the index of the coefficient of the Riemann problem that is calculated through the increment of its argument (divided by 2π) in traversing the contour anticlockwise. If the index is positive then the Riemann problem has a finite number of solutions that is proportional to the index, otherwise the BVP does not have solutions in the class of holomorphic functions. This approach can also be adopted for numerical solving complete complex SIEs [6].

DOI: 10.2495/CMEM-V6-N6-1173-1181

It is shown in Refs. [1–6] that the dominant SIE corresponds to the case of elastic half-plane regardless of the BC used. Therefore we further consider the dominant SIE only. However, in contrast to the previously used approach we deal with two real SIEs. The reason behind this is to build the successive solutions of the equations in the system for the problems that do not directly allow for the application of the technique mentioned above because of the presence of the derivatives of the unknown functions in the complex dominant SIE.

2 BOUNDARY VALUES OF STRESSES AND DISPLACEMENTS

Let us introduce the following notations based on the Kolosov formulas [8]

$$D \equiv \frac{\sigma_{22}-\sigma_{11}}{2} + i\sigma_{12} = \tau_{\max}\exp(i\alpha), \quad \alpha = \pi - 2\theta, \tag{1}$$

$$W \equiv 2G(u_1 + iu_2) = \omega \exp(i\beta). \tag{2}$$

Here D is the complex stress deviator function, σ_{11}, σ_{12}, σ_{22} are the components of the stress tensor $\tau_{max}=(\sigma_1-\sigma_2)/2\geq 0$ is the maximum shear stress, σ_1, σ_2 are major and minor principal stresses respectively, the angle $\theta=\frac{1}{2}\operatorname{arctg}\frac{2\sigma_{12}}{\sigma_{11}-\sigma_{22}}$ specifies the principal direction (the angle between σ_1 and the real axis), thus $D=-\tau_{max}\exp(-2i\theta)$; W is the complex displacement vector, G is the shear modulus, u_1, u_2 are the displacement components, $\omega\geq 0$ and β are respectively the modulus and the argument of the function W.

On the real axis, denoted as contour Γ further on, the boundary values of the functions D and W are found as follows [1, 4]

$$2D^{\pm} = \mp(g' + \bar{g}') - \mathbf{S}(g' + \bar{g}'), \tag{3}$$

$$2W^{\pm} = \pm(\kappa^{\pm} + 1)g + (\kappa^{\pm} - 1)\mathbf{S}(g). \tag{4}$$

Here the superscripts +/- stay for the boundary values for the upper/lower half-planes respectively, $\kappa^{\pm}$ are the Muskhelishvili's constants (they can be different for different half-planes) and the following notation for the singular integral was introduced

$$\mathbf{S}(g) = \frac{1}{\pi i}\int_{\Gamma}\frac{g(t)}{t-\zeta}dt = \frac{1}{\pi i}\int_{-\infty}^{\infty}\frac{g(t)}{t-\zeta}dt, \quad \zeta \in \Gamma \Leftrightarrow \operatorname{Im}\zeta = 0. \tag{5}$$

The complex operator has the following properties

$$\mathbf{S}\left[\mathbf{S}(g)\right] = g, \tag{6}$$

$$\overline{\mathbf{S}(g)} = -\mathbf{S}(\bar{g}). \tag{7}$$

The former is valid for any closed contour, while the latter is valid only for the real axis.

It should be noted that the representations in (3) and (4) are given through the only complex valued function g that is proportional to the unknown displacement jump across the contour. Meanwhile they automatically satisfy the condition of continuity of the stress vector, p, across the real axis, i.e. the condition

$$p^{+}(\zeta) - p^{-}(\zeta) = 0, \quad \operatorname{Im}(\zeta) = 0. \tag{8}$$

The following BC are used further on

$$\arg D^{\pm} = \alpha^{\pm}(\zeta), \qquad \zeta \in \Gamma, \tag{9}$$

$$\arg W^{\pm} = \beta^{\pm}(\zeta), \qquad \zeta \in \Gamma. \tag{10}$$

Taking into account the properties of the singular operator one can express (9)-(10) in terms of the sought function g by using (3)-(4) as follows

$$\left[\mp(g'+\overline{g}') - \mathbf{S}(g'+\overline{g}')\right]\exp(-i\alpha^{\pm}) - \left[\mp(g'+\overline{g}') + \mathbf{S}(g'+\overline{g}')\right]\exp(i\alpha^{\pm}) = 0, \tag{11}$$

$$\left[\pm(\kappa^{\pm}+1)g + (\kappa^{\pm}-1)\mathbf{S}(g)\right]\exp(-i\beta^{\pm}) - \left[\pm(\kappa^{\pm}+1)\overline{g} - (\kappa^{\pm}-1)\mathbf{S}(\overline{g})\right]\exp(i\beta^{\pm}) = 0. \tag{12}$$

Note that equations (11)-(12) are real valued.

It is convenient to introduce the real and imaginary parts of the sought function g by using the notations

$$g = \mathrm{M} + i\mathrm{N}, \quad g' = \mu + i\nu, \quad \mu = \mathrm{M}', \quad \nu = \mathrm{N}'. \tag{13}$$

Then (11) can be presented in the form

$$\pm i\mu\sin(i\alpha^{\pm}) - \cos(i\alpha^{\pm})\mathbf{S}(\mu) = 0. \tag{14}$$

We further consider the following BVPs:

1. Problem (α,p) assumes that boundary values of the stress orientations in (9) are known and condition (8) is satisfied.
2. Combined problem (α,p) and (β,p) assumes that BC (9)-(10) are satisfied simultaneously on upper/lower boundary and complemented by condition (8).
3. Problem (α,β) also assumes BC (8)-(10) but only for upper (or lower) half-plane alone.

It is worth to notice that these problems can be reduced to a single complex SIE. In this case problem (α,p) contains the only real valued function (next section) while other two problems would include both the complex unknown function and its derivative, which makes it impossible the direct application of the technique proposed by Gakhov [7].

3 PROBLEM (α, p) FOR JOINED HALF-PLANES

Let us consider a BVP with the BC in terms of the principal directions of the stress tensor that are specified on the x-axis that presents the interface between the upper Ω^+ and lower Ω^- half-planes.

It is assumed that the stress vector is continuous across the interface and the boundary values of the principal directions are different on the Ω^+ and Ω^- boundaries. They can be expressed as the boundary values of the argument of the complex valued function of the stress deviator $\arg D^{\pm} = \alpha^{\pm}$. The loads acting on the interface are unknown as well as displacements, body forces are absent, the stresses at infinity are uniform but unknown, or vanish.

In this case the complex dominant SIE can be expressed as follows

$$(2 - e^{4i\theta^+} - e^{4i\theta^-})\mu + (e^{4i\theta^-} - e^{4i\theta^+})\mathbf{S}(\mu) = 0. \tag{15}$$

The conditions $\tau_{max} \geq 0$ should be satisfied, which results in the following inequality

$$\sin(2\theta^{+})\sin(2\theta^{-})\geq 0. \tag{16}$$

The corresponding Riemann BVP is radically simplified to the following form

$$e^{-2i\theta^{+}}\sin(2\theta^{+})A^{+}=e^{-2i\theta^{-}}\sin(2\theta^{-})A^{-}. \tag{17}$$

Since the origin and the infinity belong to the x-axis, one should seek the solution of the Riemann BVP with the use of the linear fractional function, which leads to the following form of the BVP [7]

$$A^{+}=\left[e^{-2i(\theta^{+}-\theta^{-})}\left(\tfrac{x-i}{x+i}\right)^{-\mathrm{K}}G_{0}\right]\left(\tfrac{x-i}{x+i}\right)^{\mathrm{K}}A^{-},\quad G_{0}=\tfrac{\sin(2\theta^{-})}{\sin(2\theta^{+})}, \tag{18}$$

where K, is a non-negative index calculated in accordance with the following formula

$$\mathrm{K}=Index\left(e^{-2i(\theta^{+}-\theta^{-})}\right)=\tfrac{1}{\pi}[\theta^{+}-\theta^{-}]\big|_{\Gamma}. \tag{19}$$

As soon as the index of the linear fractional function is equal to unity and the index of G_0 is zero, then the term is square brackets in the right hand side of (18) has zero index. Therefore the solution of the Riemann BVP assumes the following form, [7]

$$A^{\pm}(z)=\frac{P_{\mathrm{K}}(z)}{(z\pm i)^{K}}e^{Y^{\pm}(z)},\quad Y^{\pm}(z)=\frac{1}{2\pi i}\int_{-\infty}^{\infty}\ln\left[\left(\tfrac{t-i}{t+i}\right)^{-\mathrm{K}}G(t)\right]\frac{1}{t-z}dt,\ \mathrm{Im}(z)\neq 0. \tag{20}$$

On the x-axis one can present the solution as follows

$$A^{\pm}=\frac{P_{\mathrm{K}}(z)}{(x^{2}+1)^{\frac{K}{2}}}\left(\frac{\sin 2\theta^{-}}{\sin 2\theta^{+}}\right)^{\pm\frac{1}{2}}e^{\mp i(\theta^{+}-\theta^{-})}e^{\frac{1}{2}\mathbf{S}\left[-\mathrm{K}\ln\left(\frac{t-i}{t+i}\right)+\ln\left(G(t)\right)\right]}. \tag{21}$$

Let us consider the complex conjugated singular operator in the solution above.

$$\begin{aligned}&\overline{\mathbf{S}\left[-\mathrm{K}\ln\left(\tfrac{t+i}{t-i}\right)+\ln\left(G(t)\right)\right]}=\\&=-\mathbf{S}\left[-\mathrm{K}\ln\left(\tfrac{t+i}{t-i}\right)+2i(\theta^{+}-\theta^{-})+\ln\left(G_{0}(t)\right)\right]=\\&=\mathbf{S}\left[-\mathrm{K}\ln\left(\tfrac{t+i}{t-i}\right)+\ln\left(G(t)\right)\right]-2\mathbf{S}\left[\ln\left(G_{0}(t)\right)\right].\end{aligned} \tag{22}$$

Here the latter term in the right hand sine is pure imagine, let us denoted it as

$$2\mathbf{S}\left[\ln\left(G_{0}(t)\right)\right]=2i\chi,\quad \mathrm{Im}\,\chi=0. \tag{23}$$

Then solution of (15) takes the form

$$\mu=-2i\sin(\theta^{+}-\theta^{-}-\chi)\frac{P_{\mathrm{K}}(z)}{(x^{2}+1)^{\frac{K}{2}}}\left(\frac{\sin 2\theta^{-}}{\sin 2\theta^{+}}\right)^{\pm\frac{1}{2}}e^{\mathbf{S}\left[-\frac{\mathrm{K}}{2}\ln\left(\frac{t-i}{t+i}\right)-i(\theta^{+}-\theta^{-})\right]}. \tag{24}$$

It follows from (23) that the exponential term in (24) is real-valued. The coefficients of the polynomial should be real to provide real-valuedness of the solution. Thus, the solution depends on (K+1) linearly independent real constants.

The solution above determines the boundary values of the complex deviator on the x-axis. However this is not sufficient for all points lying outside the x-axis. This is evident from the fact that the expression for the complex deviator function can be presented the form [1–3]

$$D(z,\bar{z}) = \frac{-1}{\pi i}\int_{-\infty}^{\infty}\frac{\mu(t)}{t-z}dt + (\bar{z}-z)\frac{1}{2\pi i}\int_{-\infty}^{\infty}\frac{\mu'(t)+i\nu'(t)}{t-z}dt\ . \tag{25}$$

Therefore we have just found a partial solution that corresponds to the real part of the sought function $g'(x)$. It can be presented as follows

$$D_{\mu}(z,\bar{z}) = -2\Phi_{\mu}(z) + (\bar{z}-z)\Phi'_{\mu}(z), \tag{26}$$

where the following notation for the holomorphic function has been introduced in terms of the Cauchy integral

$$\Phi_{\mu}(z) = \frac{1}{2\pi i}\int_{-\infty}^{\infty}\frac{\mu(t)}{t-z}dt. \tag{27}$$

This solution should be complemented by adding the terms of the form $(\bar{z}-z)\mathrm{X}^{\pm}(z)$, where $\mathrm{X}^{\pm}(z)$ are arbitrary holomorphic functions decaying at infinity faster then z^{-1}. These functions do not affect the boundary conditions on the x-axis and at infinity, and, moreover, the function D remains bi-holomorphic. Thus, the general solution is determined to within an arbitrary holomorphic function, indicating that the problem is substantially underdetermined.

Nevertheless, the analysis above makes sense for applications, since it allows one to determine such important characteristics of the stress state on the interface between the half-planes as the shear stresses and maximum shear stresses. The knowledge of these is important when assessing the onset of the propagation of shear fracture. On the other hand this solution can be combined with boundary conditions of other types, for instance, with the conditions imposed on the curvature of the stress trajectories. For example let us assume homogeneous conditions on the x-axis $\alpha^{\pm} = -2\theta_0^{\pm}$ (conditions $\tau_{max} \geq 0$ assume to be fulfilled). In this case the Riemann BVP has zero index, hence the functions $\Lambda^{\pm}(z)$ are constants as well as the sought function. Meanwhile this constant is real. Thus, the solution for the boundary values takes the form

$$D^{\pm} = Ce^{i\alpha^{\pm}}, \quad \mathrm{Im}(C) = 0. \tag{28}$$

This solution coincides with the one obtained in Ref. [1] by using a completely different approach.

Another possibility requires attracting the boundary condition in the orientations of the displacement vector.

4 COMBINED PROBLEM (a, p) AND (β, p) FOR HALF-PLANES

For simplicity, we consider the case of half-planes made of the same material and assume a pair of boundary conditions (9)-(10). The SIE of the problem takes the form (we omit the indices at the elastic constant and apply the properties of the singular operator)

$$e^{-i\beta^{\pm}}\left[\pm(\kappa+1)g+(\kappa-1)\mathbf{S}(g)\right]-e^{i\beta^{\pm}}\left[\pm(\kappa+1)\overline{g}-(\kappa-1)\mathbf{S}(\overline{g})\right]=0. \tag{29}$$

These two scalar equations can be combined into one complex equation by eliminating $\mathbf{S}(\overline{g})$ as follows

$$(\kappa+1)\left(e^{-2i\beta^{+}}+e^{-2i\beta^{-}}\right)g-2(\kappa+1)\overline{g}+(\kappa-1)\left(e^{-2i\beta^{+}}-e^{-2i\beta^{-}}\right)\mathbf{S}(g)=0\,. \tag{30}$$

We suppose that the real part of the unknown function is known from the solution of the previous problem. The complex conjugate function is further represented in terms of the integral of the function $\mu(t)$ as follows

$$\overline{g}=2\mathrm{M}-g,\quad \mathrm{M}'=\mu. \tag{31}$$

Then the SIE assumes its final form

$$\left(2+e^{-2i\beta^{+}}+e^{-2i\beta^{-}}\right)g+\frac{\kappa-1}{\kappa+1}\left(e^{-2i\beta^{+}}-e^{-2i\beta^{-}}\right)\mathbf{S}(g)=4\mathrm{M}\,. \tag{32}$$

This equation is slightly different from SIE (15) and in contrast to (15) is non-homogeneous. Its right hand side is found by integration of (24), which generates one additional arbitrary constant. The latter should be real as the function M is real. Therefore the right hand side depends on (K+2) arbitrary real constants.

The coefficient of the associated Riemann BVP is found as follows

$$G=\frac{\kappa e^{-i\beta^{-}}\cos\beta^{-}+e^{-i\beta^{+}}\cos\beta^{+}}{\kappa e^{-i\beta^{+}}\cos\beta^{+}+e^{-i\beta^{-}}\cos\beta^{-}}\,. \tag{33}$$

It is evident that *Index(G)*=0 because $\kappa\cos^2\beta^{\pm}+\cos^2\beta^{\mp}>0$, i.e. the real parts of both the numerator and the denominator of the fraction are positive. Therefore the non-homogeneous SIE in (32) has a unique solution. The general solution is found by summing the function M with the real part of the function g. Moreover, the number of arbitrary constants remains. As soon as the derivative of M is required for the stress determination, the general solution depends on up to (K+1) arbitrary constant entering into the solution for the function $\mu(t)$.

5 PROBLEM (α, β) FOR HALF-PLANE

The same method as above can be used to find the solution of the BVP formulated in terms of principal stress and displacement orientations. Let us consider the upper half-plane. Then the indices can be omitted and one can write the BC on the x-axis in the form

$$\arg D=\alpha\,, \tag{34}$$

$$\arg W=\beta\,. \tag{35}$$

We still seek solutions that provide continuity of the stress vector across the real axis, but now we use the scalar equations for the upper half-plane. Then the system of SIE assumes the form

$$\begin{aligned}&e^{-i\alpha}\left[\mu+\mathbf{S}(\mu)\right]-e^{i\alpha}\left[\mu-\mathbf{S}(\mu)\right]=0,\\&e^{-i\beta}\left[(\kappa+1)g+(\kappa-1)\mathbf{S}(g)\right]-e^{i\beta}\left[(\kappa+1)\overline{g}-(\kappa-1)\mathbf{S}(\overline{g})\right]=0.\end{aligned} \tag{36}$$

By introducing the following notations

$$g = \mathrm{M} + i\Lambda, \quad \mathrm{M}' = \mu, \tag{37}$$

we present (36) in the form

$$\begin{aligned} &\sin(\alpha)\mathrm{M}' + i\cos(\alpha)\mathbf{S}(\mathrm{M}') = 0, \\ &(\kappa+1)\cos(\beta)\Lambda - i(\kappa-1)\sin(\beta)\mathbf{S}(\Lambda) = f. \end{aligned} \tag{38}$$

Here the right hand side of the latter equation depends on the solution of the former equation. It has the form

$$f = (\kappa+1)\sin(\beta)\mathrm{M} + i(\kappa-1)\cos(\beta)\mathbf{S}(\mathrm{M}). \tag{39}$$

The system of SIE (38) can be reduced to two Riemann BVPs on the real axis. The former SIE in (38) is homogeneous; it can be reduced to the following homogeneous Riemann BVP

$$A^{+} = -e^{2i\alpha} A^{-}. \tag{40}$$

The latter equation in (38) is equivalent to the following non-homogeneous Riemann BVP

$$B^{+} = e^{2i(\beta-\varsigma)} B^{-} + \frac{e^{i\beta}}{\kappa + e^{2i\beta}} f, \qquad \varsigma = \arg\left(\kappa + e^{2i\beta}\right). \tag{41}$$

The indexes of these BVPs are determined as follows

$$2\mathrm{K} = Index(e^{2i\alpha}) = Index(e^{-4i\theta}) = -\tfrac{2}{\pi}\left[\theta\right]\Big|_{-\infty}^{\infty}, \tag{42}$$

$$2\mathrm{N} = Index(e^{2i\beta}) = \tfrac{1}{\pi}\left[\beta\right]\Big|_{-\infty}^{\infty}. \tag{43}$$

In (43) it has been taken into account that $Index(e^{i\varsigma}) = 0$ as $\mathrm{Re}(\kappa + e^{2i\beta}) > 0$.

Further on we employ the general solution from Gakhov [7] to seek solutions of (40)-(41) vanishing at infinity. These solutions obviously depend on the indexes (42)-(43) respectively. The solution for the function M is found in a similar way as (18). It has the form

$$\mathrm{M}'(x) = 2\cos\alpha(x)\frac{P_{2\mathrm{K}-1}(x)}{(x^2+1)^{\mathrm{K}}} e^{-\mathbf{S}\left[\mathrm{K}\ln\left(\frac{t-i}{t+i}\right)+i\alpha(t)\right]} \tag{44}$$

Here the exponential function is real-valued, thus the coefficients of the polynomial $P_{2\mathrm{K}-1}$ should be real as well to provide real-valuedness of the sought function. Therefore for positively defined index (42) the solution depends on 2K arbitrary real constants.

Let us solve the second SIE in (38) assuming provisionally that 2N≥0 and the function f specified by (39) is known. This solution (vanishing at infinity) is found from the general solution provided by Gakhov [7] in the form

$$\Lambda(x) = \cos(\beta(x) - \varsigma(x)) f(x) + i\sin(\beta(x) - \varsigma(x))\left[Z(x)\mathbf{S}\left(\frac{f(t)}{Z(t)}\right) - \frac{\tilde{P}_{2\mathrm{N}-1}(x)}{(x^2+1)^{\mathrm{N}}}\right]. \tag{45}$$

Here for briefness the following notation was introduced

$$Z(x) = \exp\left[-\mathbf{S}\left(\mathrm{N}\ln\tfrac{t-i}{t+i} - i\beta + i\varsigma\right)\right]. \tag{46}$$

It should be noted that this function is real-valued because the singular integral is real-valued. Then it becomes evident that the first term in the square brackets is purely imaginary, therefore, the second term should be purely imaginary as well to provide real-valuedness of the sought function Λ. Hence all the coefficients of the polynomial $\tilde{P}_{2N-1}(x)$ are purely imaginary.

The general number of arbitrary real constants entering the solution is 2K+2N+1, where we added an extra constant generated by integration of (44) in order to find the right hand side of the non-homogeneous equation in the system. This number is the maximum possible number of arbitrary constants provided that both indexes are positive. If one of them is negative, say 2N<0, then solutions exist if (-2N-1) integral conditions are imposed on the right hand side of the non-homogeneous equation of the system (see detail in Gakhov [7]). They can be always satisfied provided that 2K>-2N−1. Therefore for solvability of the system of SIE (38) it is necessary and sufficient that the cumulative index 2K+2N+1 is positive. Even more the number of solutions is an odd number because both indices determined by formulas (42) and (43) are even numbers, which is evident from the fact that the stress and displacement orientations attaint their initial values after the complete traverse of the contour.

It should be noted that the number of arbitrary constants affecting the stresses can be reduced to 2K+2N, bearing in mind that three constants can be eliminated as translation and rotation of the rigid body do not affect stresses. As soon as the displacements at infinity were zero one needs to satisfy the only additional condition on rotation, which would fix one of the coefficients of the polynomial $\tilde{P}_{2N-1}(x)$ from the condition Im(Φ(0))=0.

6 CONCLUSIONS

In this paper we investigated solvability of three non-classical boundary value problems for elastic half-plane formulated in terms of the given orientations of the displacements and stresses on the *x*-axis. This study complements the results obtained earlier (Section 3) and presents two novel solutions in the closed form (Sections 4 and 5). The latter are obtained by successive solution of the equations forming the system of SIE.

ACKNOWLEDGEMENTS

The author is thankful to the Russian Foundation for Basic Research that supported this work by the Grant 16-01-00260.

REFERENCES

[1] Galybin, A.N. & Mukhamediev, Sh.A., Plane elastic boundary value problem posed on orientation of principal stresses. *Journal of the Mechanics and Physics of Solids*, **47**(11), pp. 2381–2409, 1999. https://doi.org/10.1016/s0022-5096(99)00032-0

[2] Galybin, A.N. & Mukhamediev, Sh.A., Integral equations for elastic problems posed in principal directions: application for adjacent domains. *WIT Transactions on Modelling and Simulation*. **42**, WIT Press, Southampton, UK, pp. 51–60, 2006.

[3] Galybin, A.N., Boundary value problems posed in terms of stress orientations for plane crack systems. *Proceedings of 12th International Conference on Fracture*. Ottawa, on CD-ROM, paper T02.007 (fin 00560), 2009.

[4] Galybin, A.N., Boundary value problems of plane elasticity involving orientations of displacements and tractions. *Journal of Elasticity*, **102**, pp. 15–30, 2011. https://doi.org/10.1007/s10659-010-9259-4

[5] Galybin, A.N., Boundary integral equations for plane elastic problems posed in terms of stress orientations. *Proceeding of the 8th UK Conference on Boundary Integral Methods* (Ed. D. Lesnic), University of Leeds, UK, 2011.

[6] Galybin, A.N., On solvability of the BVP formulated in terms of displacement orientations on the interface between dissimilar elastic materials. *International Journal of Computational Methods and Experimental Measurements*, **5**(3), pp. 369–376, 2017. https://doi.org/10.2495/cmem-v5-n3-369-376

[7] Gakhov, F.D., *Boundary value problems*. Moscow, Nauka (3rd Russian edition, see also the translation of the first edition by Pergamon Press, 1966), 1977.

[8] Muskhelishvili, N.I., *Some Basic Problems of the Mathematical Theory of Elasticity*. P. Noordhoff, Groningen, The Netherlands, 1963.

SURFACE EQUIVALENCE PRINCIPLE AND SURFACE INTEGRAL EQUATION (SIE) REVISITED FOR BIOELECTROMAGNETICS APPLICATION

M. CVETKOVIĆ & D. POLJAK
University of Split, Faculty of Electrical Engineering, Mechanical Engineering and Naval Architecture, Croatia.

ABSTRACT
The paper revisits the use of a surface equivalence theorem in deriving the surface integral equation (SIE) based formulation for a homogeneous bio-electromagnetics problem. The vector analog of Green's 2nd identity is used to obtain the expression for the electric field representing the mathematical foundation of the equivalence theorem. The particular emphasis is put on the treatment of boundary integral when the observation and source points, respectively, coincide. The boundary conditions at infinity are taken into account via the Sommerfeld radiation conditions. The derived coupled SIE set can be used in problems involving biological body exposed to electromagnetic field radiation.
Keywords: Computational electromagnetics, Equivalence principle, Human exposure to electromagnetic fields, Surface integral equation,

1 INTRODUCTION

The human exposure to electromagnetic (EM) fields generated by different sources such as power and telecommunication installations has raised a number of questions regarding potentially adverse health effects. In addition to unwanted exposure to artificially generated EM fields, some biomedical applications of EM fields in medical diagnostics and for therapy purposes, has recently become of particular importance. In order to address the concern regarding the potentially harmful effects, but also to aid in the design of a more efficient medical treatments using EM radiation, the knowledge of the accurate distribution of the electromagnetic fields inside the biological tissues is required. As it is rather difficult, if possible at all, to accurately measure the induced fields within the living biological body, the exposure assessment is carried out using a sophisticated computational bioelectromagnetics models [1–3].

From the early days, bioelectromagnetics models based on differential equation approach has become the de facto standard, although the integral equation approach, using the Green integral representation, is suitable for the exact treatment of open boundary problems such as the human head or the body exposed to incident EM field [4, 5]. Regardless of the fact that the numerical methods based on the solution of integral equations in computational electromagnetics (CEM) were developed during the sixties [6, 7], only recently has this approach seen a revival in bioelectromagnetics community [2, 4, 5, 8, 9].

One interesting approach to formulate bioelectromagnetics problem is similar to a classical scattering problem [4]. If one considers the biological body of an arbitrary shape, placed in the field of an incident EM wave, an EM model based on the integral equation can be posed by using the equivalence theorem and the appropriate boundary conditions. In the case of scattering from a homogeneous dielectric body, the surface integral equation (SIE), based on the application of surface equivalence theorem is used, while in the non-homogeneous case, the appropriate volume integral equation (VIE) is derived using the volume equivalence principle.

DOI: 10.2495/CMEM-V6-N6-1182-1191

This paper aims to revisit the applications of the surface equivalence theorem and the derivation of electric field integral equation (EFIE) formulation applicable to bioelectromagnetics problem. The paper is organized as follows: The first part deals with vector analog of the Green's theorem used to derive the expression for the so called Stratton-Chu expression. This is followed by a treatment of the limiting case when observation and source points, respectively, coincide. The boundary conditions at infinity are taken into account via the Sommerfeld radiation conditions. Furthermore, the application of the equivalence principle is demonstrated to result in the coupled integral equations set. The final part presents some illustrative numerical results obtained by using SIE formulation related to bioelectromagnetics application.

2 ON THE USE OF GREEN'S SECOND IDENTITY

Let us consider the problem depicted in Fig. 1. The lossy material body V_2 of complex parameters (ε_2, μ_2), bounded by surface S, is placed in an infinite region V_1, with complex parameters (ε_1, μ_1), bounded with S_1, where $S_1 = S \cup S_\varepsilon \cup S_\infty$, with electric and magnetic source currents $(\vec{J}_1, \vec{M}_1)$. S_∞ and S_ε represent the boundary at the infinity, and of infinitesimal space V_ε required when $\vec{r} \to \vec{r}'$, respectively. $\vec{r}$ and $\vec{r}'$ represent the observation and the source point, respectively. Unit vector $\hat{n}$ is directed from V_2 to V_1, while unit vectors $\hat{n}_1$ and $\hat{n}_2$, respectively, are outward directed for each region.

If no material body V_2 is present, the field due to the currents induced by radiating source at an arbitrary point in V_1 is relatively easy to determine. However, due to the presence of a material body V_2, the field is perturbed, leading to a significantly complex problem, since it is necessary to satisfy the boundary conditions at the interface between two regions.

One way of solving such a problem is to link the field inside V_1 with sources within that region, and the field on the boundary surface of S_1. To facilitate this connection, the vector analog of Green's theorem or Green's 2nd identity is used:

$$\int_V (\vec{Q}\cdot\nabla\times\nabla\times\vec{P} - \vec{P}\cdot\nabla\times\nabla\times\vec{Q})dV = \int_S (\vec{P}\cdot\nabla\times\nabla\times\vec{Q} - \vec{Q}\cdot\nabla\times\nabla\times\vec{P})\cdot n dS \quad (1)$$

where $\vec{P}$ and $\vec{Q}$ are vector functions with continuous first and second derivatives within V_1, satisfying Helmholtz equation, selected as:

$$\vec{P} = \hat{a}G(\vec{r},\vec{r}'); \quad \vec{Q} = \vec{E}(\vec{r}) \quad (2)$$

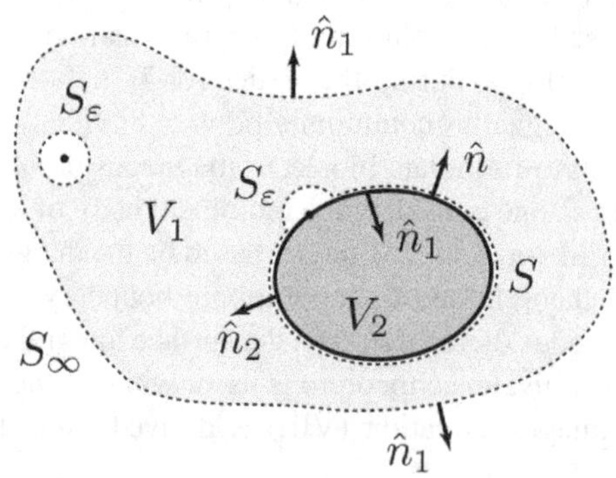

Figure 1: Problem description with denoted regions and boundaries.

where $\vec{E}$ is the electric fields, $\hat{a}$ is the consatnty unit vector, while $G(\vec{r},\vec{r}')$ denotes the free space scalar Green's function

$$G(\vec{r},\vec{r}') = \frac{e^{-jkR}}{4\pi R}; \quad R = |\vec{r} - \vec{r}'| \tag{3}$$

In (3), R is the distance from the observation point $\vec{r}$ to the source point $\vec{r}'$, and $k = \omega\sqrt{\mu\omega}$ is a wave number in free space.

2.1 Left hand side of the identity

Substituting (2) in (1), after some manipulation, the first term on the left side of (1) can be written as:

$$\nabla \times \nabla \times \vec{P} = \nabla \times \nabla \times \hat{a}G = \nabla(\hat{a} \cdot \nabla G) + k^2 G \tag{4}$$

where μ and ε are permitivity and permeability, respectively, and $\omega = 2\pi f$ is operating frequency.

Multiplying (4) with $\vec{E}$ and integrating over volume V_1, leads to the following expression

$$I = \int_{V_1} \vec{E} \cdot \left[\nabla(a \cdot \nabla G) + ak^2 G\right] dV \tag{5}$$

which, after performing some mathematical manipulations results in

$$I = \int_{S_1} \vec{E} \cdot (a \cdot \nabla G) n \, dS - \int_{V_1} (a \cdot \nabla G) \frac{\rho}{\varepsilon} dV + \int_{V_1} \vec{E} \cdot ak^2 G \, dV \tag{6}$$

where ρ represents the electric charge density.

After some rearranging, the constant vector is moved outside the integrals, leading to the following expression

$$I = \hat{a} \cdot \int_{S_1} \nabla G(\hat{n} \cdot \vec{E}) \, dS + \hat{a} \cdot \int_{V_1} \left(k^2 G \vec{E} - \frac{\rho}{\varepsilon} \nabla G \right) dV \tag{7}$$

Repeating the same procedure, the 2nd term on the left side of (1) can be written as:

$$\vec{P} \cdot \nabla \times \nabla \times \vec{Q} = \hat{a} \cdot G\left[k^2 \vec{E} - j\omega\mu\vec{J} - \nabla \times \vec{M}\right] \tag{8}$$

where the following substitution has been used:

$$\nabla \times \nabla \times \vec{E} = k^2 \vec{E} - j\omega\mu\vec{J} - \nabla \times \vec{M} \tag{9}$$

Integrating (8) over volume V_1, after some rearranging, the following expression is obtained:

$$II = \hat{a} \cdot \int_{V_1} G\left(k^2 \vec{E} - j\omega\mu\vec{J}\right) dV - \hat{a} \cdot \int_{V_1} G(\nabla \times \vec{M}) dV \tag{10}$$

Partial integration of the last term in (10), followed by application of vector form of Stokes theorem, leads to:

$$II = \hat{a} \cdot \int_{V_1} G\left(k^2 \vec{E} - j\omega\mu\vec{J}\right) dV - \hat{a} \cdot \int_{S_1} \hat{n} \times (G\vec{M}) dS + \hat{a} \cdot \int_{V_1} \nabla G \times \vec{M} \, dV \tag{11}$$

Finally, subtracting (11) from (7), left side of Green's second identity (1) can be written as follows:

$$\begin{aligned} \text{L.H.S.} \equiv a\cdot\int_{S_1} \nabla G(n\cdot\vec{E})\, dS + a\cdot\int_{V_1}\left(k^2 G\vec{E} - \frac{\rho}{\varepsilon}\nabla G\right)dV \\ -a\cdot\int_{V_1} G\left(k^2\vec{E} - j\omega\mu\vec{J}\right)dV - \hat{a}\cdot\int_{S_1}\hat{n}\times(G\vec{M})\,dS \\ +\hat{a}\cdot\int_{V_1}\nabla G\times\vec{M}\,dV \end{aligned} \tag{12}$$

which, after some rearrangement, results in:

$$\int_{V_1}\left(j\omega\mu\vec{J}G - \frac{\rho}{\varepsilon}\nabla G - \nabla G\times\vec{M}\right)dV + \int_{S_1}\left[\nabla G(n\cdot\vec{E}) + n\times(G\vec{M})\right]dS \tag{13}$$

2.2 Right hand side of the identity

Treatment of the right hand side of (1) is practically identical. The first term can be written as:

$$\vec{P}\times\nabla\times\vec{Q} = \hat{a}G\times\nabla\times\vec{E} \tag{14}$$

Where the use of $\nabla\times\vec{E} = -j\omega\mu\vec{H} - \vec{M}$ will lead to:

$$\left(\vec{P}\times\nabla\times\vec{Q}\right)\cdot\hat{n} = \hat{a}\cdot\left[\hat{n}\times\left(j\omega\mu G\vec{H} + G\vec{M}\right)\right] \tag{15}$$

Repeating the same procedure, the second term from the right hand side of (1) can be written as

$$\left(\vec{Q}\times\nabla\times\vec{P}\right)\cdot\hat{n} = \left[\vec{E}\times\nabla\times(\hat{a}G)\right]\cdot\hat{n} = \hat{a}\cdot\left(\hat{n}\cdot\vec{E}\right)\times\nabla G \tag{16}$$

Integrating (15) and (16), respectively, over the surface S_1, followed by a subtraction, the following expression is obtained:

$$\begin{aligned} \text{R.H.S.} \equiv \int_{S_1}\hat{a}\cdot\left[\hat{n}\times\left(j\omega\mu G\vec{H} + G\vec{M}\right)\right]dS - \int_{S_1}\hat{a}\cdot\left(\hat{n}\cdot\vec{E}\right)\times\nabla G\, dS = \\ = \hat{a}\cdot\left[-\left(\hat{n}\times\vec{E}\right)\times\nabla G + j\omega\mu G(\hat{n}\times H) + \hat{n}\times\left(G\vec{M}\right)\right]dS \end{aligned} \tag{17}$$

Finally, by equating (13) and (17), after some additional mathematical manipulations, results in the following formula:

$$\begin{aligned} \int_{V_1}\left[j\omega\mu\vec{J}G + \vec{M}\times\nabla' G - \frac{\rho}{\varepsilon}\nabla' G\right]dV' = \\ = \int_{S_1}\left[j\omega\mu G(n\times H) - \left(n\times\vec{E}\right)\times\nabla' G - \left(n\cdot\vec{E}\right)\nabla' G\right]dS' \end{aligned} \tag{18}$$

In (18), additional substitution between the source and the observation points, has been carried out.

The expression (18) is in a suitable form, as the left side contains all the sources $(\vec{J},\vec{M},\rho)$ within volume V_1, while the right side includes electric and magnetic field, respectively, on the boundary $S_1 = S \cup S_\varepsilon \cup S\infty$ of the domain. Next, the behavior of the surface integral needs to be considered separately for each part of the boundary.

3 REGION AROUND SINGULARITY

In the limiting case when $\vec{r} \to \vec{r}'$, it is necessary to determine surface integral of (18) for two special cases: when observation point is inside V_1, and when on the boundary surface S_1, respectively.

When former is the case, a small spherical region S_ε, radius of ε, is excluded around the observation point, as shown on Fig. 2a).

Inserting (3) in (18), the first integral on the right side of (18) vanishes, when R → 0:

$$\lim_{R\to 0}\int_{S_\varepsilon} j\omega\mu \frac{e^{-jkR}}{4\pi R}\left(n\times\vec{H}\right)\varepsilon^2 \sin\theta \; d\theta \; d\varphi = 0 \tag{19}$$

Leaving

$$\lim_{R\to 0}\int_{S_\varepsilon}\left[-\left(n\times\vec{E}\right)\times\nabla' G-\left(n\cdot\vec{E}\right)\nabla' G\right] dS \tag{20}$$

where the Green's function gradient is inserted in the following form

$$\nabla G(\vec{r},\vec{r}') = -(1+jkR)\frac{G(\vec{r},\vec{r}')}{4\pi R^2}\vec{R} \tag{21}$$

Leading to:

$$\begin{aligned}&\lim_{R\to 0}\int_{S_\varepsilon}\left[\vec{E}(n\cdot n)-\left(n\cdot\vec{E}\right)n+\left(n\cdot\vec{E}\right)n\right](1+jkR)\frac{e^{-jkR}}{4\pi R^2}=\\&\lim_{R\to 0}\int_{S_\varepsilon}\left[\vec{E}\right](1+jkR)\frac{e^{-jkR}}{4\pi R^2}dS=\frac{\vec{E}}{4\pi}\int_{\theta=0}^{2\pi}\int_{\varphi=0}^{\pi}\sin\theta \; d\theta \; d\varphi=\vec{E}(\vec{r})\end{aligned} \tag{22}$$

In the limiting case when $\vec{r} \to \vec{r}'$, and the observation point is on the boundary surface S, the hemispherical region around the observation point needs to be excluded as shown in Fig. 2b, resulting in the following value for the above integral: $\frac{1}{2}\vec{E}(\vec{r})$.

4 BOUNDARY CONDITIONS AT INFINITY

The boundary conditions on the surface S_∞, where R→∞, represents the so called Sommerfeld radiating condition. The surface integral of (18) vanishes on this surface, as easily showed if

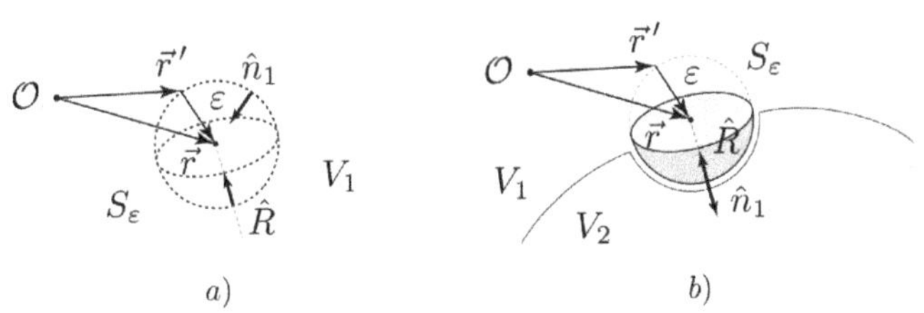

Figure 2: Special case when observation point is: a) in V_1, b) on S_1.

the surface of infinite radius R is selected, with center placed at the source point $\vec{r}'$, and radius vector $\hat{R}$ pointing outwards from V_1:

$$\lim_{R\to\infty}\int_{S_\infty}\left\{j\omega\mu G\left(R\times\vec{H}\right)+\left(\frac{1}{R}+jk\right)\underbrace{\left[\left(R\times\vec{E}\right)\times R+\left(R\cdot\vec{E}\right)R\right]}_{\vec{E}}\frac{e^{-jkR}}{4\pi R}\right\}R^2\sin\theta\,d\theta\,d\varphi= \tag{23}$$

$$=\lim_{R\to\infty}\int_{S_\infty}\left[jkR\left(\eta R\times\vec{H}+\vec{E}\right)\frac{e^{-jkR}}{4\pi}\right]\sin\theta\,d\theta\,d\varphi=0$$

Where $\eta=\sqrt{\mu/\varepsilon}$ is the wave impedance of free space.

5 STRATTON–CHU EXPRESSION

Inserting (22) in (18), while replacing the surface integral designation (S_1 changing to S), yields the integral relation for the electric field in V_1:

$$\begin{aligned}\vec{E}(\vec{r})=&\iiint_{V_1}\left[j\omega\mu\vec{J}G+\vec{M}\times\nabla'G-\frac{\rho}{\varepsilon}\nabla'G\right]dV'+\\&+\iint_S\left[-j\omega\mu G(n\times H)+(n\times\vec{E})\times\nabla'G+\left(n\cdot\vec{E}\right)\nabla'G\right]dS'\end{aligned} \tag{24}$$

The integral expression (24) is the so called Stratton–Chu [10] solution for the electric field. The first term from (24) represents the electric field due to the sources placed inside the region V_1, while the second one is the integral representation of the sources placed outside this region expressed in terms of the field over the boundary surface S.

Thus, (24) enables one to consider the contribution of the sources placed outside the region of interest via their field components or the equivalent sources on the boundary surface, and represents the mathematical form of the equivalence principle [7].

6 ON THE USE OF THE EQUIVALENCE PRINCIPLE

Introducing the equivalent electric and magnetic current densities $\vec{J}_S$ and $\vec{M}_S$, respectively, on the surface S

$$\vec{J}_S=\hat{n}\times\vec{H}_1=-\hat{n}_1\times\vec{H}_1;\quad \vec{M}_S=-\hat{n}\times\vec{E}_1=\hat{n}_1\times\vec{E}_1 \tag{25}$$

the electric field in region V_1, can be expressed the following way:

$$\begin{aligned}\vec{E}_1(\vec{r})-=&\int_{V_1}\left[j\omega\mu_1\vec{J}_1G_1+\vec{M}_1\times\nabla'G_1-\frac{\rho_1}{\varepsilon_1}\nabla'G_1\right]dV'+\\&\int_S\left[-j\omega\mu_1\vec{J}_SG_1+\vec{M}_S\times\nabla'G_1-\frac{\rho_1}{\varepsilon_1}\nabla'G_1\right]dS'\end{aligned} \tag{26}$$

The last term of (26) featured the use of a continuity equation and the following substitution:

$$\hat{n}_1\cdot\vec{E}_1=\frac{j}{\omega\varepsilon_1}\nabla\cdot\left(\hat{n}_1\times\vec{H}_1\right) \tag{27}$$

According to (26), the electric field in V_1 is due to the sources inside this volume $\left(\vec{J}_1, \vec{M}_1\right)$, as well as due to the equivalent surface currents $\left(\vec{J}_S, \vec{M}_S\right)$, introduced to the surface S. Since the sources outside V_1 are omitted in (27), the entire space is homogenized, facilitating the use of a free space Green's function G_1.

The left hand side of (26) results from integral (18) evaluated over the infinitesimal surface S_ε, while it is equal to zero when (18) is evaluated at S_∞. When the observation point is on the boundary S, the expression needs to be modified, resulting in:

$$\begin{aligned}\frac{1}{2}\vec{E}_1(\vec{r}) = \int_{V_1}\left[j\omega\mu_1\vec{J}_1G_1 + \vec{M}_1\times\nabla'G_1 - \frac{\rho_1}{\varepsilon_1}\nabla'G_1\right]dV' + \\ + \int_S\left[-j\omega\mu_1\vec{J}_SG_1 - \vec{M}_S\times\nabla'G_1 + \frac{\rho S}{\varepsilon_1}\nabla'\frac{\rho S}{\varepsilon_1}\nabla'G_1\right]dS'\end{aligned} \tag{28}$$

Finally, when the observation point is outside V_1, i.e. inside V_2, it can be readily shown that the left side of (26) vanishes [11]. Namely, when the source point and the observation point are in V_1 and V_2, respectively, it is not required to exclude the infinitesimal region around singularity, leading to:

$$\begin{aligned}\int_{V_1}\left[j\omega\mu_1\vec{J}_1G_1 + \vec{M}_1\times\nabla'G_1 - \frac{\rho_1}{\varepsilon_1}\nabla'G_1\right]dV' = \\ = -\int_S\left[-j\omega\mu_1\vec{J}_SG_1 - \vec{M}_S\times\nabla'G_1 + \frac{\rho_S}{\varepsilon_1}\nabla'G_1\right]dS'\end{aligned} \tag{29}$$

Subsequently inserting (29) in (26), it follows $\vec{E}_1(\vec{r}) = 0$.

On the other and, electric field in V_2 can be expressed as:

$$\begin{aligned}\vec{E}_2(\vec{r}) - \int_{V_2}\left[j\omega\mu_2\vec{J}_2G_2 + \vec{M}_2\times\nabla'G_2 - \frac{\rho_2}{\varepsilon_2}\nabla'G_2\right]dV' + \\ + \int_S\left[-j\omega\mu_2\vec{J}_2G_2 - \vec{M}_S\times\nabla'G_2 + \frac{\rho S}{\varepsilon_2}\nabla'G_2\right]dS'\end{aligned} \tag{30}$$

where equivalent electric and magnetic currents are introduced on the surface S

$$\vec{J}_S = \hat{n}\times\vec{H}_2 = \hat{n}_2\times\vec{H}_2; \quad \vec{M}_S = -\hat{n}\times\vec{E}_2 = -\hat{n}_2\times\vec{E}_2 \tag{31}$$

and using $\hat{n}_1 = -\hat{n}_2$.

Again, (30) states that electric field in V_2 is due to sources inside this region $\left(\vec{J}_2, \vec{M}_2\right)$, as well as some external sources taken into account via equivalent surface currents. Moreover, the whole region is homogenized again.

Similar to region 1, it is easily shown that left hand side of (30) vanishes when observation point is outside region 2 [11], while the observation point is on surface S, it follows:

$$\begin{aligned}\frac{1}{2}\vec{E}_2(\vec{r}) - = \int_{V_2}\left[j\omega\mu_2\vec{J}_2G_2 + \vec{M}_2\times\nabla'G_2 - \frac{\rho_2}{\varepsilon_2}\nabla'G_2\right]dV' + \\ + \int_S\left[-j\omega\mu_2\vec{J}_SG_2 - \vec{M}_S\times\nabla'G_2 + \frac{\rho S}{\varepsilon_2}\nabla'G_2\right]dS'\end{aligned} \tag{32}$$

The boundary conditions for the electric field on the surface *S* require the tangential components of vector $\vec{E}$ to be continuous. One can assume there are no field sources inside region 2, while sources inside region 1 (volume integral over V_1) can be taken into account via incident field vector $\vec{E}^{inc}$. The electric field at point $\vec{r}$, on the boundary surface *S*, from the viewpoint of region 1 and expression (28) becomes

$$\frac{1}{2}\vec{E}_1(\vec{r}) = \vec{E}^{inc} + \int_S \left[-j\omega\mu_1 \vec{J}_S G_1 - \vec{M}_S \times \nabla' G_1 + \frac{\rho S}{\varepsilon_1} \nabla' G_1 \right] dS' \tag{33}$$

The electric field at the same point $\vec{r}$, on the boundary surface *S*, from the viewpoint of region 2 and expression (30) is:

$$\frac{1}{2}\vec{E}_2(\vec{r}) = \int_S \left[-j\omega\mu_2 \vec{J}_S G_2 - \vec{M}_S \times \nabla' G_2 + \frac{\rho S}{\varepsilon_2} \nabla' G_2 \right] dS' \tag{34}$$

Equating the left sides of (33) and (34), i.e. satisfying the boundary conditions at surface S, $\hat{n} \times \vec{E}_1 = \hat{n} \times \vec{E}_2$, the right hand sides are

$$\begin{aligned} -\hat{n} \times \int_S \left[-j\omega\mu_1 \vec{J}_S G_1 - \vec{M}_S \times \nabla' G_1 + \frac{\rho S}{\varepsilon_1} \nabla' G_1 \right] dS' &= \hat{n} \times \vec{E}^{inc} \\ -\hat{n} \times \int_S \left[-j\omega\mu_2 \vec{J}_S G_2 - \vec{M}_S \times \nabla' G_2 + \frac{\rho S}{\varepsilon_2} \nabla' G_2 \right] dS' &= 0 \end{aligned} \tag{35}$$

Furthermore, substituting the continuity equation into (35), yields the coupled set of SIEs, representing the EFIE formulation, which can be solved by a Moment Method (MoM) scheme, as reported in [4]:

$$\begin{aligned} \vec{E}^{inc} &= \int_S j\omega\mu_1 \vec{J} G_1 dS - \frac{j}{\omega\varepsilon_1} \int_S \nabla'_S \cdot \vec{J} \nabla' G_1 \, dS + \int_S \vec{M} \times \nabla' G_1 \, dS \\ 0 &= \int_S j\omega\mu_2 \vec{J} G_2 dS - \frac{j}{\omega\varepsilon_2} \int_S \nabla'_S \cdot \vec{J} \nabla' G_2 \, dS + \int_S \vec{M} \times \nabla' G_2 \, dS \end{aligned} \tag{36}$$

7 BIOELECTROMAGNETICS APPLICATION

Some computational examples obtained using SIE formulation (36) related to the assessment of the induced field in the human brain are presented in this section.

The first example is related to the induced electric field in the brain of adult, 10-year old, and 5-year old child, respectively, exposed to horizontally polarized incident plane wave of 900 MHz, as shown in Fig. 3 [12].

It can be seen that at horizontally polarized incident wave, the higher values of induced electric field are obtained in smaller brain models in the frontal lobe areas.

The second example gives the comparison of the induced electric field in the same models, due to a transcranial magnetic stimulation (TMS) coil positioned 1 cm over primary motor cortex [13], as shown on Fig. 4.

Although the rapid decrease in the induced electric field is evident in all brain models, Fig. 4 shows that for smaller ones, the place where electric field falls to half its maximum value is moved closer to the surface.

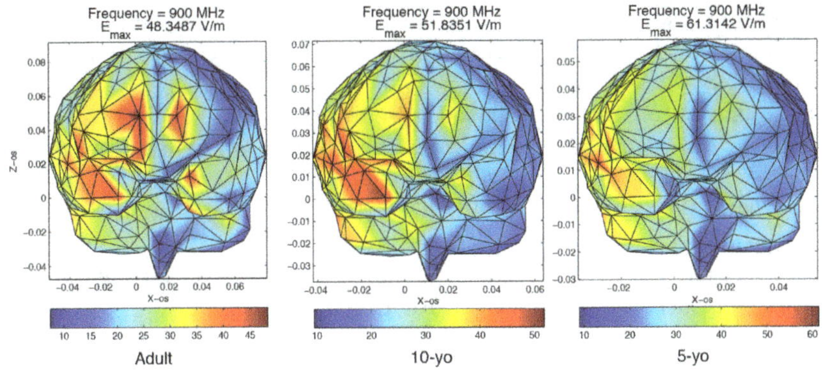

Figure 3: Comparison of the induced electric field due to plane EM wave.

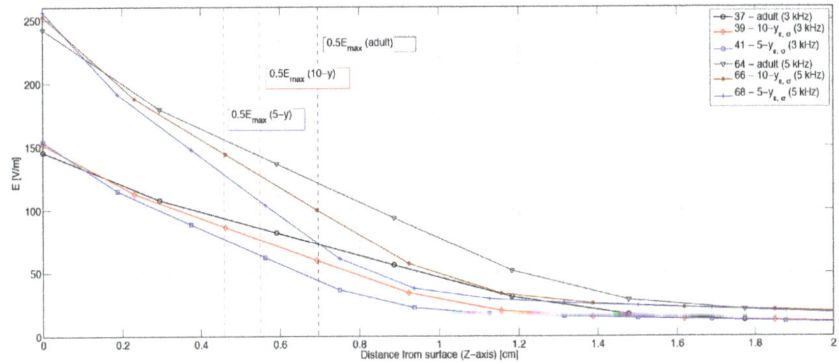

Figure 4: Comparison of the induced electric field due to TMS coil.

8 CONCLUSION

The paper revisits the application of surface equivalence theorem in the derivation of the EFIE formulation. The resulting coupled SIE set, although known for some problems arising in the CEM community, has only recently begun to appear in the bioelectromagnetics, offering a powerful, accurate and efficient alternative to the de facto standardly used differential equation based methods. Some computational examples obtained using the SIE based method are presented in the paper. The proposed method could be found useful in the initial assessment of the human brain exposed to EM radiation, while more realistic scenarios would require the use of a non-homogeneous models, resulting in the appropriate volume integral equation based formulation and the application of the volume equivalence principle.

REFERENCES

[1] Hand, J., Modelling the interaction of electromagnetic fields (10 MHz– 10 GHz) with the human body: methods and applications. *Physics in Medicine and Biology*, **53**(16), pp. R243–R286, 2008.
https://doi.org/10.1088/0031-9155/53/16/r01

[2] Poljak, D., Cavka, D., Dodig, H., Peratta, C. & Peratta, A., On the use of the boundary element analysis in bioelectromagnetics. *Engineering Analysis with Boundary Elements*, **49**, pp. 2–14, 2014.
https://doi.org/10.1016/j.enganabound.2014.02.008
[3] Poljak, D., *Human Exposure to Electromagnetic Fields*. WIT Press, Boston, 2003.
[4] Cvetković, M. & Poljak, D., *An efficient integral equation based dosimetry model of the human brain*. Proceedings of EMC EUROPE 2014, Gothenburg, Sweden, 1–4 September 2014, pp. 375–380, 2014.
[5] Poljak, D., Cvetković, M., Peratta, A., Peratta, C., Dodig, H. & Hirata, A., *On some integral approaches in EM dosimetry*. BIOEM 2016, 2016.
[6] Poggio, A.J. & Miller, E.K., *Integral equation solutions of three– dimensional scattering problems. Computer Techniques for Electromagnetics*, 2.E., ed. R. Mittra, pp. 159–264, 1987.
[7] Chew, W.C., Tong, M.S. & Hu, B., *Integral Equation Methods for Electromagnetic and Elastic Waves*. Morgan & Claypol Publishers, San Rafael, CA, 2009.
[8] Cvetković, M., Poljak, D. & Haueisen, J., Analysis of transcranial magnetic stimulation based on the SIE formulation. *IEEE Transactions on Biomedical Engineering*, **62**(6), pp. 1535–1545, 2015.
https://doi.org/10.1109/tbme.2015.2393557
[9] Wei, F. & Yilmaz, A., A more scalable and Efficient parallelization of the adaptive integral method – Part II: BIOEM Application. *IEEE Transactions on Antennas and Propagation*, **62**(2), pp. 727–738, 2014.
https://doi.org/10.1109/tap.2013.2291564
[10] Stratton, J.A. & Chu, L.J., Diffraction theory of electromagnetic waves. *Physical Review*, **56**(1), pp. 99–107, 1939.
https://doi.org/10.1103/physrev.56.99
[11] Chen, K.M., A mathematical formulation of the equivalence principle. *IEEE Transactions on Microwave Theory and Techniques*, **37**(10), pp. 1576–1581, 1989.
https://doi.org/10.1109/22.41004
[12] Cvetković, M. & Poljak, D., Electromagneticthermal dosimetry comparison of the homogeneous adult and child brain models based on the SIE approach. *Journal of Electromagnetic Waves and Applications*, **29**(17), pp. 2365–2379, 2015.
https://doi.org/10.1080/09205071.2015.1092896
[13] Cvetković, M., Poljak, D., Rogić Vidaković, M. & Đogaš, Z., TMS induced fields in different brain models. *Journal of Electromagnetic Waves and Applications*, **30**(14), pp. 1820–1835, 2016.
https://doi.org/10.1080/09205071.2016.1216807

Author index

www.ingramcontent.com/pod-product-compliance
Lightning Source LLC
LaVergne TN
LVHW062233200926
848484LV00006B/1528